Advances in

Heterocyclic Chemistry

Volume 50

Advances in

HETEROCYCLIC CHEMISTRY

Edited by

ALAN R. KATRITZKY, FRS

Kenan Professor of Chemistry
Department of Chemistry
University of Florida
Gainesville, Florida

Volume 50

ACADEMIC PRESS, INC.
Harcourt Brace Jovanovich, Publishers
San Diego   New York   Boston
London   Sydney   Tokyo   Toronto

ACADEMIC PRESS, INC.
San Diego, California 92101

United Kingdom Edition published by
ACADEMIC PRESS LIMITED
24-28 Oval Road, London NW1 7DX

LIBRARY OF CONGRESS CATALOG CARD NUMBER: 62-13037

ISBN   0-12-020650-1   (alk. paper)

PRINTED IN THE UNITED STATES OF AMERICA
90 91 92 93   9 8 7 6 5 4 3 2 1

Contents

Thiadiazines with Adjacent Sulfur and Nitrogen Ring Atoms
REGINALD E. BUSBY

Preface

Volume 50 of *Advances in Heterocyclic Chemistry* comprises four chapters. Howard D. Perlmutter has contributed a survey of 1,2- and 1,3-diazocines. This is the fourth in a series of reviews by Perlmutter, whose previous surveys include azocines (Volume 31, 1982), 1,4-diazocines (Volume 45, 1989), and 1,5-diazocines (Volume 46, 1989). The present chapter completes this treatment of eight-membered heterocycles containing nitrogen.

R. E. Busby has written a chapter on thiadiazines containing adjacent sulfur and nitrogen ring atoms. This chapter complements one in Volume 44 which dealt with sulfamides of this type. H. Quiniou and O. Guilloton of Nantes have covered the chemistry of monocyclic 1,3-thiazines, a group that, surprisingly, has not been reviewed comprehensively for many years. Finally, E. V. Kuznetsov and I. V. Shcherbakova of Rostov collaborate with A. T. Balaban of Bucharest in a summary of the chemistry of benzo[c]pyrylium salts. This review complements the review by a group of authors under the leadership of Balaban who covered the chemistry of pyrylium salts in Supplement 2 of this series, which appeared in 1982.

ALAN R. KATRITZKY

1,2-Diazocines, 1,3-Diazocines, Triazocines, and Tetrazocines

HOWARD D. PERLMUTTER

*Department of Chemical Engineering, Chemistry and
Environmental Science,
New Jersey Institute of Technology
Newark, New Jersey 07102*

I. Introduction, Scope, and Nomenclature

This chapter covers eight-membered heterocycles containing (a) two nitrogens in a 1,2 (**1**) and 1,3 (**2**) position to each other, (b) three nitrogens (e.g., **3**), and (c) four nitrogens (e.g., **4**). Excluded from coverage are compounds where *all* nitrogens are members of another fused ring (e.g., **5**) and where *nonadjacent* carbons of the eight-membered ring form part of another ring (e.g., **6**).

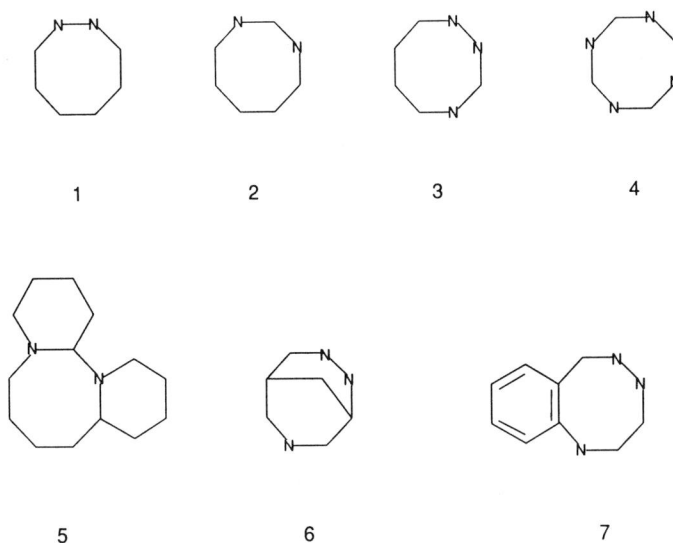

In this chapter, all compounds are classified according to the actual relationship of the eight-membered ring nitrogens to each other, regardless of formal nomenclature. Thus, 1,4,5-benzotriazocine **7** would be covered in the section on 1,2,5-triazocines. For other aspects of the nomenclature of these compounds, refer to previous reviews of azocines (82AHC115), 1,4-diazocines (89AHC185), and 1,5-diazocines [89AHC(46)1]. This chapter completes the review of eight-membered heterocycles containing nitrogen.

II. 1,2-Diazocines

A. Preparative Methods

1. *Via Cyclization*

Formation of a 1,2-diazocine by cyclization or ring closure has been almost exclusively accomplished by formation of the N=N or N–N bond by coupling of open-chain dinitro or diamino compounds. Reaction of *o,o'*-dinitrodiphenylethane with zinc accompanied by hydrogen chloride (09CR401) or barium hydroxide (75MI4) yielded **8** or its *N*-oxide, respectively. The latter compound and the *N,N'*-dioxide of **8** were formed by oxidative and reductive coupling of *o,o'*-diamino- and *o,o'*-dinitrodiphenylethane, respectively [73AC(P)329; 80JOC4597]. Reduced compound **9** was produced, using barium hydroxide and zinc, from either **8** (10BSF727; 65USP3170929; 69JOC3237) or *o,o'*-dinitrodiphenylethane under nitrogen (75MI4). The reverse **9** → **8** conversion was accomplished with mercuric oxide (69JOC3237), refluxing acetone, or methyl sulfate/butyllithium (75MI4) [the last reaction also yielded a tetrazine incorporating two diazocines (see Section II,C,1).

Small amounts of the *N*-oxide of **8** accompanied the formation of *o,o'*-dinitrodiphenylethane by the reaction of *o*–nitrotoluene with potassium *tert*-butoxide [73AC(P)329; 75MI4]. Formation of **8** and its *N*-oxide also resulted from the controlled-potential electrolysis of *o*-nitrobenzyl thiocyanate, a reaction proceeding by dimerization of the initially formed *o*-nitrobenzyl radicals followed by simultaneous or subsequent reductive coupling of the nitro groups. In strongly acidic medium, the reaction took a different course (85CCC33).

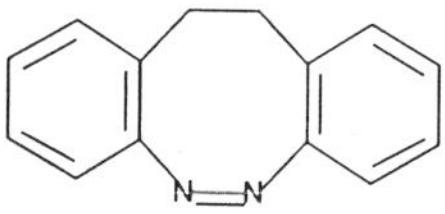

8

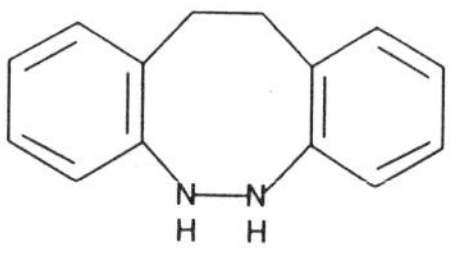

9

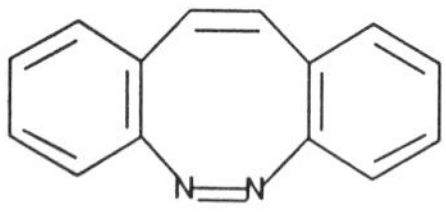

10

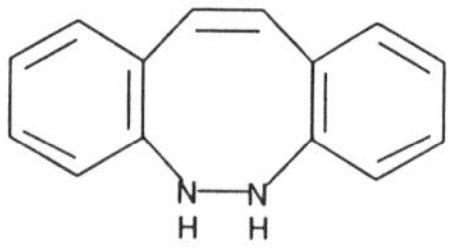

11

Bromination of **8** followed by dehydrobromination afforded completely unsaturated compound **10**. Barium hydroxide and zinc reduction of **10** yielded **11** (69JOC3237). (*E*)-1,2-Diazacyclooctene **12** (R^1 – R^4 = Me) was synthesized by oxidative cyclization of 2,7-dimethyl-2,7–octanediamine with iodine pentafluoride (78CB596). [For preparation of the stereoisomeric "*Z*" compounds, see Section II,A,3,a.] Stetter and Koch reacted bis(*N*-bromoamino)biadamantyl **13** with potassium *tert*-butoxide and isolated a molecule containing a 1,2-diazocine sandwiched between two adamantane moieties (**14**) [see also Section II,A,3,c] (75LA1357).

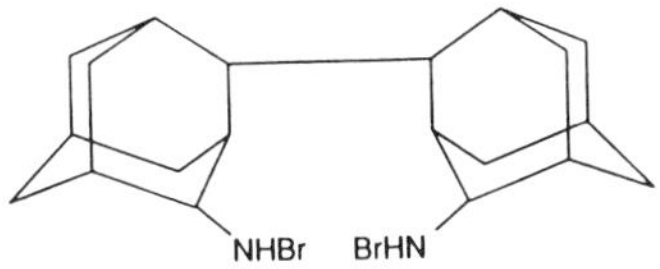

13

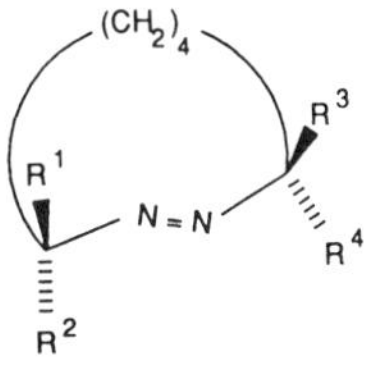

12

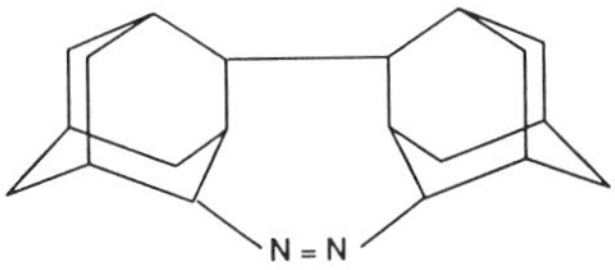

14

In an uncommon example of 1,2-diazocine synthesis by C–N bond formation, Wittig and Grolig reacted hydrazobenzene with adipic anhydride to give the monohydrazide of adipic acid (**15**). Ring closure of **15** using thionyl chloride afforded diazocine **16** (X = O) (61CB2148). Huang *et al.* condensed benzyl 2-phthalimido-6-bromohexanoate with a trisubstituted hydrazine to obtain **17** (R = CH$_2$Ph, R′ = CO$_2$CH$_2$Ph). Repeated hydrogenolysis of the latter compound gave **17** (R = R′ = H), cyclization of which yielded **18**. (Gabriel reaction of **18** followed by reductive alkylation and decarboxylation afforded a compound of some medicinal value; see Section II,D) (84USP4465679).

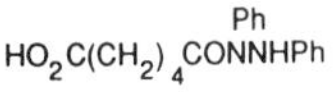

HO$_2$C(CH$_2$)$_4$CONNHPh

15

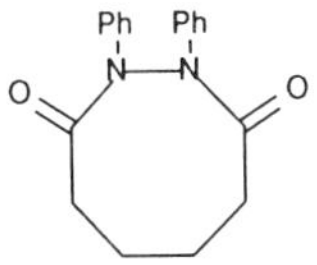

16

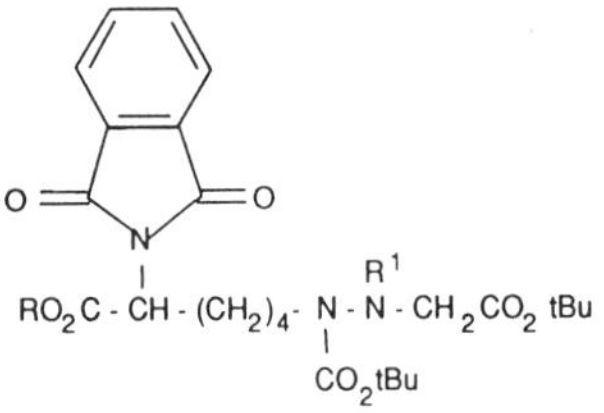

17

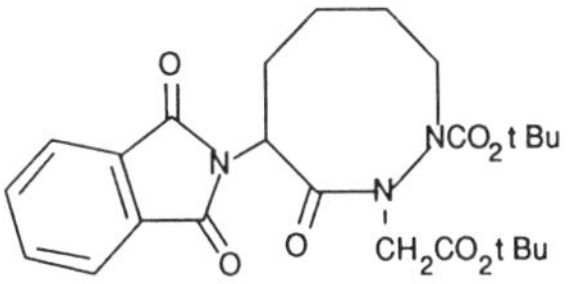

18

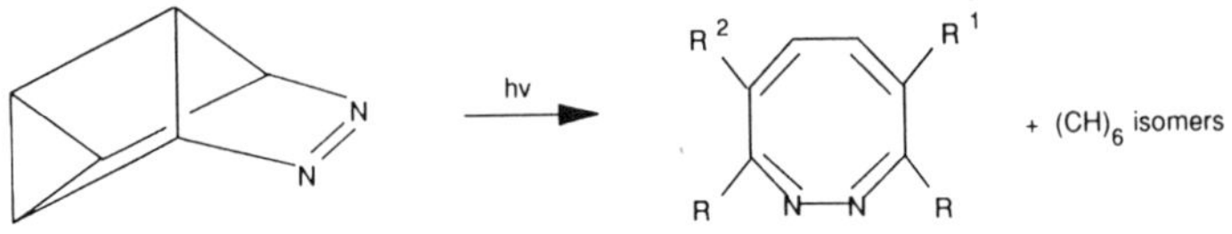

19   20

SCHEME 1

2. Via Ring Expansion

Totally conjugated 1,2-diazocine **20** ($R = R^1 = R^2 = H$) was prepared by direct photochemical excitation of polycyclic azoalkane **19** (Scheme 1). The ratio of $(CH)_6$ isomers (benzvalene, Dewar benzene, prismane, and benzene) to **20** ($R = R^1 = R^2 = H$) is strongly dependent on both solvent and temperature. For example, at ambient temperature the ratio of $(CH)_6$ isomers to **20** ($R = R^1 = R^2 = H$) is 92/8 in pentane and 38/68 in dimethyl sulfoxide (DMSO). Photolysis at $-78°C$ in a variety of solvents gives mainly **20** ($R = R^1 = R^2 = H$). The same results were obtained using direct triplet photosensitized excitation of **19**. Spectroscopic analysis showed the diazocine to be **20** ($R = R^1 = R^2 = H$) and not isomeric **21** (71JA5573; 73JA2738; 76JA4320; 79JOC1264; 79RTC173). For discussion of another preparation, the structure, and the chemistry of conjugated 1,2-diazocines **20,** see Sections II,A,3,c; II,B; and II,C,2.

21

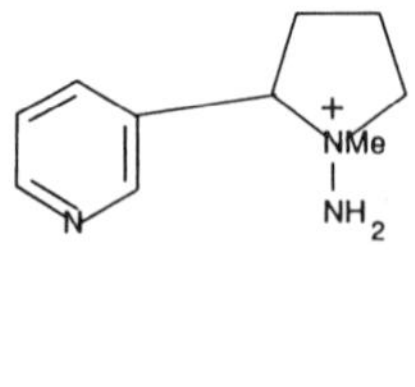

22 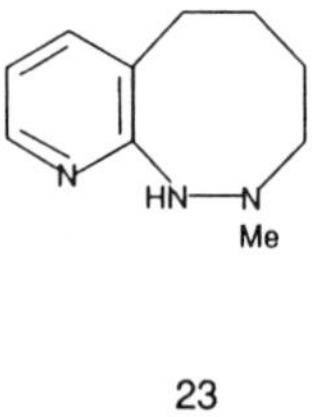   23

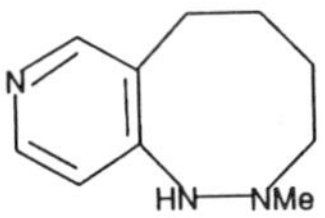

24

In a base-catalyzed ring expansion route to 1,2-diazocines, Sommelet–Hauser rearrangement of the *N*-amino derivative of nicotine (**22**) using sodium in ammonia afforded pyrido-1,2-diazocine **23**. The reaction was regiospecific, i.e., no isomeric **24** was detected (82H1595).

There have been three reported syntheses of 1,2-diazocines by means of ring-expansive cycloaddition reactions. In the first case, Sasaki and co-workers reacted 4,4-dimethyl-3,5-diphenylisopyrazole **25** with diphenylcyclopropenone to obtain diazocine **26** (73SC249). The second cycloaddition was achieved by Haddadin *et al.*, who condensed cyclobutanone with tetrazine **27** (Ar = Ph) to afford diazocine **28** (Ar = Ph, X = O) or **29**, depending on whether methanolic base or diethylamine was used as catalyst (84TL2577). A similar reaction was used to prepare a triazocine; see Section IV,A,1.

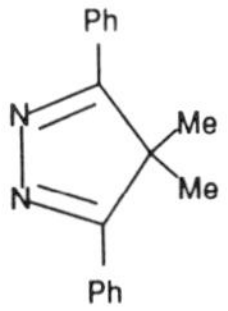

25

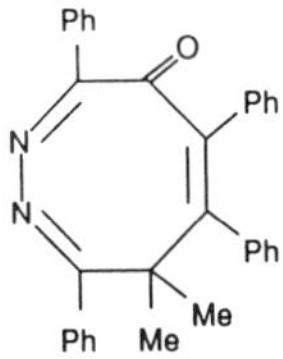

26

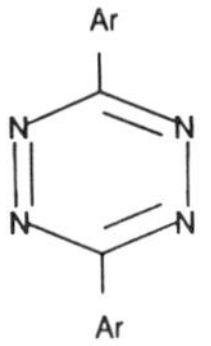

27

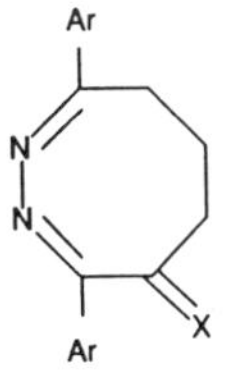

28

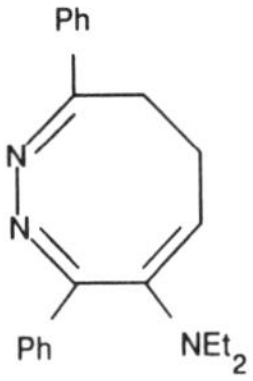

29

In the last case, 1-oxido-3-phenylphthalazinium ion (**30**) underwent cycloaddition with unsaturated molecules to give 8-phenyl-3,4-benzo-1,8-diazabicyclo[3.2.1]octane-2-ones, 8-phenyl-3,4-benzo-1,8-diazabicyclo-

[3.2.1]-6-octene-2-ones and/or benzodiazocinones **31**; the product composition shows an interesting dependence on solvent and substrate. For example, whereas reaction of **30** with styrene or diphenylacetylene (or with dimethyl acetylenedicarboxylate (DMAD) in xylene) and with phenylacetylene in *o*-dichlorobenzene afforded only the diazabicyclic compounds, reaction of **30** with phenylacetylene in xylene gave mainly **31** (R = Ph, R' = H) plus a small amount of the diazabicyclics. Condensation of **30** with DMAD in chloroform yielded only diazocine **31** (R = R' = CO₂Me) [76JCS(P1)2281; 76TL1569].

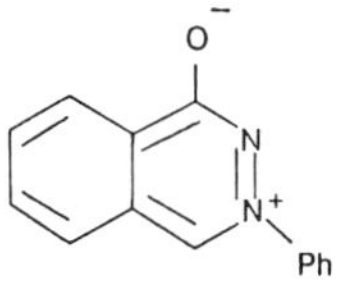

30 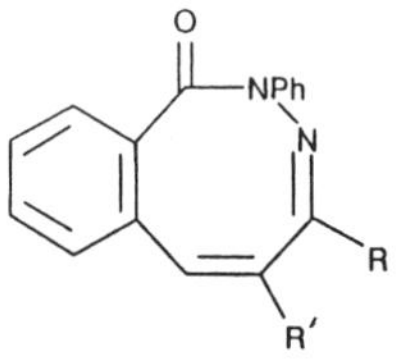   31

3. Via Condensations

By far the most frequently used preparation method of 1,2-diazocines has been condensation reactions using hydrazine, substituted hydrazines, or their alkali metal salts to produce eight-membered cyclic hydrazines (**32**), hydrazides (**33**), and azines (**34**). These fall into three categories, as described in the following paragraphs.

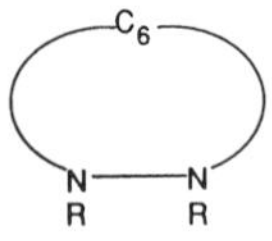

32 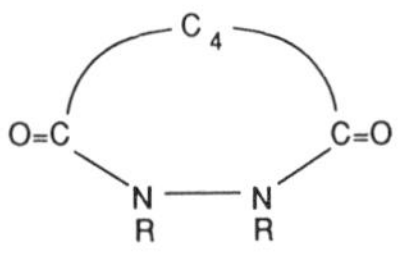   33

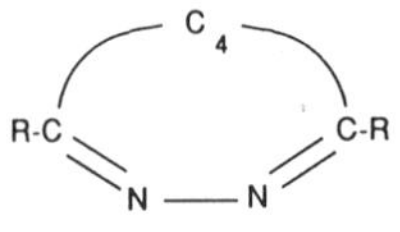

34

a. *Reaction of Hydrazine and Derivatives with Six-Carbon Saturated Electrophiles to form Hydrazines* **32.** Wittig and Grolig reacted dilithiohydrazobenzene (prepared from hydrazobenzene and methyllithium) with 1,6-hexaneditosylate to produce diazocine **16** ($X = H_2$) (61CB2148). Carpino treated 2,2'-bis(bromomethyl)biphenyl with *tert*-butyl hydrazodiformate (**35,** R = *tert*-butyl) in the presence of potassium *tert*-butoxide. Removal of the carbalkoxy groups with hydrogen chloride afforded the hydrochloride of 5,8-dihydrodibenzo (*d,f*)-1,2-diazocine (**36**). Free base **36**

$RO_2C\text{-}NH\text{-}NH\text{-}CO_2R$

35

36

37

12

was liberated from the salt with base. Mild oxidation of **36** yielded the very sensitive azo compound **37,** which detonated on heating (63JA2144). In a similar manner, Carpino and Masaracchia reacted **35** (R = *tert*-butyl) with (Z)-1,6-dibromo-3-hexene and 1,6-dibromohexane to obtain **38** and **39,** respectively. [An attempt to prepare the triply unsaturated 1,2-diaza-1,4,6-cyclooctatriene was unsuccessful; bromination of **38** followed by base treatment resulted in ring contraction. See Section II,C,3 (72JOC1851).]

As part of an extensive study of cyclic azo compounds, Overberger and co-workers condensed 1,6-hexadiyl ditosylate and 2,7-octadiyl ditosylate with **35** (R = Et) in the presence of sodium hydroxide. Removing the ethoxycarbonyl group with potassium hydroxide followed by mercury(II) oxide oxidation afforded (Z)-1,2-diaza-1-cyclooctene (**40,** R = H) and the *cis*- and *trans*-dimethyl isomers **40** (R = Me), respectively (70JA4922). Quinkert and co-workers prepared **40** (R = H) in the same manner and also

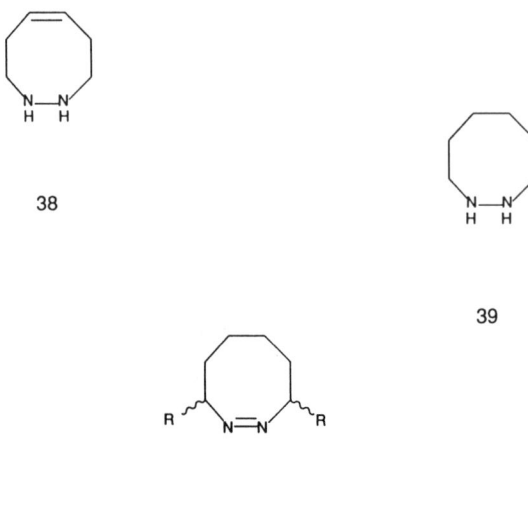

38

39

40

found that oxidation of **39** with molecular O_2 produced an E/Z mixture of **12** ($R^1 - R^4 = H$) and **40** (R = H) (76CB518). In subsequent work, the 1,2-diaza-1-cyclooctenes could be separated into the (Z) (**40**) and (E) (**12**) double-bond isomers (72TL4565; 81JOC303). (E)-Isomers **12** could be prepared more efficiently by photoisomerization of **40** (R = H or Me) (72TL4565; 76CB518; 81JOC303; 85CJC1829). This and other photochemistry of 1,2-diaza-1-cyclooctene and derivatives is discussed in Section II,C,2.

b. *Reaction of Hydrazine and Derivatives with Hexanedioic (Adipic) Acid Derivatives of Form Hydrazides* **33.**   A very common example of this type of reaction has been the condensation of 2,2′-biaryldicarboxylic acid or acid derivatives with hydrazine or substituted hydrazines. For example, Vasserman and Miklukhin prepared hydrazide **41** (R = R′ = X = H) by reacting biphenyl-2,2′-dicarboxylic acid or its diethyl ester with hydrazine hydrochloride or hydrazine hydrate, respectively (39JGU606). Similarly, Bhargava and Heidelberger synthesized hydrazide **42** starting with 2-phenylphenanthrene via either the anhydride or the diacid chloride (56JA3671). Roldan and co-workers isolated substituted **41** (R = Me, Et, Pr, Bu; R′ = X = H) by treating diphenic anhydride with appropriately substituted hydrazines (74GEP2323554; 75 MI1). Wasfi had similarly found that diphenic anhydride reacted with phenylhydrazine to afford **41** (R = Ph, R′ = X = H). However, when hydrazine hydrate was substituted for phenylhydrazine, N-aminodiphenimide **43** was isolated (66JIC723).

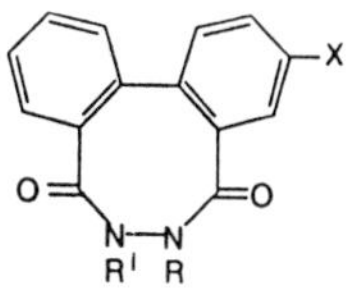

41

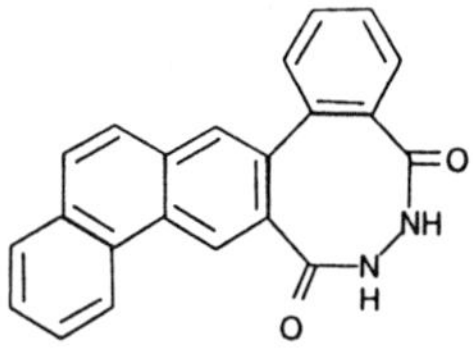

42

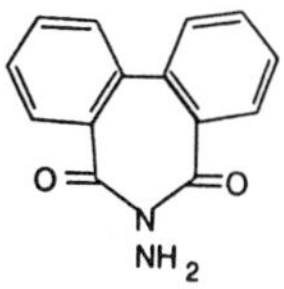

43

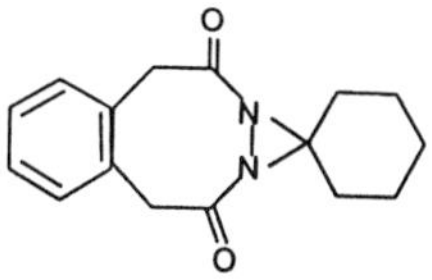

44

As part of a study of 1,2-diaroyldiaziridines, Heine and co-workers synthesized hydrazide **44** by condensing *o*-phenylenediacetyl chloride with 3,3-pentamethylenediaziridine. The chemistry of **44** was discussed, including an interesting ring contraction (Section II,C) (76JOC3229). A slightly different approach was taken by Wittig and co-workers, who treated dilithiohydrazobenzene (PhNLi)$_2$ with adipoyl chloride to yield **45** (R' = Ph, R$_8$ = H$_8$), in addition to a 16-membered cyclic bishydrazide and azobenzene (66LA109). In a novel perfluorination reaction, Ogden reacted **46** with adipoyl fluoride in the presence of cesium fluoride and obtained the perfluorinated 1,2-diazocine **45** (R' = CF$_3$, R$_8$ = F$_8$) [71JCS(C)2920] (use of a different acid halide, oxalyl fluoride, under similar conditions afforded a tetrazocine; see Section V,B,1,d).

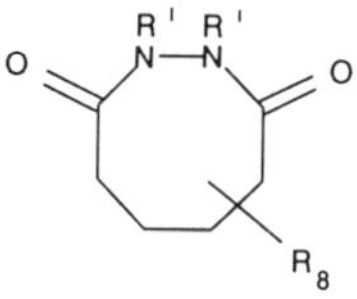

$CF_2=N-N=CF_2$

46

45

c. *Reaction of Hydrazine with Aldehydes and Ketones to Form Azines* **34**. This type of condensation, i.e., the reaction of hydrazine with a diketone to give an azine, has been the most frequently used method of preparing 1,2-diazocines. A common example of this is the synthesis of diazocine-fused biphenyls (see also Section II,A,3,a and b). Hall and co-workers condensed 2,2'-diacetylbiphenyl with hydrazine hydrate to obtain diazocine **47** (R = Me) [plus a small amount of 9,10-disubstituted phenanthrene (56JCS3475)]. Similarly, Bacon and Lindsay reacted 2,2'-diacetyl-, 2,2'-dipropionyl-, or 2,2'-dibenzoylbiphenyl with hydrazine to obtain **47** (R = Me, Et, or Ph) with varying amounts of 9,10-disubstituted phenanthrenes. However, when hydrazine was treated with 2,2'-biphenyldicarboxaldehydes, only phenanthrenes were obtained. It was shown that the phenanthrenes did *not* result from ring contraction of azines **47** [56CI(L)1479; 58JCS1375]. Benzodiazocines **48** and related quin-oxalinodiazocines **49** have been prepared by reacting hydrazine with α,α'-diaroyl-o-xylenes (60JOC1509) and 2,3-bis(aroylmethyl)quinoxa-lines. (77MI1, 77MI3), respectively.

47

48

49

(Ar=Ph,2,4-xylyl)   (Ar=Ph, p=tolyl)

Nonfused ring 1,2-diazocines **28** (Ar = Ph)(59JA217; 63CR929), *p*-anisyl (56RTC1159), and *p*-ethylphenyl (57RTC519) (X = H$_2$) have resulted from the reaction of 1,6-diaryl-1,6-hexanediones with hydrazine. Interestingly, when 1,6-dialkyl-1,6-hexanediones were used, no diazocine was iso-lated, but instead, an unstable diazabicyclo[3.3.0]octene was obtained (63CR929). Compound **28** (Ar = Ph, X = H$_2$) has been used as the starting material for the synthesis of totally conjugated 1,2-diazocines. Allylic 4,7-dihalogenation and 4,4,7,7-tetrahalogenation of **28** (Ar = Ph, X = H$_2$) followed by dehydrohalogenation afforded **20** (R = Ph, R^1 = R^2 = H) and **20** (R = Ph, R^1 = R^2 = Cl) (82CL1579; 86BCJ1087). The chemistry of these last compounds, including formation of other substituted derivatives of **20**, is discussed in Section II,C,2.

Compounds **28** were also reduced to 1,2-diaza-1-cyclooctenes. Overberger and co-workers reported that **28** (Ar = Ph, X = H$_2$) was hydrogenated to give the two stereoisomeric *cis* and *trans*-diphenyl compounds **40** (R = Ph) (59JA217; 69JA3226). However, subsequent repetition of this reaction by Quinkert and co-workers demonstrated the latter compound was not as previously reported, being instead (*E*)-isomer **12** (R^1 = R^4 = Ph, R^2 = R^3 = H) (see Section IIB). Quinkert's group proceeded to photolyze *cis*-diphenyl isomer **40** (R = Ph) and compound **12** (R^1 = R^4 = Ph, R^2 = R^3 = H) either directly or using photosensitization to afford, respectively, two new compounds isomeric at the azo bond, ie., **12** (R^1 = R^3 = Ph, R^2 = R^4 = H) and *trans*-diphenyl compound **40** (R = Ph) (76CB518). For other synthetic and structural aspects of compounds **12** and **40** see Section II,A,1; II,A,3,a; II,B; and II,C,2.

Diazocine rings have been placed in unusual places via carbonyl-hydrazine condensations. Both Roth and Dvorak (63AP510) and Jena and co-workers [77IJC(B)867] reacted hydroxyketone **50** (X = CH$_2$) with hydrazine to yield ketazine **51** (X = CH$_2$). In a closely related reaction, Nagai and co-workers combined hydrazine with hydroxybis(piperidine) **50** (X = NCH$_2$Ph) to afford **51** (X = NCH$_2$Ph) (77CPB1911). Stetter and Koch inserted a 1,2-diazocine ring between two adamantane moieties. Reaction of 2a,2a'-biadamantyl-4,4'-dione (**52**) with hydrazine afforded azine **53**, hydrogenation of which yielded 4,4'-azo-2,2'-biadamantyl (**14**). For the preparation of **14** by ring closure, see Section II,A,1 (75LA1357).

50   51   14

52   53

Bicyclic dihydrazines **54** (R = Me, Et; $R^1 = R^2 = OH$) were prepared by reacting tetraketodiacids **55** (R = Me, Et) with sodium cyanide and hydrazine sulfate. Compounds **54** were in turn oxidized to bisazo compounds **56** (R = Me, Et; $R^1 = R^2 = OH$). Carboxy acid azides **56** (R = Me, Et;

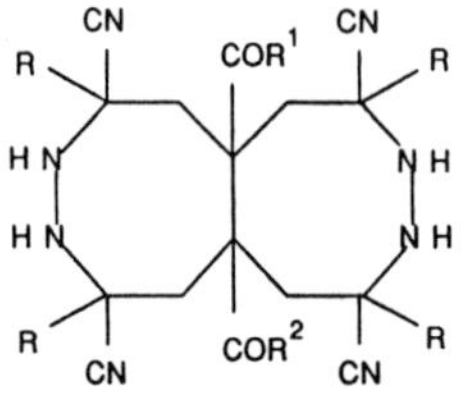

54

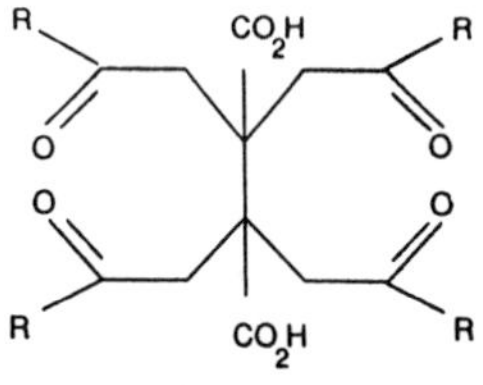

55

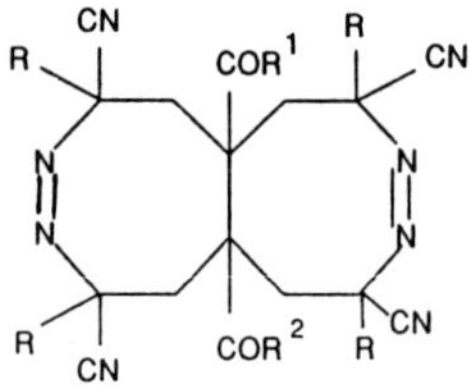

56

$R^1 = N_3$; $R^2 = OH$) were useful compounds (see Section II,D). The starting materials **55** had been synthesized via Diels–Alder reaction of acetylene-dicarboxylic acid with 2,3-dialkyl-1,3-butadienes, followed by oxidation (85USP4532301; 85USP4556512). Yoshida *et al.* condensed the doubly cross-conjugated cyclopentadiene **57** with hydrazine to obtain 1,2-diazocine **58** (84TL4223) (For structural studies of **58**, see Section II,B.)

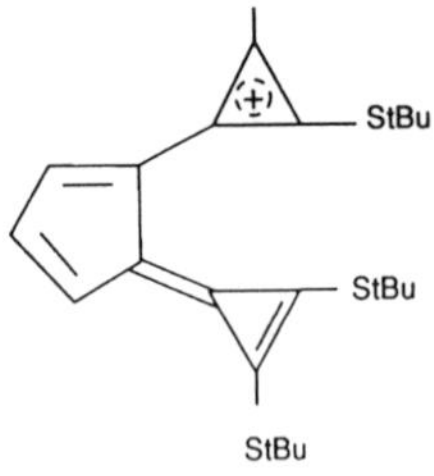

57

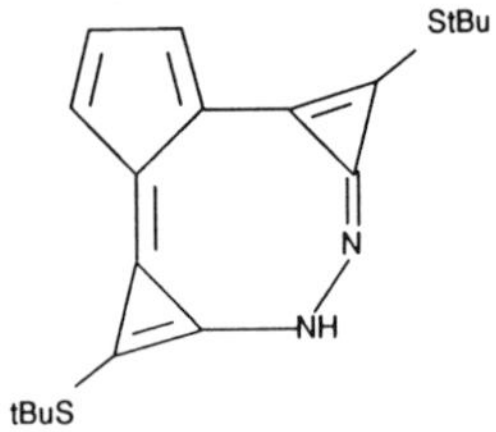

58

In an interesting replacement of a steroid B-ring with a diazocine ring, Lettre and Schelling ozonized cholesterol derivatives **59**. Reduction of the ozonide followed by treatment of the product with phenylhydrazine gave compounds **60**. The latter hydrazones were further hydrogenated to yield various 3β-substituted-6-amino-5,β-hydroxy-5,6-*seco*cholestanes (see Section II,C,4) (63LA160).

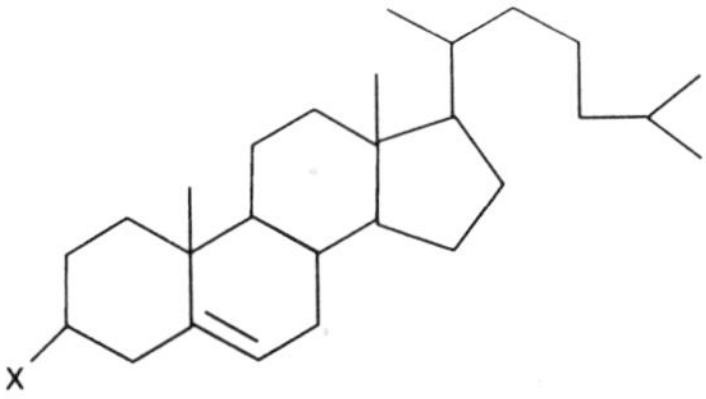

59

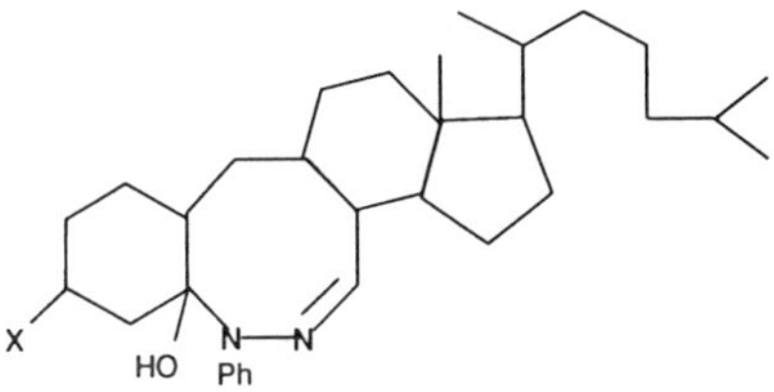

60

d. *Mixed-Type Condensations.*   As an example of the replacement of steroidal A-ring with a diazocine, Rodewald and Olejniczak synthesized diazocinone **61** by reacting ketoester **62** with hydrazine (80MI2). Leclerc condensed δ-benzoylvaleric acid with methylhydrazine or morpholino-ethylhydrazine to produce **63** [R = Me or $(CH_2)_2$–(C_4H_8NO)]. Reaction of 1-benzoyl-5-chloropentane with methylhydrazine afforded **64** (R = Ph, R^1 = Me, R^2 = H). This can also be synthesized by hydride reduction of **63** (73BSF2029).

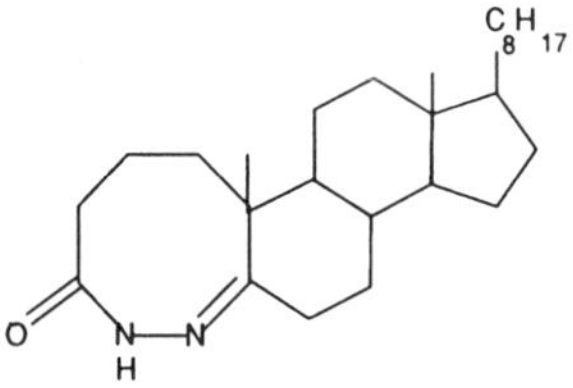

61

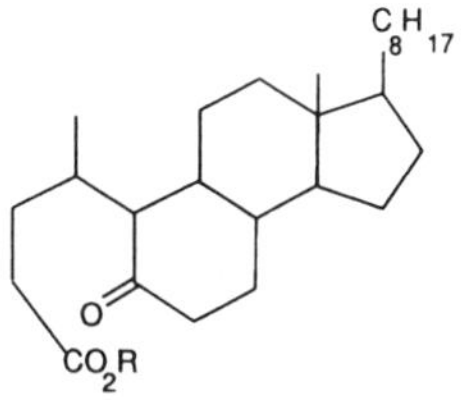

62

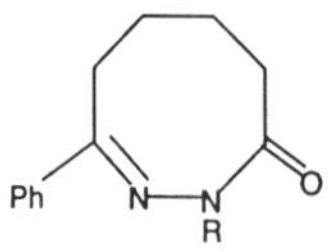

63

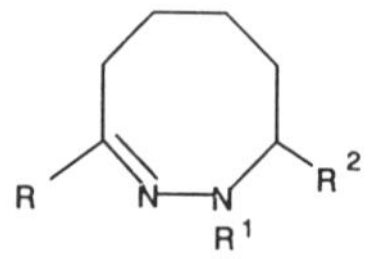

64

B. Theory and Structural Aspects

Allinger and Youngdale used molecular orbital calculations to obtain resonance energies of the 10π-electron diazocines **65** and **66** (R = H). Although these values were large compared to benzene, the calculational results were nevertheless regarded as consistent with the fact that benzodiazocines **48** (Ar = Ph; 2,4-xylyl) (See Section II,A,3,c) failed to isomerize to **66** (R = Ph; 2,4-xylyl) (60JOC1509). Ponce and co-workers also reported quantum mechanical calculations on 1,2-diazocines **65** and **66** (R = H). Bond lengths, electron densities, free valences, superdelocalizabilities, delocalization energies, steric energies, and the Julg/Lebarre–Gallais-Julg degrees of aromaticity were calculated. The results showed that bond alternation was not important in **65** and **66** (R = H), and that these 1,2-diazocines were much more aromatic than the 1,4-diazocines studied [75MI8; 89AHC(45)185].

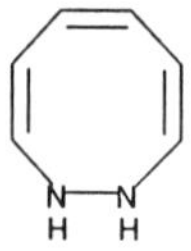

65

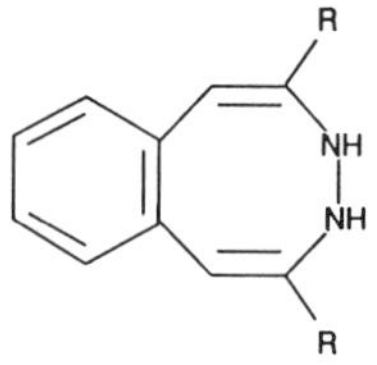

66

Michl and co-workers calculated magnetic circular dichroism (MCD) spectra for diazocine **65** and the isomeric 10π-electron 1,4-dihydro-1,4-diazocine. See Perlmutter [89AHC(45)185] for a brief description of MCD relevant to the latter compound; for more details, see Michl (77MI4) and Waluk *et al.* (81JOC3306). Compound **65** was predicted to approach being a "negative-hard" MCD chromophore, while the 1,4-isomer was more distinctly a "single-soft" one (81JOC3306).

Trost and co-workers used proton and carbon NMR to show that the

1,2-diazocine they synthesized (see Section II,A,2) had structure **20** ($R = R^1 = R^2 = H$), not **21**, and consisted almost exclusively of monocyclic, as opposed to diazabicyclo[4.2.0]octatriene (see **86**, Section II,C,3), valence tautomers. This latter observation is also supported by estimated enthalpies of isomerization to bicyclic tautomers, which are less favorable than in the carbocyclic series (79JOC1264).

Cyclic azo compounds **40** (R = Alk, Ar, H) and their (*E*)-isomers **12** have been the subject of much study, both synthetic [see Section II,A,3,a] and structural. The major work has been by Overberger and co-workers, who, over a number of years, have synthesized [see Section II,A,3,a] and studied both **12** and **40** and their 3,8-dialkyl and diaryl derivatives. The (*Z*)-isomer **40** and its derivatives were the subject of earlier work. Evidence for the (*Z*)-azo linkage was found in the dipole moment, UV, IR, and NMR spectra, all of which were more consistent with properties of known seven- and six-membered ring homologs [of known (*Z*)-azo structures] than of typical (*E*)-acyclic azo compounds (63JA2752; 69JA3226; 72TL4565). In the case of the 3,8-diaryl derivatives of **40**, it was shown that the two isomeric compounds prepared from diazacyclooctadienes **28** (X = H$_2$) (see Section II,A,2) differed in their stereochemistry at carbons 3 and 8 and not about the azo bond (69JA3226). Also the rate of double-bond isomerization of **40** (R = Ar) to cyclic hydrazone **64** ($R = R^2 = Ar$, $R^1 = H$) was more consistent with a (*Z*)- than an (*E*)-azo linkage.

When the (*E*)-isomer (**12**; $R^1 - R^4 = H$) was finally prepared (Section II,A,3,a) its dipole moment, IR, UV, and NMR spectra were distinguishable from (*Z*)-isomer **40** (R = H) (72TL4565). NMR studies were particularly relevant. Shift reagent experiments showed the europium complex coordinated more strongly with (*Z*)-compound **40** (R = H) than with (*E*)-isomer **12** ($R^1 - R^4 = H$) (72TL4565). Also, the (*E*)-isomer and its 3,8-disubstituted derivatives (**12**) showed nonequivalent methylene and methine protons, in contrast to (*Z*)-isomers **40** (81JOC303).

Quinkert and co-workers, using X-ray diffraction, showed that (*E*)-compound **12** ($R^1 = R^4 = Ph$, $R^2 = R^3 = H$) had a slightly distorted crown shape. The configuration of epimeric **12** ($R^1 = R^3 = Ph$, $R^2 = R^4 = H$) was established by nonempirical NMR. *cis* and *trans*-Diphenyl (*Z*)-isomers **40** (R = Ph) were adequately characterized as photoisomers of corresponding (*E*)-compounds **12** (76CB518) (see Section II,A,3,c).

Kao and Huang employed molecular mechanics and *ab initio* molecular orbital (MO) theory to determine the structure, energies, and conformations of the (*Z*)- (**40**, R = H) and (*E*)- (**12**, $R^1 - R^4 = H$) isomers. For the (*Z*)-isomer, an unsymmetrical conformation (**67**) is favored by at least 6 kcal over symmetrical chair, boat, and twist forms, which suffer from ethane-type and other H−H repulsions. In the case of (*E*)-isomer, the twist

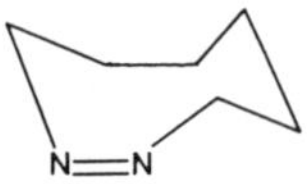

67 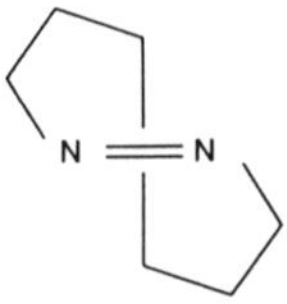   68

form (**68**) was favored by almost 6 kcal over the chair and was in agreement with the aforementioned X-ray diffraction results (76CB518). These findings were very similar to those for the isomeric cyclooctenes (79JA5546).

Dibenzo-1,2-diazocine **10** was found to have an NMR spectrum consistent with a tub-shaped structure similar to the carbocyclic analog. Similarities were also found in the UV spectra. 10π-Electron compound **11** evinced enhanced conjugation based upon a comparison of its UV and NMR spectra with those of **9** and **10** and upon the comparison of the spectra of **11** with its N-methyl and N,N'-dimethyl derivatives (69JOC3237).

Heilbroner and co-workers found the UV spectra of N-protonated 1,2-diazocine **8** and N-protonated (Z)-azobenzene were very similar. This and other spectroscopic and chemical evidence indicated that upon protonation (Z)- and (E)-azobenzene retain their configuration and are classical rather than bridged or nonclassical cations (60HCA1890). Haselbach and co-workers studied protonated **8** (and other azo compounds) using electron spectroscopy for chemical analysis (ESCA) and concluded that the cation has classical structure **69** as opposed to a nonclassical structure (**70**) (72HCA705).

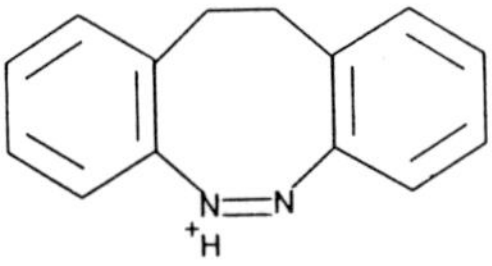

69 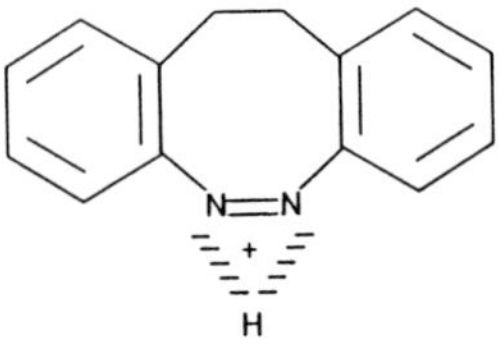   70

Diazocine **8** was found to form complexes very slowly or not at all with palladium chloride or silver nitrate. (E)-Azobenzene behaved similarly, and since both compounds have similar pK_a values, the ease of complex formation with the ligand was concluded to be a function of its electron-pair donating (basic) character. The N-oxide of **8** seemed to behave similarly to **8** itself, perhaps indicating lack of any coordination to oxygen (73JHC43). Compounds **12** ($R^1 - R^4 = H$) and **40** ($R = H$) were among a number of cyclic azo compounds that gave a reliable and selective color test on thin-layer chromatograms with copper(I) chloride in acetonitrile (87CB251).

The simulated electron spin resonance (ESR) spectrum of the radical cation generated from **71** indicated there was considerable twist about the N-aryl bond, as compared to other compounds such as a dibenzocinnoline derivative, and thus little delocalization of the unpaired electron into the benzene rings (78JPC1152). Proton and ^{13}C-NMR spectra of the novel peripheral 14π-electron bis(cyclopropeno)cyclopentadieno-1,2-diazocine **58** (see Section II,A,3,c) indicated that important ground-state contributions were made by polarized structure **72** (84TL4223).

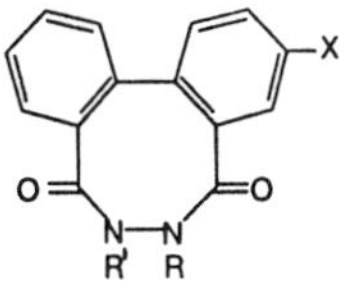

41

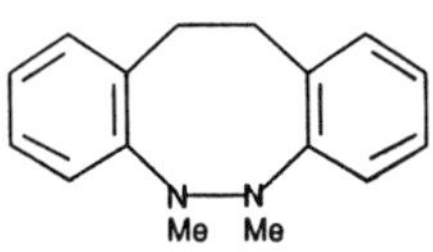

71

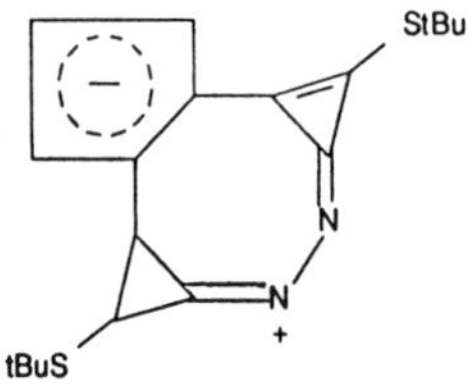

72

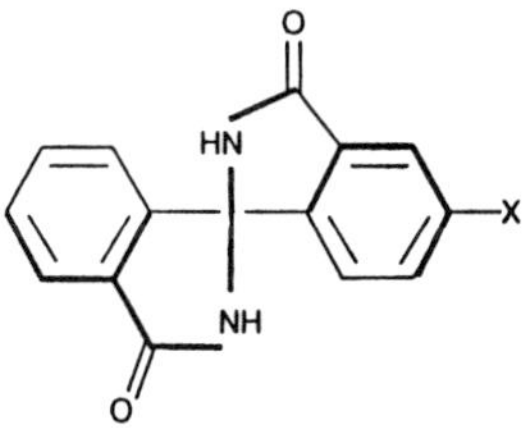

73

Beaven and Johnson compared the UV spectra of bridged biphenyl **47** (R = Me) with those of the acyclic benzylidene azine and benzylidene imine and thus demonstrated a serious inhibition of any biphenyl-type conjugation or of any similar interaction between benzene rings and the imine bond or through the azine bridge (57JCS651). Riggs and Verma found that at temperatures up to 140°C, the methyl singlet of **41** (R = R' = Ac, X = H) and the methylene AB quartet of **41** (R =

R' = Ch$_2$Ph, X = H) remained sharp, thus indicating R and R' equivalency. This was consistent with a virtually strainless conformation (**73**) (perhaps similar to that for **47**), where the two benzene rings make a dihedral angle of 60° with each other, and the two amidic moieties are in a near perpendicular arrangement. As expected, benzene substitution (**41**, **73**, X = NO$_2$) rendered the *N*-substituents magnetically nonequivalent (70AJC1913). Vasserman and Miklukhin found that **41** (R = R' = X = H) exhibited chemiluminescence, but not as much as the cyclic hydrazides of 4,5- and 1,2-dihydrophthalic acids (40JGU202).

Wawzonek and Shradel reinvestigated the NMR spectrum of **74** (R = H) [previously prepared by Leclerc from **63** or **64** (R = Ph, R^1 = Me, R^2 = H) (73BSF2029); see Section II,A,3,d]. A singlet was obtained for the *N*-methyl protons of **74** (R = H) instead of the previously reported doublet (73BSF2029), thus indicating a single conformation for the compound at room temperature. However, *N*-acetyl compound **74** (R = Ac)[prepared by catalytic reduction of **64** (R = Ph, R^1 = Me, R^2 = H) in the presence of acetic anhydride] evinced proton NMR peaks indicative of the presence of three stable conformations. Although a total of eight possible crown and chair forms are possible, the one thought to predominate was crown conformation **75,** where all three substituents assumed equatorial positions (80JOC5216).

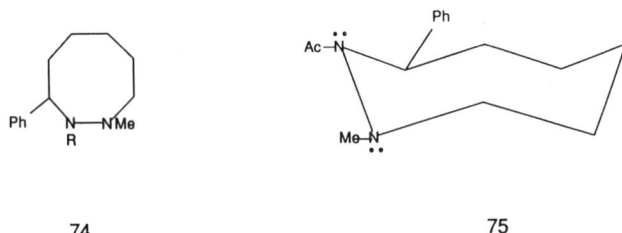

<table>
<tr><td>74</td><td>75</td></tr>
</table>

The molecular structure of diazocine **31** (R = R' = H) (see Section II,A,2) was determined by X-ray diffraction to be a tub conformation with no extensive π-conjugation and a very short C–H---O intramolecular distance [76AX(B)2314]. The mass spectra of a number of diaryl-1,2-diazocines **28** (X = H$_2$) indicated no migration of aryl groups (79MI1).

C. REACTIONS

1. *Eight-Membered Ring Preserved: Annelations*

Monothiocarbamoyldiazocines **76** were converted into diazocino-dithiourazoles **77** by reaction with carbon disulfide (76JAP76-86489). Re-

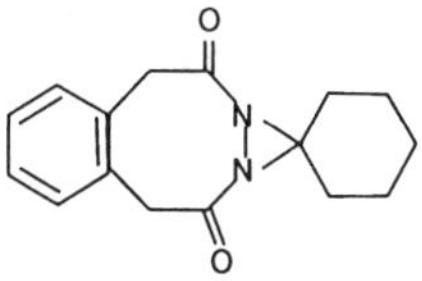

44 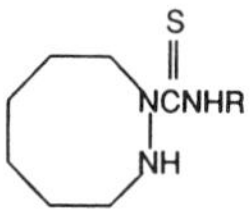   76 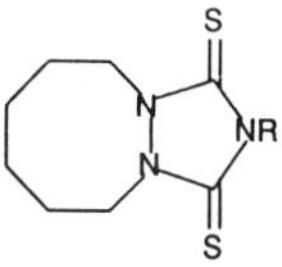   77

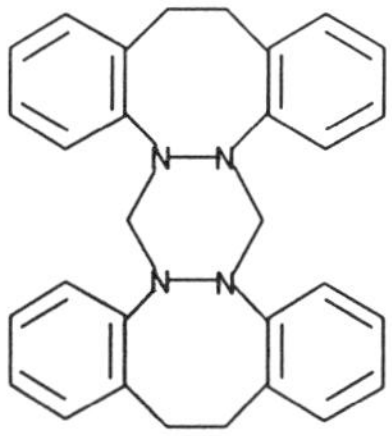

78 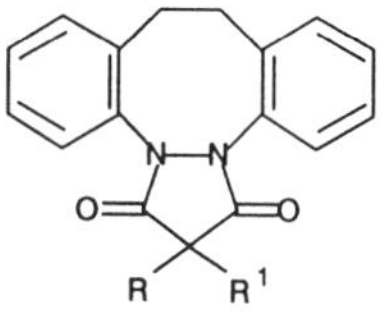   79 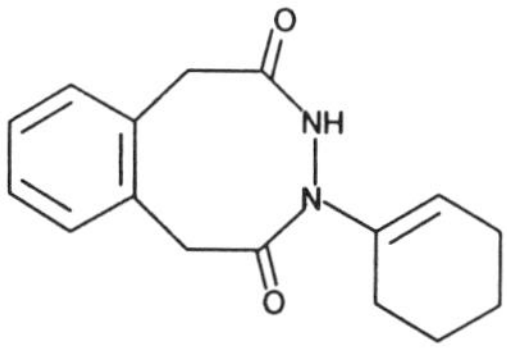   80

action of dibenzodiazocine **9** with *n*-butyllithium and methyl sulfate resulted in formation of dimer **78** (75MI4). Condensation of **9** with malonyl dichlorides afforded hydrazides **79** (65USP3170929).

2. *Eight-Membered Ring Preserved: Other Reactions*

When spirodiazirinodiazocine **44** (see Section II,A,3,b) was heated in benzene containing triethylamine, cyclohexenyldiazocine **80** resulted (76JOC3229). There have been a number of reports of photochemical (*Z*)- **(40)** and (*E*)- **(12)** isomerizations in 1,2-diaza-1-cyclooctenes (72TL4565; 76CB518; 81JOC303; 85CJC1829).

Eamples of reductions in 1,2-diazocines are reduction of 1,2-diaza-2,8-cyclooctadienes **28** to 1,2-diaza-1-cyclooctenes **40** (56RTC1159; 59JA217; 76CB518) and to 1,2-diazacyclooctane **39** (56RTC1159); reduction of **40** to **39** (69JOC3237); and lactam–amine transformation (73BSF2029). Oxidations include N—N to N=N conversions using mercury(II) oxide (56RTC1159; 63JA2144; 69JOC3237; 72TL4565; 76CB518; 81JOC303; 85CJC1829) and oxidation of an azo monoxide to an *N,N'*-azo dioxide (80JOC4597).

Acylations at nitrogen include *N*-acetylation using acetic anhydride (82H1595) and *N*-mesylation (84TL4223), as well as the *N*-carboxyalkylation of **81** and reaction of **82** with isocyanates or isothiocyanates to afford **83** (Scheme 2) (76JAP76-65757). Alkylations include methylation to yield *N*-methyl and *N,N'*-dimethyl-1,2-diazocines (69JOC3237). An in-

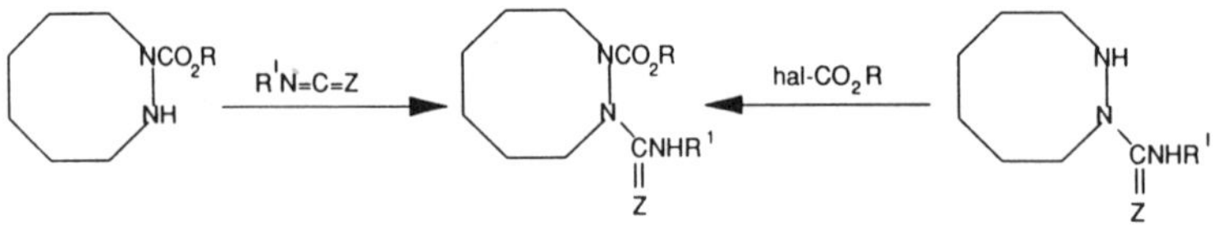

SCHEME 2

teresting deacylation–oxidation sequence beginning with **84** resulted in copper complex **85** (Scheme 3) (72JOC1851). For a discussion of other 1,2-diazocine metal complexes, see Section II,B. Elimination reactions include the conversion of **28** (X = H_2) via **86** ($R^1 = R^2 = H$ or Cl) (see Section II,A,3,c) to the completely conjugated, 1,2-diazocine **20** (R = Ph, $R^1 = R^2 = H$ or Cl) (82CL1579).

Nucleophilic substitution on diazocinyl carbon can be accomplished on **20** ($R^1 = R^2 = Cl$, R = Ph) by reaction with various metal carboxylates (86BCJ1087), arylthiols (87BCJ335), and benzenesulfinate (87BCJ343) to afford **20** [R = Ph; $R^1 = R^2 = $ acyloxy (86BCJ1087), $R^1 = R^2 = $ ArS (87BCJ335), $R^1 = R^2 = PhSO_2$ (87BCJ343)]. Other bissulfonyl derivatives of **20** were prepared by oxidizing the arylthio derivatives (87BCJ343). Also, **20** (R = Ph, $R^1 = R^2 = $ Cl) was condensed with various imides. For example, use of phthalimide gave **20** (R = Ph, $R^1 = R^2 = $ phthalimido) and **20** (R = Ph, $R^1 = R^2 = $ succinimido). Also prepared were **20** (R = Ph, $R^1 = $ AcO, $PhCO_2$, SPh; $R^2 = $ phthalimido) (87BCJ731).

3. Ring Contractions

Thermolysis of **20** (R = $R^1 = R^2 = $ H) afforded benzene and pyridine, whereas photolysis of the molecule gave only benzene (79JOC1264). Diazocines **20** (R = Me, Ph; $R^1 = R^2 = $ H, Me), although not isolated, were probably intermediates in the thermal isomerization of 3,4-diazabicyclo-[4.2.0]octatrienes **87** (R = Me, Ph; $R^1 = R^2 = $ H, Me) to substituted ben-

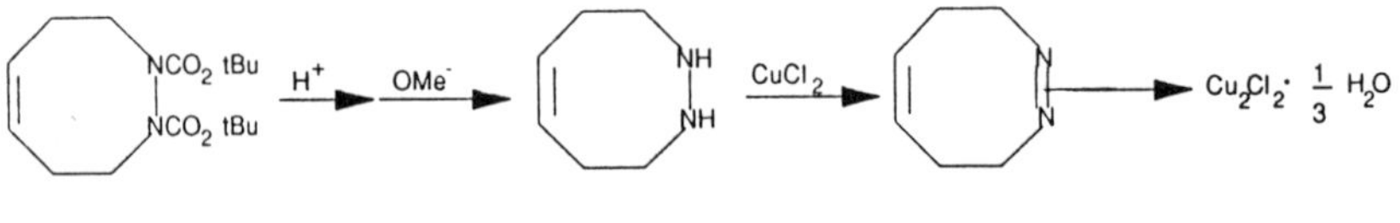

SCHEME 3

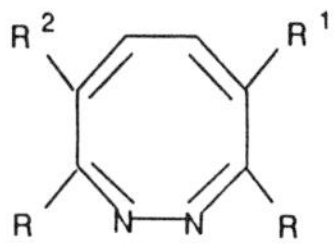

20 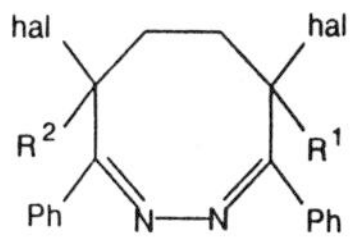   86

zenes. Dehydrobromination of **86** ($R^1 = R^2 = H$) afforded, in addition to diazocine **20** ($R = Ph$, $R^1 = R^2 = H$), the ring-contracted cyclobutapyridazine **88**. When **20** ($R = Ph$, $R^1 = R^2 = H$) and **20** ($R = Ph$, $R^1 = R^2 = Cl$) were heated in refluxing toluene, 2-phenylpyridine [plus a trace of o-terphenyl (**89**, $R = H$)] and 2-phenyl-3,6-dichloropyridine were formed. If the refluxing toluene was wet, an additional compound, 2-benzoyl-5-chloro-6-phenylpyridine, was isolated. These results, plus the fact that when **20** ($R = Ph$, $R^1 = R^2 = Cl$) was heated in refluxing benzene containing six equivalents of silver acetate, acetoxy-substituted diazocines **20** ($R = Ph$, $R^1 = Cl$, $R^2 = OAc$) and **20** ($R = Ph$, $R^1 = R^2 = OAc$) were formed, strongly supporting the postulated ring contraction of the diazocines to **90** and **91**, which, although not isolated, would be rationally expected to produce the aforementioned diazocine thermolysis products. On the other hand, irradiation of diazocines **20** ($R = Ph$, $R^1 = R^2 = H$ or Cl) afforded o-terphenyls **89** ($R = H$ and Cl) as the only products (82CL1579).

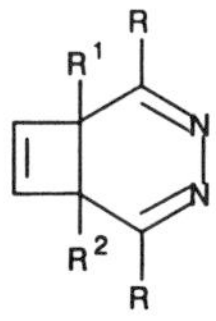

87 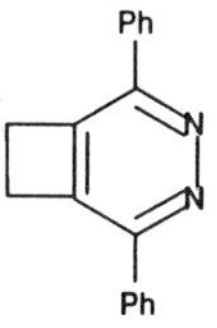   88 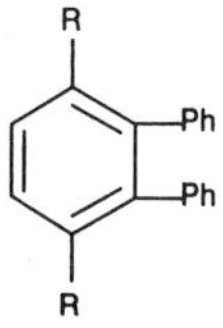   89

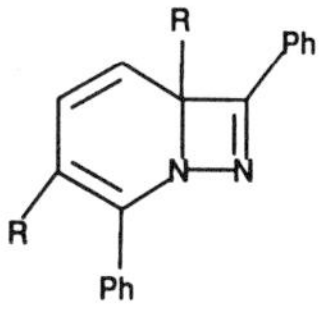

90 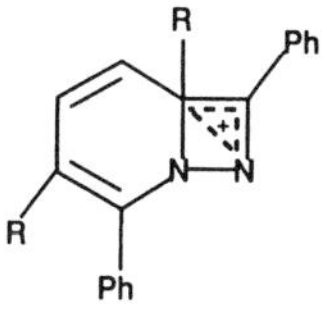   91

The same group thermolyzed **20** ($R = Ph$, $R^1 = R^2 = Cl$) (86BCJ1087), **20** ($R = Ph$, $R^1 = Cl$, ArS; $R^2 = ArS$) (87BCJ335), and **20** ($R = Ph$, $R^1 = Cl$, $R^2 = SO_2Ar$) (87BCJ343) in refluxing toluene to produce pyridines **92**

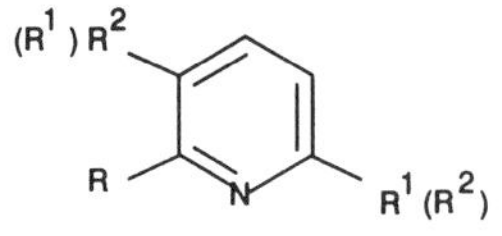

92 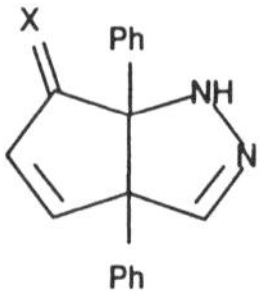   93

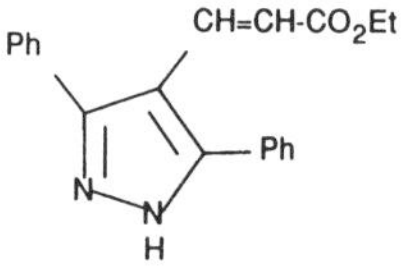

94

(86BCJ1087; 87BCJ335; 87BCJ343) with extrusion of benzonitrile (86BCJ1087; 87BCJ335). Thermolysis of **20** (R = Ph; R^1 = imido; R^2 = H, imido, OAc, O_2CPh, SPh) proceeded similarly but at higher temperatures. Reaction of **20** (R = Ph, R^1 = H, R^2 = imido) afforded **20** (R = Ph, $R^1 = R^2$ = imido) in addition to pyridines (87BCJ731). On the other hand, hydrolysis of **20** (R = Ph, R^1 = Cl, SPh; R^2 = OAc, phthalimido) gave ring-closed bicyclic intermediate **93** (X = O, NH), which then cleaved to form pyrazole **94** (87CL157).

Attempted dehydrobromination of dibromodiazocine **95** with sodium methoxide resulted in transannular reaction and ring contraction to give **96**. Use of the more bulky potassium *tert*-butoxide instead of methoxide resulted in debromination of **95** to yield **84** (72JOC1851). Treatment of dibenzodiazocine **9** with dilute acids afforded spirotetrahydroisoquinoline **97**, representing the first isolated stable *o*-semidine rearrangement intermediate (73JHC423). Sublimation of benzodiazocine **31** gave tricyclic **98** (76TL1569). Although heating of diaziridinodiazocine **44** in benzene containing a base left the diazocine ring intact (see Section II,C,2), heating in the absence of base yielded benzazepine **99** (76JOC3229).

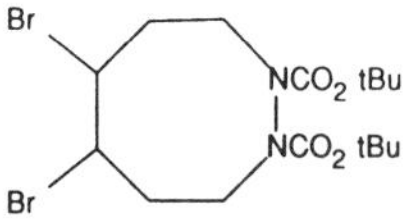

95 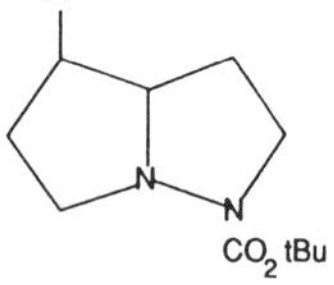   96

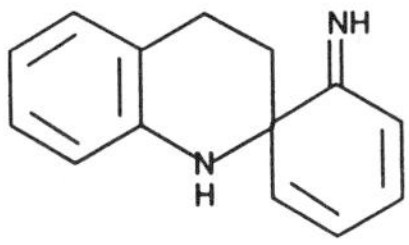

4. Ring Openings

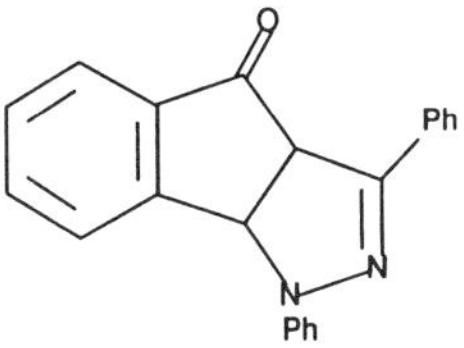

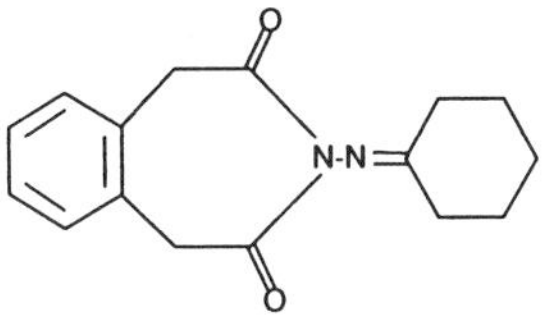

Two reductive ring-openings of 1,2-diazocines have been reported. The cholestane analog containing a diazocine B-ring present as a cyclic hydrazone (60) was hydrogenated to afford *seco*cholestane 100 (63LA160). Wolff–Kishner reduction of the azine moiety in diazocinol 51 (X = H) yielded a mixture of unsaturated hydrocarbons (101) [77IJC(B)867].

The direct irradiation of 1,2-diazocines, 40 [R = H, Me (isom. mix., Ph-(*trans*)] produced disproportionation (102) and coupled (103) products. The ratio of 102/103 increased with increasing amounts of nonketonic photosensitizer, indicating the intermediacy of a relatively long-lived 1,6-diradical in which more time is allowed for intramolecular hydrogen transfer (70JA4922).

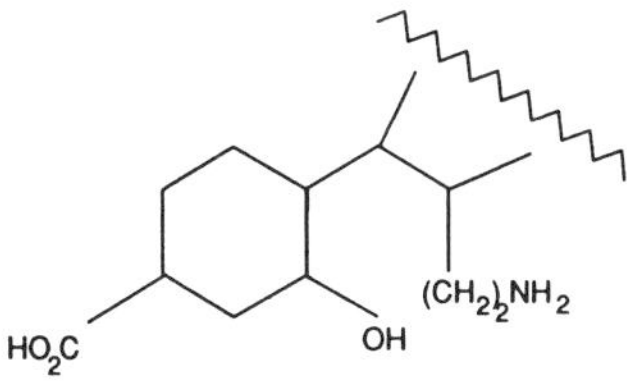

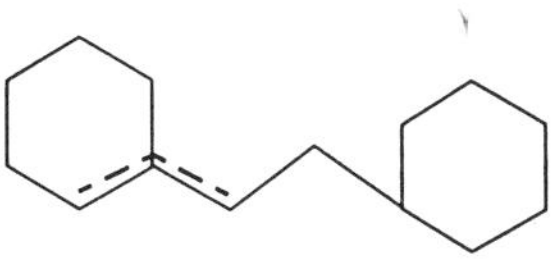

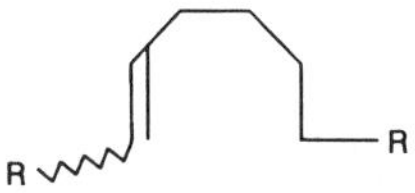

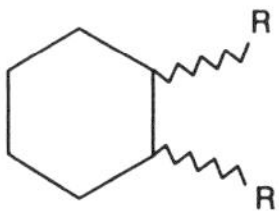

D. Applications

A number of N,N'-diacyl and thioacyl derivatives of 1,2-diazocines [e.g., **77** (76JAP76-86489) and **83** (76JAP76-38425, 76JAP76-65757; 77JAP77-83552)] have been found to be herbicides. 1,2-Diazocine-3-one **18** (see Section II,A,1), as well as lower and higher ring homologs, were useful as antihypertensives (84USP4465679). Dibenzodiazocine **41** tested as an analgesic (75MI1), while malonylhydrazides **79** were reported to have antibacterial properties (65USP3170929). The parent compound (**79,** R = R' = H) has unspecified pharmacological activity (63BRP940165).

Diazocine **12** (R = R' = Me) was among a number of nonaromatic azo compounds that, because they underwent an exothermic reaction upon application of heat, were useful in providing heat amplification in thermal transfer printing (84EUP113017). Bisazo compounds **56** (R = Me, Et; $R^1 = N_3$; $R^2 = OH$) (see Section III,A,3,c) were used as initiators in the preparation of radial block copolymers having four or more arms (these compounds were useful as impact modifiers for thermodeformable plastics) (85USP4532301, 85USP4556512).

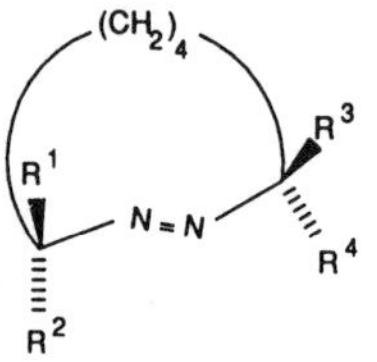

12

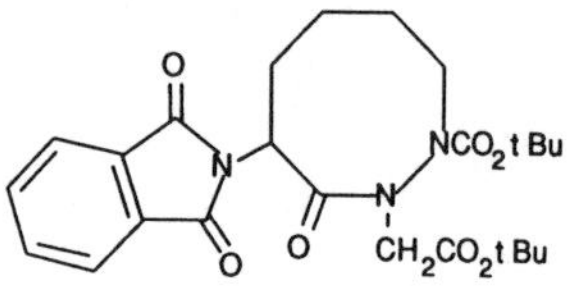

18

III. 1,3-Diazocines

A. Cyclic Ureas, Thioureas, and Derivatives; Carbodiimides

Since the most common group of 1,3-diazocines is related to cyclic urea **104** (called, for example, perhydro-1,3-diazocine-2-one or N,N'-pentamethyleneurea), a separate section is devoted to the synthesis, chemistry, and applications of **104,** substituted **104,** and related molecules such as thiourea **105,** guanidines **106,** and carbodiimide **107.**

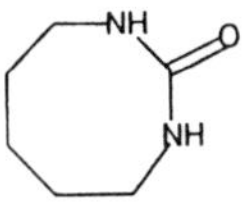

104 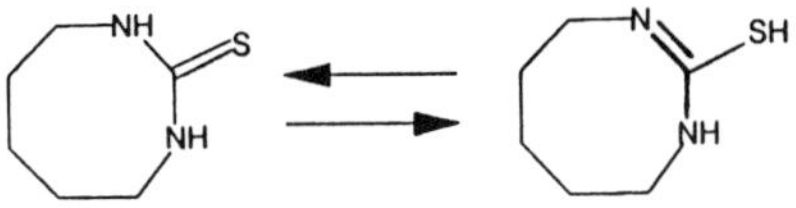   105

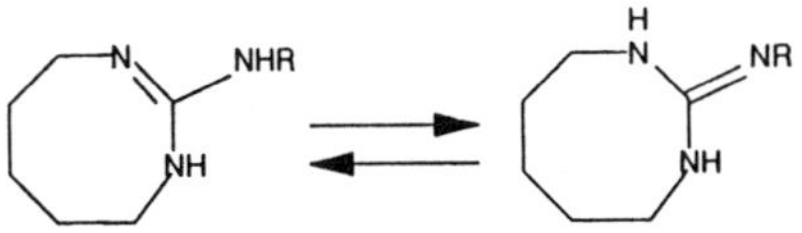

106 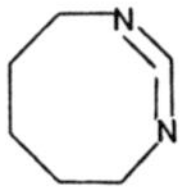   107

1. Preparative Methods

a. *Via Cyclization.*   Klein *et al.* prepared ω-guanidino and α-ureido-α-ketoacids **108** (X = NH, O; R = H, Me) from the corresponding α-amino-acids via oxazolidines. Compound **108** (R = Me, X = NH) was then cyclized to 1,3-diazocine **109**, whereas **108** (R = H, X = NH, O) was converted into diazabicyclic compounds (83LA1623). Compound **104** itself

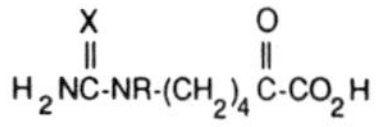

108

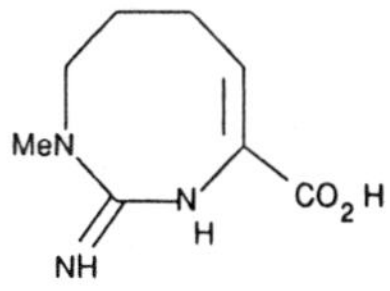

109

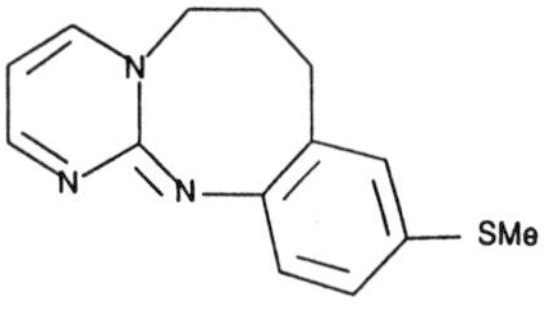

110

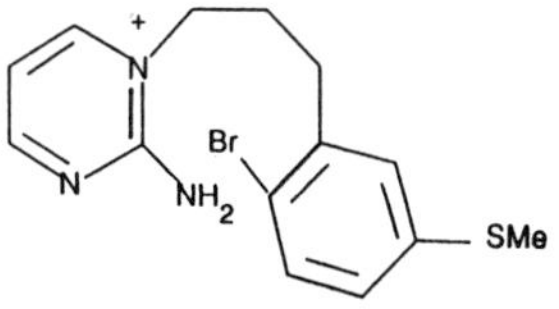

111

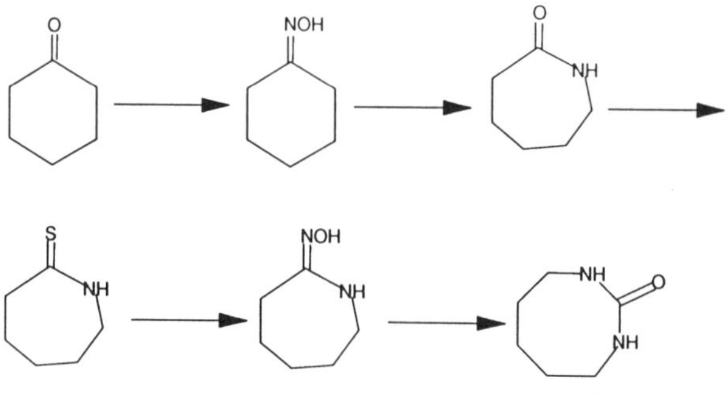

112   104

SCHEME 4

has been prepared by the action of water on pentamethylenediisocyanate (57JA4358, 57NKZ1416). Fused pyrimido compound **110** has been made by treating **111** with potassium carbonate and copper bronze (77USP4001233).

b. *Via Ring Expansion.* Schmidt-type reactions of lower ring homologs have been common methods used to prepare eight-membered cyclic ureas. Behringer and Meier, beginning with cyclohexanone, employed two ring expansions to prepare **104** via caprolactam oxime **112** (Scheme 4) (57LA67). Le Berre and co-workers also prepared **104** by aqueous base treatment of the *O*-tosylate of **112**. The same tosylate, when reacted with alkoxides, yields alkoxyisoureas **113** (R = Me, Et, *n*-Bu, *n*-hex, *n*-oct, *n*-dodecyl) (71BSF3245). Ulrich and co-workers found that Lossen rearrangement of the *O*-mesylate of **112** in base gave **104** in yields better than these found in previous syntheses (78JOC1544).

Schmidt reaction of tetralone **114** gave, in addition to the expected isomeric benzazepinones, an unusual product of further ring expansion: the eight-membered cyclic urea **115** (86JHC975). Cartier and co-workers treated the *E*-homovincamone derivative **116** with thionyl chloride to isolate **117** and **118** as major and minor products, respectively (76BSF1962).

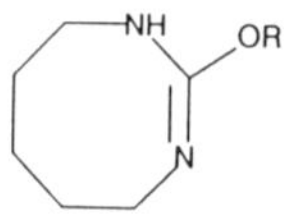

113

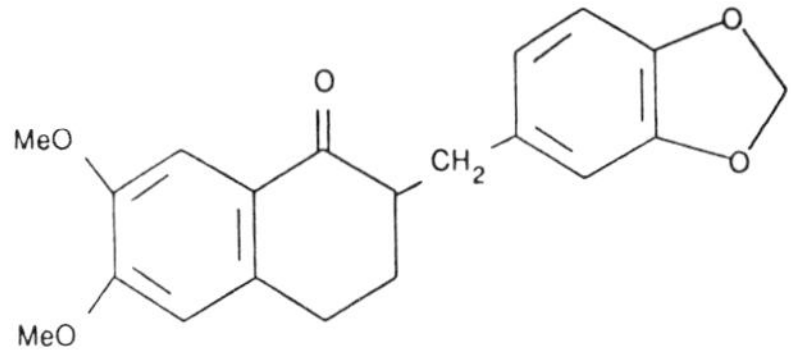

114

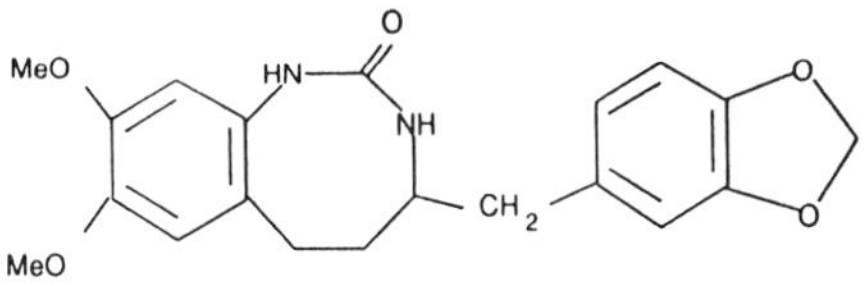

115

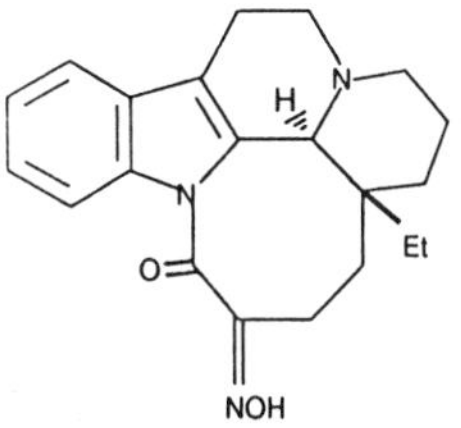

116

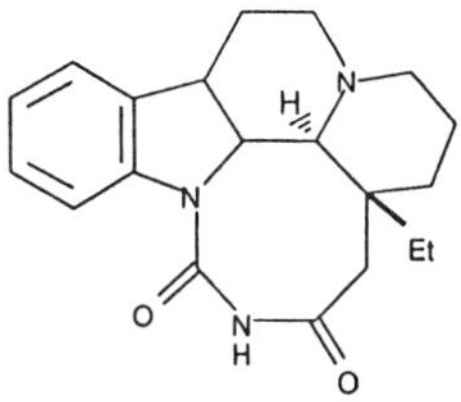

117

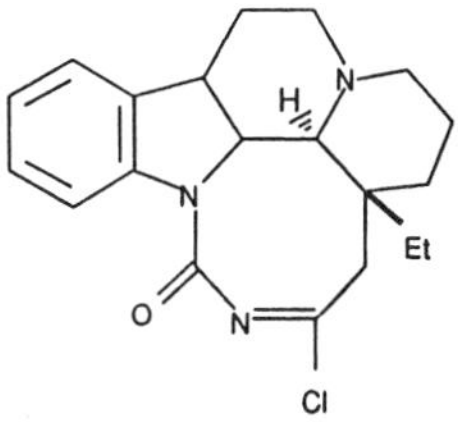

118

Rose-Bengal sensitized photooxygenation of cyclic tryptophan derivative **119** gave 1,3-diazocine-2,6-dione **120** (76H53). Murato and coworkers reductively cleaved β-carboline derivative **121** and isolated the novel diazocinoindole **122** (77CPB1559).

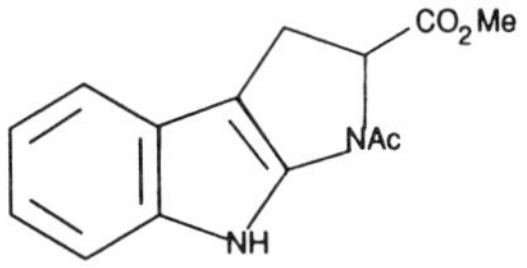

119

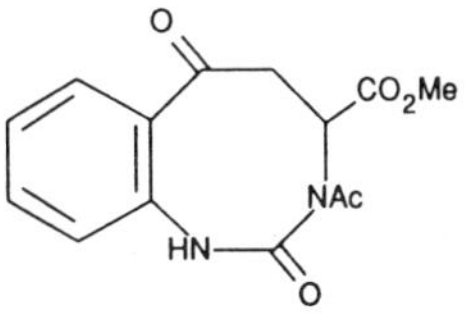

120

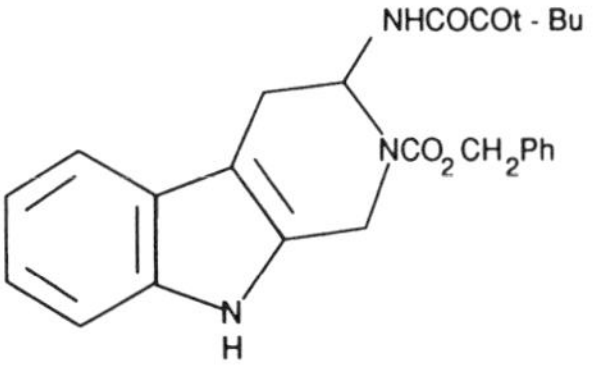

121

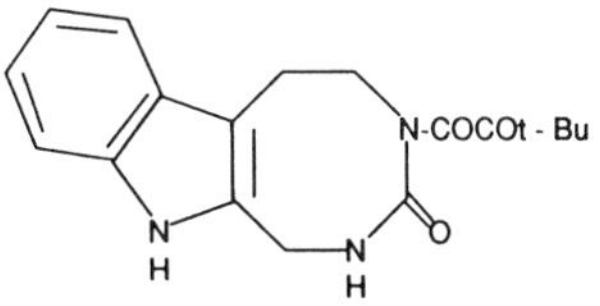

122

Swenton and co-workers showed that uracyl-alkyne photoadducts **123** [R^1 = H, R^2 = n–C$_5$H$_{11}$ (82TL4207); R^1 = F, R^2 = n–C$_3$H$_7$ (83JOC2337); R^1 = SiMe$_3$, R^2 = n–C$_3$H$_7$ (83JOC2337)] yield the corresponding diazocines **124** on treatment with trimethylsilyl chloride/hexamethyldisilazine followed by silica gel. The latter compounds afforded pyridones **125** on mixing with potassium *tert*-butoxide. These findings were taken as strong evidence for the intermediacy of diazocines **124** in the base-catalyzed fragmentation of **123** to give **125**, a mechanism that was strongly suggested by ^{15}N and ^{13}C labeling and kinetics studies and ruled out a Dewar pyridone-type intermediate (82TL4207; 83JOC2337).

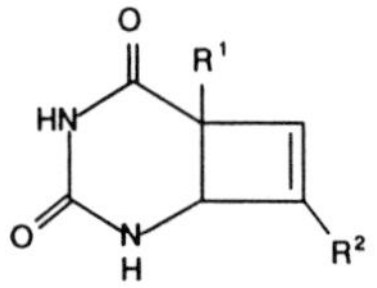

123

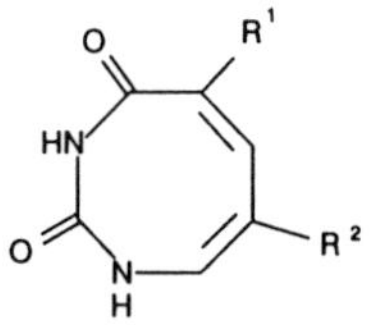

124

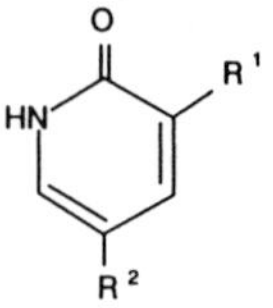

125

c. *Via Condensations.* Berke and Rosen reported the 1 : 1 reaction of glutaraldehyde and urea in alkaline solution afforded diazocine **126,** although no structure proof was offered. This material exhibited phar-

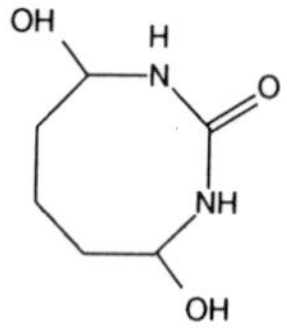

126

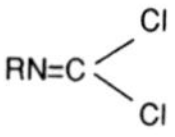

127

macological activity (See Section III,A,2,a) (84USP4454133). Condensation of dichloromethyleneimino compounds **127** (R = PhCO) or **127** (R = 2,6-Cl$_2$C$_6$H$_3$) with pentamethylenediamine afforded diazocines **106** (R = PhCO) or **106** (R = 2,6-Cl$_2$C$_6$H$_4$), respectively (67CB2569). Pentamethylenediamine and its *N* and *C*-substituted derivatives, when treated with carbon disulfide, afforded thiourea **105** [61USP2988478; 76IJC(B)773] as well as *N* and *C*-substituted derivatives (61USP2988478).

Dipyrazolodiazocines **128** (R = R^2 = H, R^1 = Ph or PhCH$_2$) were obtained by treating bisurea **129** (R = H) with pyrazoles **130** (R = Ph or PhCH$_2$, R^1 = NH$_2$, R^2 = H). Similarly, reaction of **129** (R = Ph) with **130** (R = Ph, R^1 = H, R^2 = NH$_2$) produced **128** (R = R^2 = Ph, R^1 = H) (74KGS233). A similar reaction occurred when bisthiazoles **131** (R = SMe, OMe) were condensed with carbon disulfide to give **132**. Reaction of **131** (R = SMe) with triethyl orthoformate, acetate, and propionate yielded 1,3-diazocines **133** (R = SMe; R' = H, Me, Et) (86JHC1435).

128	RCH(NHCONH$_2$)$_2$ **129**	**130**

131	**132**	**133**

Kempter and co-workers prepared benzodiazocines **134** (R = R^1 = H; R^2 = H, Me) from **135** (R = H, Me) by reduction and subsequent cyclocondensation with the appropriate carbonyl compound (75MI7; 87ZC36). For example, these workers condensed 3-(*o*-aminophenyl)propylamine with either phosgene or alkyl chloroformate to obtain **134** (R = R^1 = R^2 = H, X = O). Reaction of the same diamine (77MI5; 87ZC36) and substituted derivatives (77MI5) with carbon disulfide afforded thioureas **134** (X = S; R = H; R^1 = H, Me; R^2 = H, Me, Ph), while the same reaction using 3-(*o*-aminophenyl)propylureas gave **134** [X = S, R = R^1 = H, R^2 = CONHAlk(Ar)] (86EGP235452).

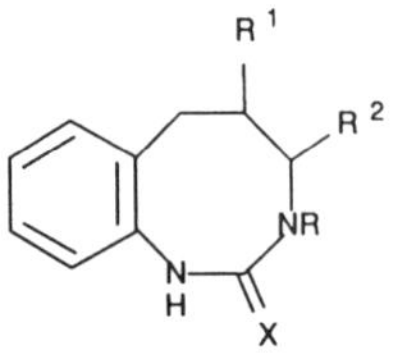

R^1

R^2

NR

N
H   X

134

(CH$_2$)$_2$C-R
           ||
           NOH

NO$_2$

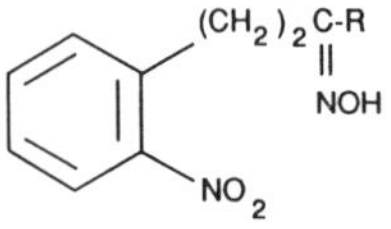

135

2. Chemistry and Applications

a. *Eight-Membered Ring Preserved: Substitution at Nitrogen, Sulfur, and Carbon.* Dihydroxyurea **126** evinced antibacterial activity (84USP4454133). Benzo-fused ureas and thioureas **134** (X = O, S) were useful as potential herbicides (75MI7), while compounds **134** [X = S, R = R^1 = H, R^2 = CONHAlk(Ar)] specifically were used as intermediates for bioactive compounds (86EGP235452).

Danchev and co-workers, by treating, *N,N'*-pentamethyleneurea (**104**) with sodium nitrite and acid followed by reduction with zinc, obtained azomethine **136** (Ar = *p*-O$_2$NC$_6$H$_4$, 3-OMe, 4-OHC$_6$H$_3$, 5-O$_2$N-2-furyl). The nitrofuryl derivative was found to be active against *Escherichia coli* (70MI1). A number of *N*-aroyl and *N,N'*-diaroyl derivatives of **104,** obtained by acylation of **104** with ArCOCl [Ar = Ph, 2-AcOC$_6$H$_4$, 4–O$_2$NC$_6$H$_4$, and 3,4,5–(OMe)$_3$C$_6$H$_2$], were found to exhibit sedative and muscle-relaxant activities (74AP673).

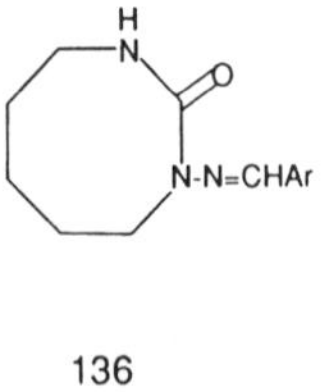

H
N   O

N-N=CHAr

136

HOCH$_2$ O N

NH
O

HO   OH

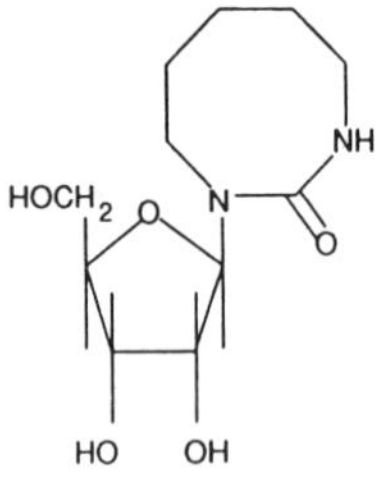

137

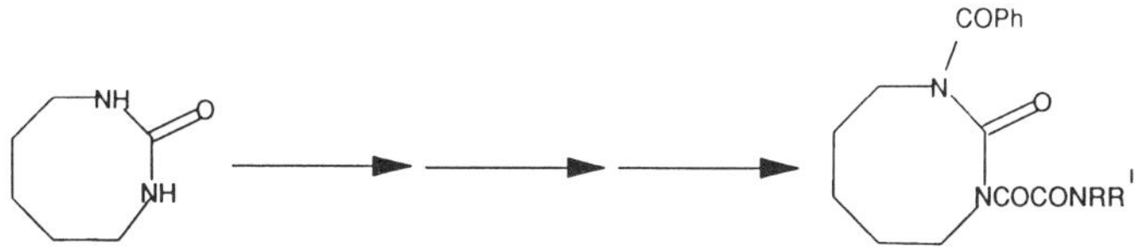

104   138

SCHEME 5

Liu and co-workers treated the N,N'-bis(trimethylsilyl) derivatives of **104** with 2,3,5-tris-O-benzoyl-D-ribosyl bromide. Debenzoylation of the intermediate product afforded β-D-ribofuranosyl-N,N'-pentamethyleneurea nucleoside **137**, found to be a cytidine deaminase (CDA) inhibitor against mouse kidney enzyme (although less effective than the seven- or six-membered ring homologs) (81JMC662).

Richter *et al.* prepared N-benzoyl-N'-(N-aryloxamoyl) ureas **138** (R = Ph; R' = H, Me) by using a sequence of reactions involving benzoylation, reaction with oxalyl chloride, and, finally, reaction with anilines (Scheme 5) (78JOC4150). Karparov *et al.* found, in broad-spectrum antiviral screening, that Mannich bases **139** (m = 4, 5) showed some activity, but that the parent urea (**104**) and thiourea (**105**) did not (84AF9). However, a later study revealed the two Mannich bases were inactive toward alphavirus models (86MI1).

Reaction of thioether **140** (R = Me) with amines gave **106.** Reaction of isothiourea (thiourea) **140** (R = H) (also **105**) with alkyl halides yielded **140** (R = Alk). Both the latter compounds (61USP2988478) and **106** (65USP3219522) and their N and C ring-substituted derivatives are useful in controlling plant and animal phytopathogenic agents; (**106**) is also an anthelmintic agent (65USP3219522). Substitution of this type is very common, another example being the conversion of **132** to **133** (R' = SMe, NHR) (86JHC1435).

Reaction of isocyanates with diazocines **106** (R = H) afforded amidinoureas **141** (R = Me, Et). These compounds, along with others that have 2,6-disubstituted benzene residues, evinced pharmacological activity

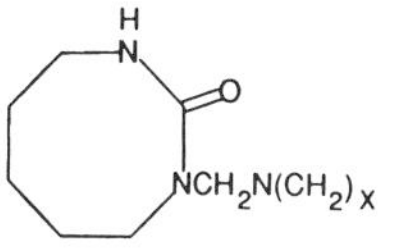

139 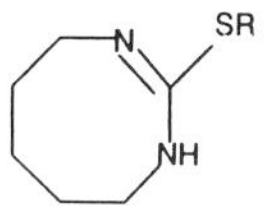   140

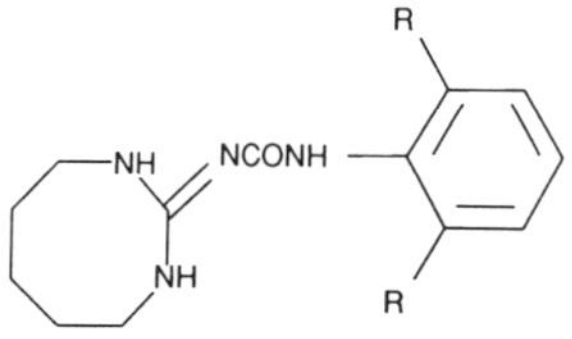

141

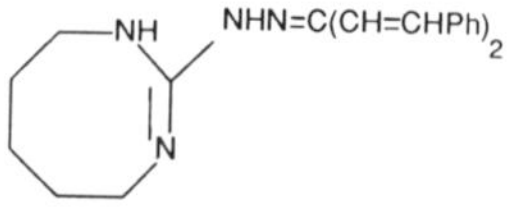

142

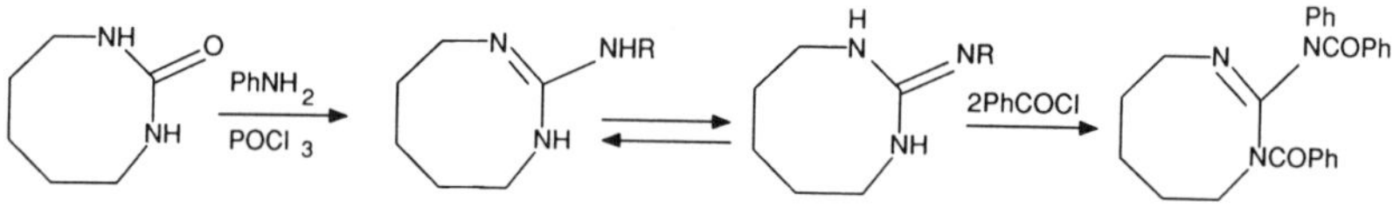

104

106
(R=Ph)

143

SCHEME 6

upon the gastrointestinal tract (78AF1435). Compounds **142** (benzene rings are variously H-, Alk-, OAlk-, hal-, and SAlk-substituted) found use as insecticides (87JAP6216921). Richter *et al.*, in a reaction sequence starting with **104** (Scheme 6) prepared guanidine **106** (R = Ph) using phosphoryl chloride and aniline. This intermediate underwent benzoylation to afford *N,N'*-dibenzoylated guanidine **143** (78JOC4150).

Treatment of a mixture of 2,6-di-*tert*-butylphenol and trimethylamine with a product prepared *in situ* from **104** and oxalyl chloride in acetonitrile gave quinonemethide **144** (84CZP212600). 2,6-Dichlorophenylamino-

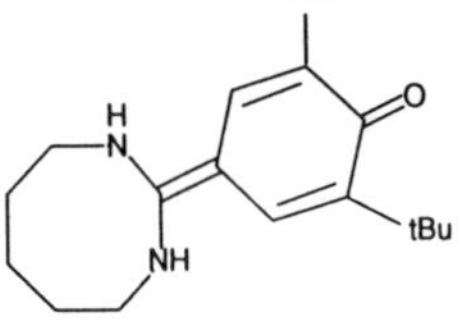

144

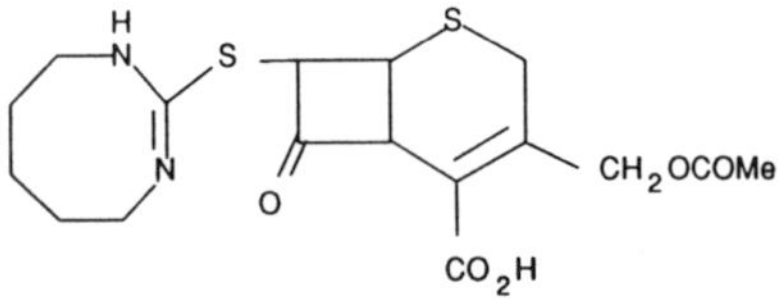

145

diazocine **106** (R = 2,6–C_6H_3) (see Section III,A,1,c) was found to have antihypertensive, sedative, vasoconstrictive, and/or secretion-inhibiting properties (68SAP6706503). Indolyl derivative **140** [R = 3-(1,2-dimethylindolyl)], prepared by oxidative coupling of 1,2-dimethylindole with **105** using iodine [81IJC(B)672], exhibited vasoconstrictive, hypotensive, and antihypertensive properties, the vasoconstrictive activity being the most potent (81MI5). Substituted cephalosporanic acid **145** was found to have antibacterial activity (72AAC54). Some of the guanidinothiazole derivatives (**106**) (R = 2-thiazolyl plus derivatives substituted on diazocine and/or thiazole rings) demonstrated anticholesteremic and antithrombotic activity (83GEP3220118).

b. *Eight-Membered Ring Preserved: Annelated Compounds.* Pyrimidobenzodiazocine **110** was found to be useful as a central nervous system (CNS) stimulant and muscle relaxant (77USP4001233). Reaction of **105** with the appropriate α-haloketone afforded thiazolodiazocines **146** (R = Alk, Ar; R^1 = Ar, CHAr, CArAr') (81USP4283334; 82USP4340734) and thiazino compound **147** [76IJC(B)773]. Coumpounds **146** were found to be diuretics (81USP4283334; 82USP4340734). Dehydrated compounds

146

147

148

149

148 [R^1 = Me, R^2 = CO_2Me; R^1R^2 = $(CH_2)_4$; R^1R^2 = $(CH_2)_2SCH_2$; R^1R^2 = $(CH_2)_2NAcCH_2$] were similarly prepared [76IJC(B)773]. Also prepared were compounds **148** [(R^1 = $(CH_2)_{0-2}$—Ar; (Ar = Ph or F, Cl, Br, Alk, NMe_2 subst. Ph); R_2 = Ph, or F, Cl, Br, Alk subst. Ph]. These compounds have diuretic properties (81USP4283334; 82USP4340734; 82USP4347248). Analogously prepared were **148** (R^1 = Ph, pClC$_6$H$_4$; R^2 = H); these compounds were evaluated for bacteriocidal and fungicidal properties (82JIC1170). These same properties were evaluated in thiazolidinones **149** (R = R' = H) and their salts, which are prepared by condensing **105** with chloroacetic acid and ethyl chloroacetate, respectively. Also evaluated was sulfamidophenylazo derivative **149** [R = H, R' = $N_2C_6H_4SO_2NH_2(p)$], 2-arylidene derivatives (**149**) (RR'are =CHPh and =CHthienyl), and the products of bromine addition to the latter two compounds. These compounds were found to be more active than higher-ring homologs, but less active than lower-ring ones(83JIC970).

Kempter and co-workers condensed thiourea **134** (X = S, $R^1 = R^2 = H$) with various α-halocarbonyl compounds to obtain benzothiazolodiazocines **150** [R^1 = H, Me, Et, n-Pr, n-C_7H_{15}, Ph; R^2 = Me, Ch_2Ph, Ph, var. aryl; R^1R^2 = —(CH$_2$)$_3$—, —(CH$_2$)$_4$—] (77MI5). The reaction of **134** (X = S, $R^1 = R^2 = H$) with maleic anhydride and DMAD afforded **151** (R = CH$_2$-CO$_2$H, R' = H) and **151** (RR' is =CHCO$_2$Me), respectively (87ZC36).

The S-methyl compound (**140**) (R = Me), when treated with aminoacetal **152**, yielded imidazodiazocine **153** (R = H). If, in addition, various electrophiles are added, such as aroyl isothiocyanates and aroyl chlorides, N-substituted compounds **153** (R = CONHC$_6$H$_3$R$_2^2$ (R^2 = Cl, Me); 3,4,5-(OMe)$_3$C$_6$H$_2$CO, and 5-NO$_2$-2-furyl] are obatined [76IJC(B)773]. Reaction of isoureas **113** (R = Me, Et) or isothioureas **140** (R = Me, Et) with hydrazine gave **106** (R = NH$_2$). The latter compound was condensed with pyruvate or 2-ketophenylacetate esters to afford diazocino-1,2,4-triazinones **154** (R = Me, Ph) (72LA112).

150

151

H$_2$ NCH$_2$ CH(OEt)$_2$

152

153

154

c. *Ring Contractions and Ring Openings; Polymerizations.* When **138** was heated in methanol, ring contraction occurred to yield parabanic acid derivative **155** (78JOC4150). As part of a study of the utility of cyclic ureas as masked isocyanates for potential polyurethane formation, the same workers heated the *N*-benzoyl derivative of **104,** upon which ring opening occurred to give novel benzamidoisocyanates **156** (R = NCO). Heating of this material in methanol solution afforded carbamate **156** (R = NHCO$_2$Me) (78JOC1544). Reaction of **104** with isophthaloyl chloride produced bisurea **157** which, upon heating, produced bisisocyanate **158** (R = NCO) (not isolated). This material could be trapped with methanol as the biscarbamate (**158**) (R = NHCO$_2$Me) (78JOC1544). However, **157** has not been employed to make polyurethanes (86PC1), although the next lower homolog, the bis(*N,N'*-tetramethylene)urea, has been so used (81MI1).

Diazocinyl-substituted phthalic anhydride **159** (R is inert, m = 0–3), prepared from **104** and the appropriately substituted acid chloride, underwent self or copolymerization to give polyamide–polyimides (83USP4391751).

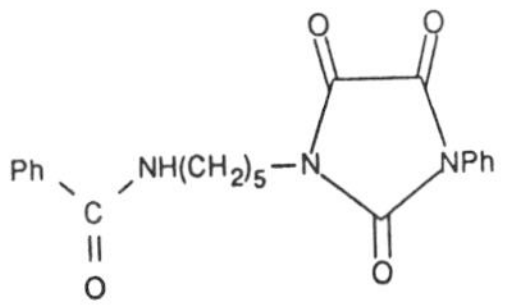

155 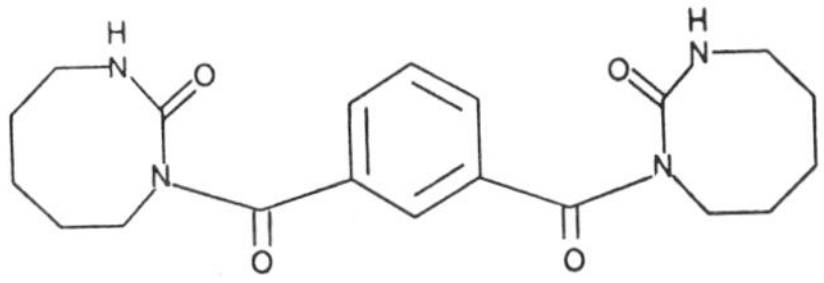   157

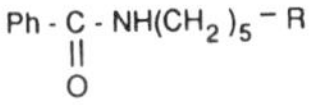

156

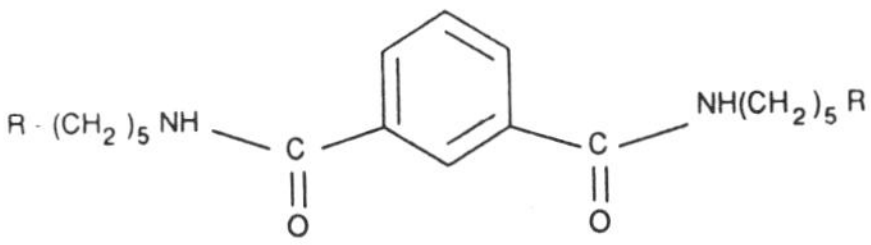

158 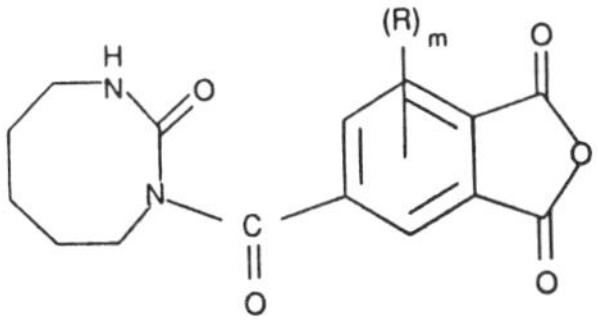   159

3. *N,N'-Pentamethylenecarbodiimide*

Behringer and Meier found that treatment of thiourea **105** with yellow mercury(II) oxide gave traces of *N,N'*-pentamethylenecarbodiimide (**107**), the smallest known cyclic carbodiimide. Compound **107** was reported as unstable and able to polymerize readily (larger ring homologs were, expectedly, more stable) (57LA67). An improved route to **107** has been developed. Dehydrosulfurization of **105** with mercury(II) oxide in the presence of sodium sulfate as a dehydrating agent afforded **107** in high yield and free of side products. Compound **107** could also be prepared by the Tiemann rearrangement of the *O*-mesylate of **112,** followed by treatment with aqueous sodium hydroxide (83JOC1694).

Compound **107** formed cycloadduct **160** on reaction with aryl isocyanates (83JOC1694). The reaction of **107** and diphenylketene occurred readily to give cycloaddition product **161** (R = R' = Ph). The condensation of **107** with diphenylketene or phenylethylketene in a 1 : 2 molar ratio occurred slowly to produce the 1 : 2 cycloadducts **162** (R = Ph or Et). (Interestingly, the same 1 : 2 reaction using the 14-membered homolog of **107** yielded only 1 : 1 cycloadduct, presumably due to ring-strain effects.) The *in situ* generation of dichloroketene in the presence of **107** afforded crude cycloadduct **161** (R = R' = Cl). However, attempted purification of this material using aqueous alcohol resulted in isolation of β-lactam-opened *N*-acylurea **163** (R = Cl). The aforementioned reaction that produced **161** (R = R' = Ph) was quenched very early in water, and **163** (R = Ph) was formed. This compound was *not* formed from **161** (R = Ph) under these conditions. Thus, the intermediacy of dipolar molecule **164** was postulated in the condensation of **107** with ketenes (85JHC357).

160   161   162

163   164   165

In another study of the chemistry of **107,** the reaction of this compound with 2 mol of oxalyl chloride produced bis-adduct **165.** Once again the

14-membered homolog of **107** behaved differently, yielding an imidazo-lidinedione resulting from 1,3 addition across the carbodiimide moiety (86JOC417).

B. OTHER 1,3-DIAZOCINES

1. *Preparation*

a. *Via Cyclization.* Cyclization of *N*-aryl-*N'*-aroyl-1,5-pentane-diamine in ethyl polyphosphate afforded diazocine **166** (Ar = *p*-$O_2NC_6H_4$, Ar' = Ph, *p*-$O_2NC_6H_4$) [77JCS(P2)2068]. The relative basicities and hydro-lysis rates of **166** and lower-ring homologous amidines were investigated (see Section III,B,2). The bis-Schiff bases, *N,N'*-diisopropylidene-1,5-diaminopentane and *N,N'*-dibenzylidene-1,5-diaminopentane, underwent benzoylation–cyclization to produce 1,3-diazocines **167** (R = *i*-Pr and Ph, R' = PhCO) (85IZV1612). Pyridobenzodiazocine **168** was prepared by cy-clizing **169** in the presence of copper bronze (74GEP2365309).

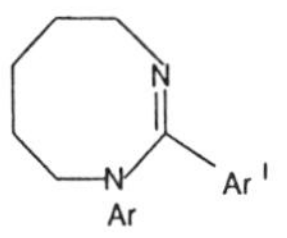

166

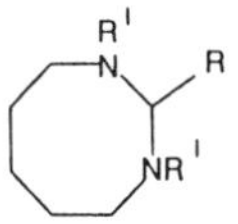

167

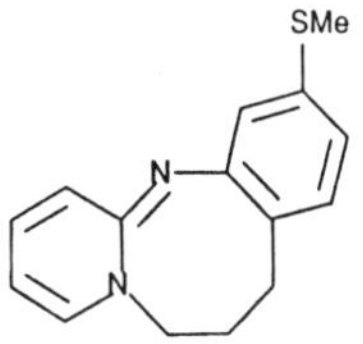

168

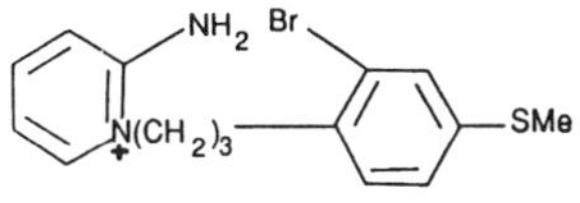

169

In a study of the diastereoselective hydrogenation of one of the chiral precursors of lysine, i.e., **170,** this compound was hydrogenated in acetic anhydride and potassium hydroxide using Raney nickel to produce diace-tyllysineamide **171,** *N,N'*-diacetyl-1,5-diaminopentane, and, if the amount of alkali is increased, imidazolono-1,3-diazocine **172.** It was concluded that **172** arose from ring closure of **171** (87JOU306).

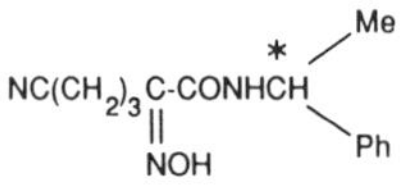

170

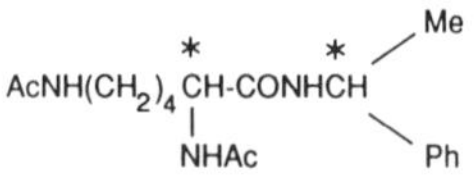

171

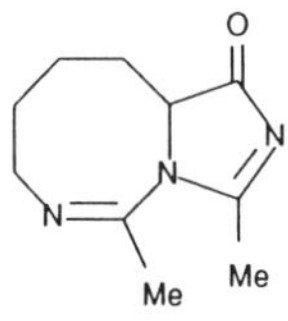

172

   Daidone and co-workers cyclized the diazonium salt derived from *N*-pyrazolylbenzamide **173** to obtain, in addition to other products, a small amount of pyrazolodiazocine, **174** (R = Ph) (80JHC1409). Rahman condensed 3-chloro-4-aryl-1,2,4-triazoles with *N*-methylanthranilic acid to afford triazoles **175** (R = H, Me; R′ = Me). Conversion of this compound to the acid chloride, followed by cyclization with aluminum chloride,

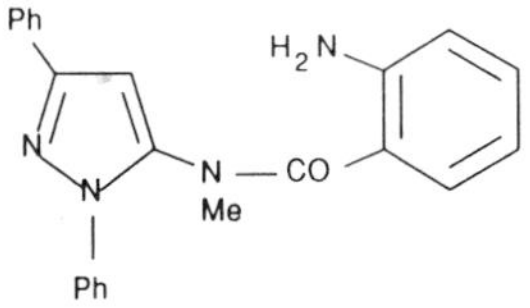

173

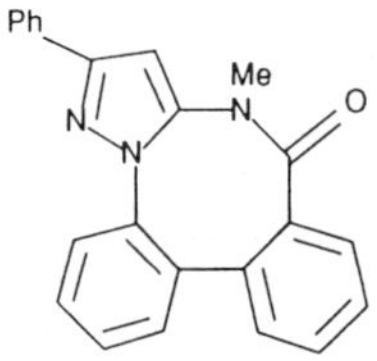

174

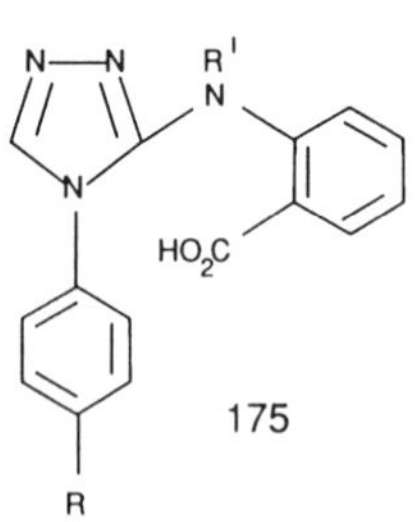

175

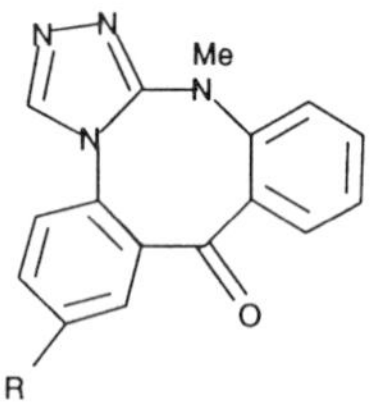

176

yielded diazocine **176** (R = H, Me). These diazocines were also prepared from 3-amino-4-phenyl-1,2,4-triazoles and *o*-iodobenzoic acid via **175** (R = H Me; R' = H), the latter compounds being ring-closed and *N*-methylated (86BSB141)

b. *Via Ring Expansion.* Pyrroloimidazolium iodide **177** underwent transannular ring-opening with sodium hydride or lithium/ammonia to afford 1,3-diazocine **178** or its saturated analog, respectively (74JOC1710). Van der Plas and co-workers synthesized the novel azetodiazocine **181** (R^1 = var., R^2 = H) by a series of cycloadditions of 5-nitropyrimidines **179** (R^1 = var., $R^2 = R^3$ = H) with an excess of 1-(diethylamino)-propyne. They envisaged **181** being formed in a series of reactions shown in Scheme 7 (85JOC270).

In a further study of this reaction, pyrimidines with different combinations of methoxy and methyl substituents were reacted with the ynamine. It was found that when **179** (R^1 = H, $R^2 = R^3$ = OMe), **179** ($R^1 = R^2$ = OMe, R^3 = H) and **179** ($R^1 = R^3$ = H, R^2 = OMe) were used, no diazocine could be isolated, but instead they produced pyridine, a pyridylcarbamoylnitrile,

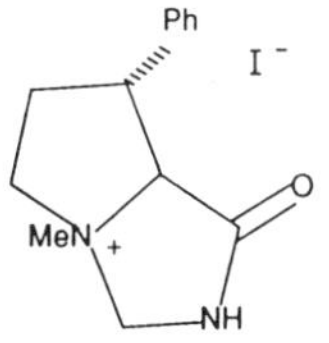

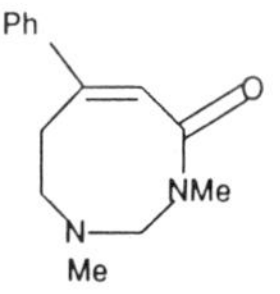

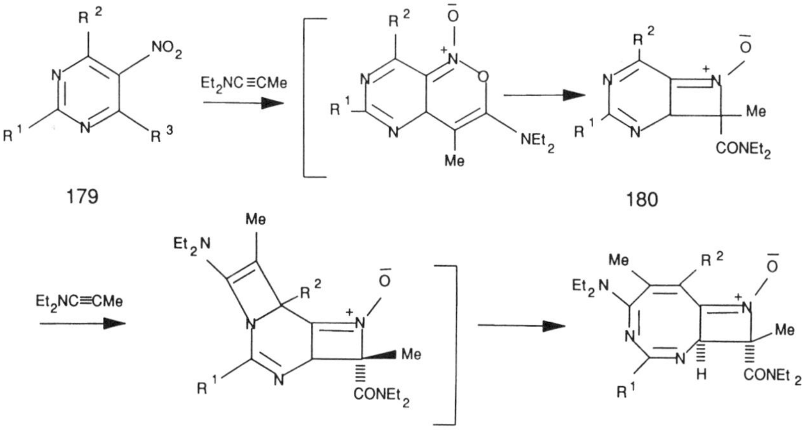

SCHEME 7

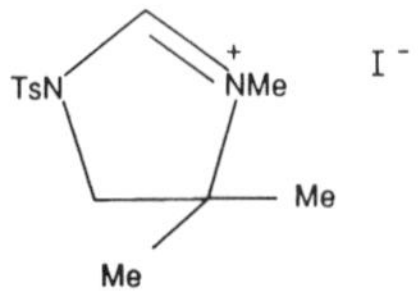

182 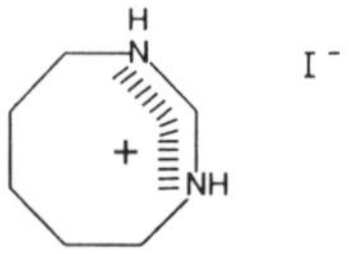   183

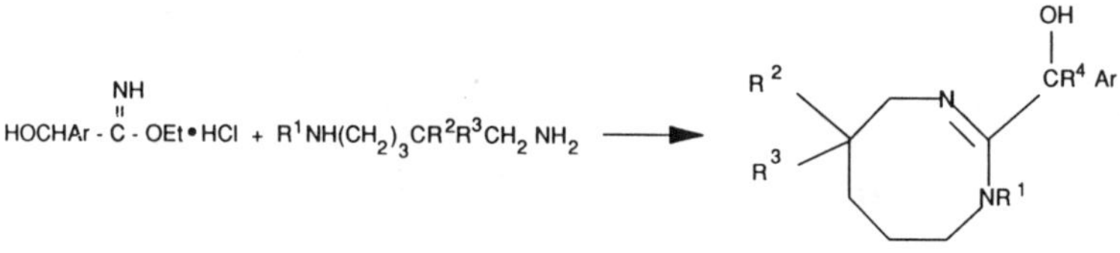

SCHEME 8   184

and a mixture of these two, respectively. The absence of **181** and the presence of other products was explained by invoking a combination of steric and electronic effects, some involving ring opening of **181** (Scheme 7). However, when **179** ($R^1 = R^3 = H$, $R^2 = Me$) was used, diazocine **181** ($R^1 = H$, $R^2 = Me$) was formed in addition to two pyridines and, probably, a carboxynitrone. In this case, the methyl, being less electron-donating than the methoxy, does not cause as extreme an electronic effect (86JOC67).

c. *Via Condensations.* Krakowiak *et al.* condensed the sodium salts of N,N'-ditosyl and N,N'-dibenzenesulfonyl-1,5-diaminopentanes with an excess of dibromomethane to obtain diazocine **167** [R = H; R' = Ts (83MI3), Bs (84MI2)]. Using a triamine instead of a diamine yielded a triazocine (see Section IV,A). Imidazolium salts **182,** described as models for tetrahydrofolate coenzymes, underwent carbon transfer to bifunctional nucleophiles such as 1,5-diaminopentane to yield amidinium salt **183.** [Higher and lower ring homologs could also be prepared (83T3971).] 1,3-Diazocines **184** (Ar = Ph, 4-ClC$_5$H$_4$, 2-BrC$_6$H$_4$; R^1 = H, Me, Ac; R^2 = H, Me, OH; R^3 = H, Me; R^4 = H) were obtained by reacting iminoesters with diamines (Scheme 8) (75USP3926994). Kempter and co-workers

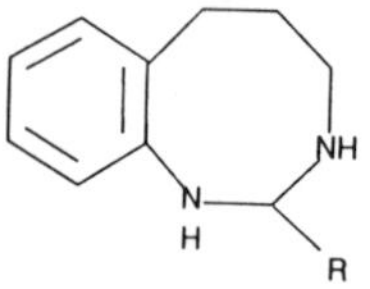

185

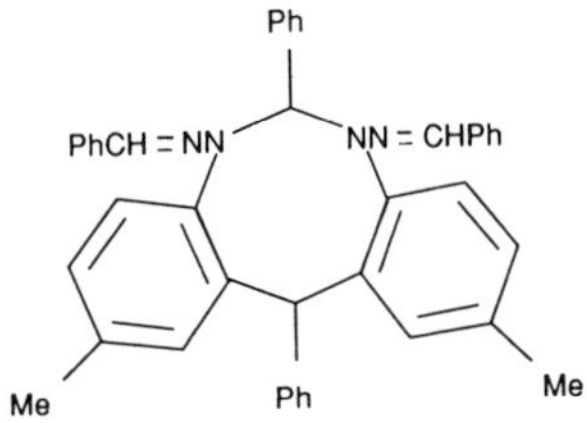

186 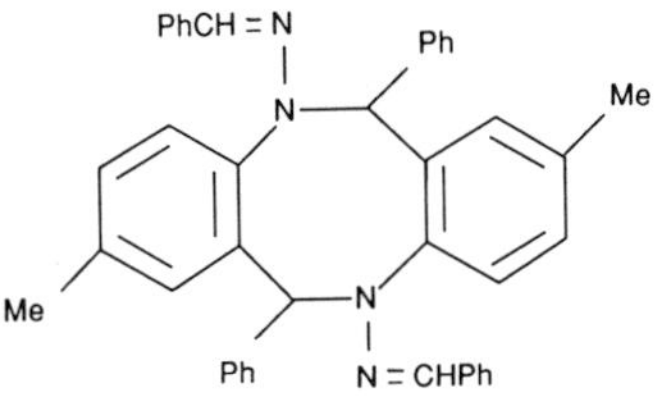   187

treated 3-(*o*-aminophenyl)-propylamine with aldehydes to get benzodiazocines **185** (R = H, Me, Ph) (77MI4).

Giacalone condensed benzaldehyde with the *p*-tolylhydrazone of benzaldehyde in the presence of zinc chloride to isolate a compound, $C_{42}H_{36}N_4$, whose structure was postulated as either 1,3-diazocine **186** or 1,5-diazocine **187** (32G20). To establish the structure of the compound, **187** was synthesized by condensing 1 mol of benzaldehyde with 2 mol of *p*-toluidine. The condensation product (**188**) was then converted to **186** by the route shown in Scheme 9. The product of this sequence was shown to be different from the condensation product of benzaldehyde and benzal tolylhydrazone (*vide supra*). The product of Scheme 9 was 1,3-diazocine **186,** while the first compound of formula $C_{42}H_{36}N_4$ was claimed to be 1,5-diazocine **187** [33G764; 89AHC(46)1]. When the methylenebis(quino-

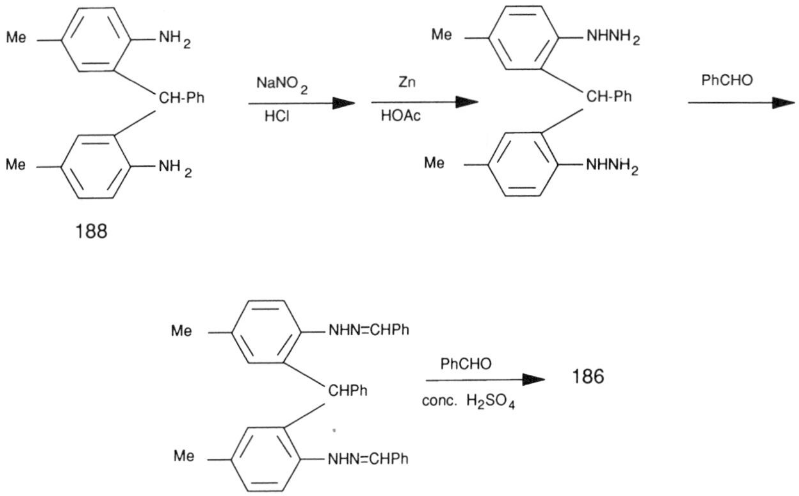

188

SCHEME 9

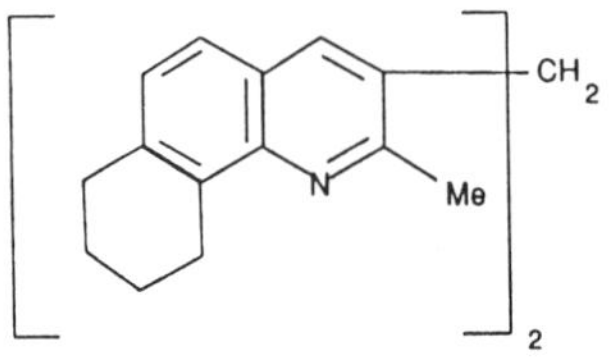

189

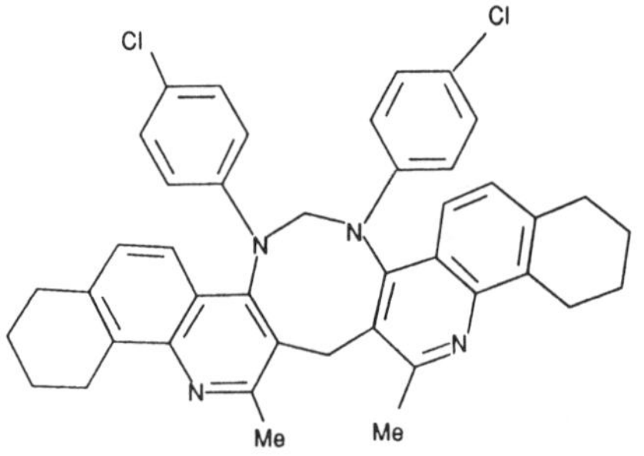

190

line) derivative (**189**) was heated with paraformaldehyde, diquinolino-diazocine **190** was obtained (78JHC649).

2. Theory and Structure

Hückel MO calculations on 1,3-diazocine (**191**) showed a resonance energy (RE) roughly equal to other alternant systems such as azocine (82AHC115), 1,4-diazocine [89AHC(45)185], and 1,5-diazocine [89-AHC(46)1]. Nitrogen substitution did not appreciably alter the RE's when compared with the parent carbocycle cyclooctatetraene (COT), in contrast to nonalternant compounds such as pentalene. However, the RE of **191**, although similar to 1,5-diazocine, was half as negative as 1,4-diazocine (75T295). Molecular orbital calculations using the Pariser–Parr–Pople (PPP) method gave RE's for **191**, its 1,4- and 1,5-isomers, and azocine which were almost identical to the aforementioned values (75T295). The calculations on **191** produced a single solution corresponding to a C_{2v} π-delocalized structure, in contrast to the results for 1,4-diazocine [89AHC(45)185]. It was concluded that **191** as well as its 1,5-isomer and azocine) have less antiaromatic character than 1,4-diazocine or COT, and that in **191** (and 1,5-diazocine), isomerization by π-bond alternation may compete with ring inversion (83MI4).

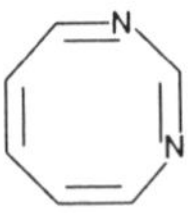

191

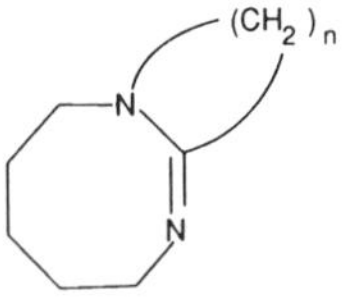

192

Perillo *et al.* determined the basicities of **166** (Ar = Ph and p-$O_2NC_6H_4$, Ar′ = p-$O_2NC_6H_4$) and similarly substituted diazepines. Comparison of the pK_a of **166** (Ar = Ph, Ar′ = p-$O_2NC_6H_4$) with seven-, six-, and five-membered homologous cyclic amidines showed the order of basicity to be 6 > 7 > 8 > 5. The relatively low basicity of the diazocine was explained on the basis of molecular models, which indicated unfavorable transannular interaction between C-4 and C-8 protons and also showed some eclipsing of the C-5, C-6, and C-7 methylene protons. A protonated **166**, *i.e.*, amidinium ion, would be prevented from being planar, with a concomitant loss of basicity in free **166**. Less easily explained was the finding that **166** (Ar = Ph, Ar′ = p-$O_2NC_6H_4$) was more resistant to alkaline hydrolysis than all lower homologs. It was speculated that this was due in part to steric hindrance of the attack of hydroxide at C-2 [77JCS(P2)2068].

3. *Chemistry and Applications*

1,3-Diazocinylarylcarbinols **184** (R^4 = H) (see Section III,B,1,c) were oxidized to the corresponding ketones, and these ketones reacted with aryl Grignard reagents to afford 1,3-diazocinyldiarylcarbinols **184** (R^1 = H, Me, Ac; R^2 = H, Me, OH; R^3 = H, Me; R^4 = Ph, 4-ClC_6H_4, 4-$MeOC_6H_4$, 2,6-$Me_2C_6H_3$, 1-napthyl; Ar = Ph, 4-ClC_6H_4, 2BrC_6H_4) (see Section III,B,1,c). These compounds were active as diuretics and hypoglycemics in rats (75USP3926994).

Diazocines **192** and alkyl-substituted derivatives were among lower and higher ring homologs used as catalysts for making epoxy resin hardeners and materials for elastomers, polyurethanes, paints, etc. (86JAP6133158).

IV. Triazocines

A. PREPARATIVE METHODS AND STRUCTURE

1. *1,2,4-Triazocines*

Sodium amalgam reduction of the semicarbazide of β-benzylpropionic acid was reported to yield 1,2,4-triazocine **193** (47CR876; 48CR818). The same product was formed by hydrogenation of the semicarbazide of β-benzylacrylic acid, followed by sulfuric acid-catalyzed cyclization (49CR1952). The previously reported condensation of aminoguanidine with 2,5-hexanedione to give a triazocine was shown to be in error; the correct product structure contained a five-membered heterocycle (60-

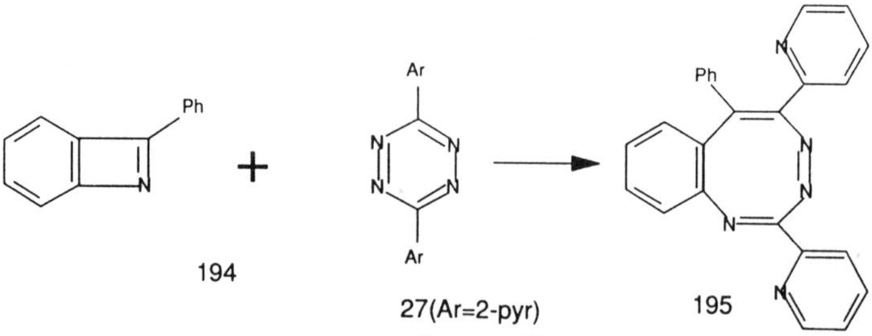

194   +   27(Ar=2-pyr)   195

SCHEME 10

LA150). A similar report of triazocine formation by cyclization of phenacyl bromide guanylhydrazone hydrobromide with base (56AG374; 57-LA50) was also shown to be incorrect, the product being instead an imidotriazine (68TL789).

A ring enlargement to a triazocine occurred when 2-phenylbenzazete (**194**) reacted with dipyridyltetrazine **27** (Ar = 2-pyr) to yield **195** (Scheme 10). In this reaction, Diels–Alder addition is accompanied by spontaneous extrusion of molecular nitrogen and σ-bond cleavage [75JCS(P1)45]. (A similar reaction was used to prepare, 1,2-diazocines; see Section II,A,2.) Condensation of phthalazine derivative **196** with phthalaldehydic acid in aqueous medium afforded 1,2,4-triazocinone **197** and a phthalazine-containing molecule. However, **197** was not formed if the reaction was run in ethanol (81JHC1625).

2. 1,2,5-Triazocines

O-Aminophenylhydrazine hydrochloride was condensed with dialkyl malonate in the presence of sodium acetate, resulting in triazocinedione

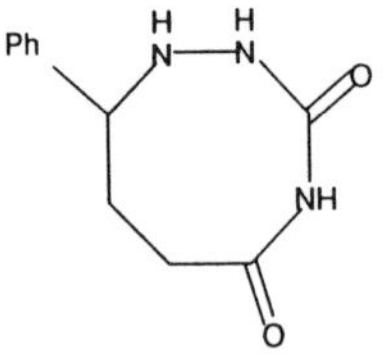

193

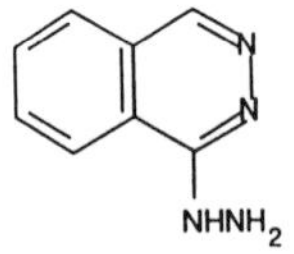

196

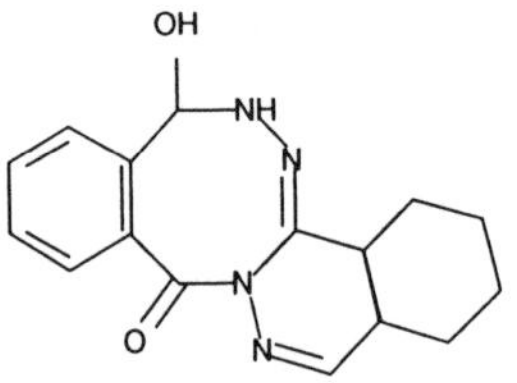

197

198 (26MI1). Triazocine **199** (R = R' = Ts) was obtained by reacting 1,2-dibromoethane with **200** (obtained from 2-chloronicotinic acid) (85MI3).

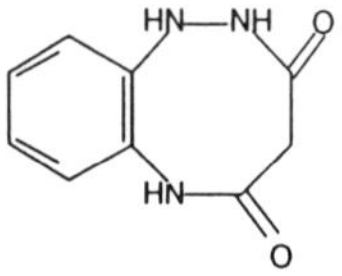

198

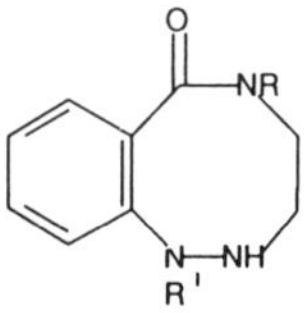

199

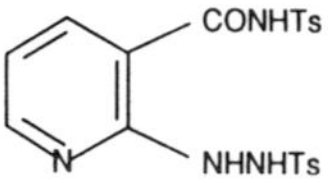

200

Natsukari and co-workers were the authors of several reports involving ring closure of 2-chloroacetamidobenzophenone hydrazones **201** (hal = Cl) to give 1,4,5-benzotriazocinium salts **202** and benzotriazocines **203** (73JAP73-33760; 73JAP73-35276; 79CPB2084) (for an interesting ring contraction of these compounds, see Section IV, B,3). Shindo and co-workers

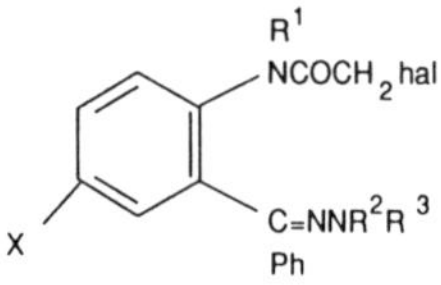

201

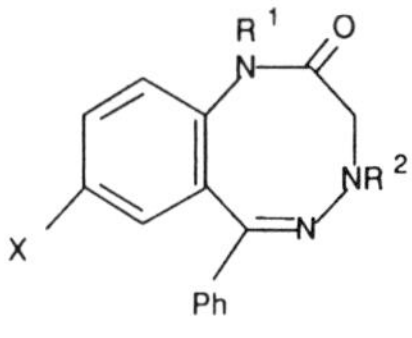

203

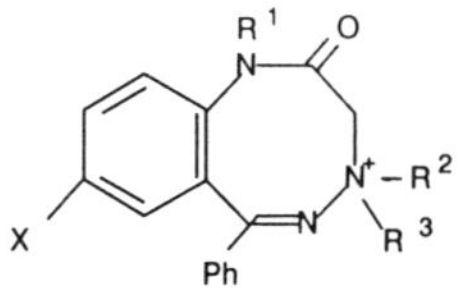

202

similarly cyclized **201** (hal = Br, $R^1 = R^2 = H$, $R^3 = COCH_2Br$, X = hal, alk) with sodium hydroxide to produce **203** ($R^1 = H$, $R^2 = COCH_2Br$, X = hal, alk) (72GEP2216837; 73GEP2308064; 74JAP74-95989; 81JAP81-03349, 81JAP81-05745). For reactions and pharmaceutical uses of these compounds, see Sections IV,B and IV,C, respectively.

Andronati and co-workers reported a detailed study of the *N*-basicity of compounds **203** (X = H, Me; $R^1 = H$; $R^2 = H$, Br, Cl, Me) which contain amide, imine, and amine nitrogens in one molecule. The basicities, obtained from the half-neutralization potentials in potentiometric titrations, showed **203** to be monoacidic bases; the basicity varied predictably with substitution and fell between the stronger quinoxaline-2-ones and the weaker 1,4-benzodiazepine-2-ones. Ultraviolet spectroscopic studies demonstrated that the amino, not the imino, nitrogen was *N*-protonated (83CHE337).

The 1,2,5-triazocine ring in **203** (X = Me, $R^1 = H$, $R^2 = Ts$) was found by X-ray diffraction to be a deep, somewhat puckered boat with an *S-cis* amide configuration (80MI1). Dynamic proton NMR showed that **203** (X = Br, $R^1 = H$, $R^2 = Ac$) interconverted between two enantiomeric boat conformations (Scheme 11) with a $\Delta G^{\ddagger}$ of 85 kJ/mol (84CHE451, 84KGS552).

N-Oxalylbenzoate **204** (R = H, Cl) was amidated with methylhydrazine and the product cyclized with sodium hydroxide in dimethylformamide (DMF) to give triazocinedione **205** (R = H, Cl) (78JHC1309). Kamiya *et al.*, on the basis of X-ray diffraction analysis, suggested that the compound previously reported by Derieg *et al.* [68JCS(C)1103] as the product of the reaction of **206** with hydrazine is not a benzodiazepine, but rather benzotriazocine **207** (R = H, R' = OH) (73CPB1520). The same reaction was reported by Meguro and Kuwada (73CPB2375) [for an annelation reaction of **207** (R = R' = H), see Section IV,B,1].

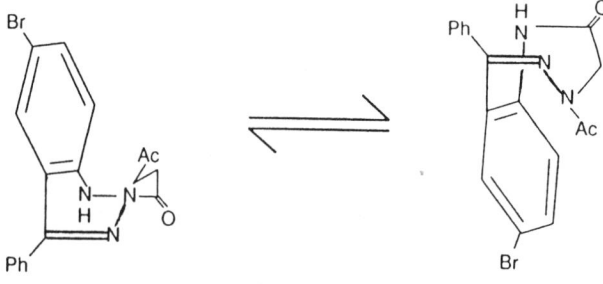

SCHEME 11

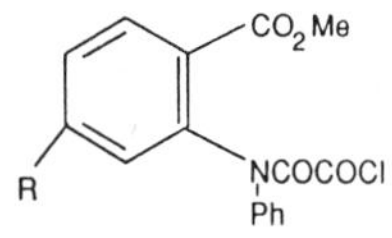

204

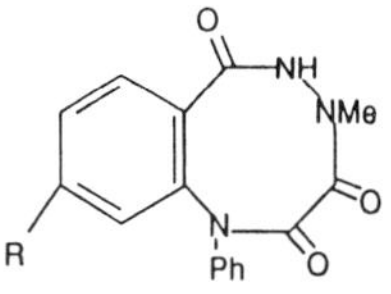

205

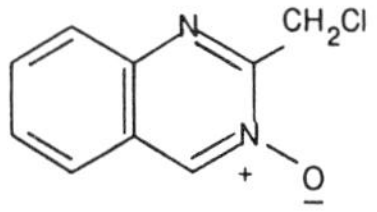

206

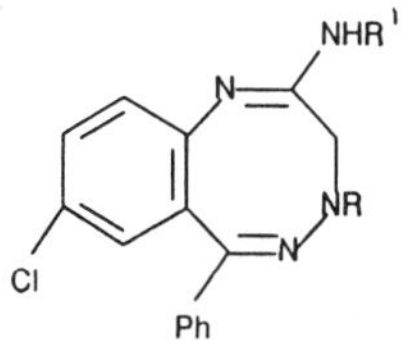

207

Condensation of piperonal derivative **208** with arylhydrazines afforded 1,4,5-benzotriazocines **209** (83M1231). The appropriately substituted triazolylbenzophenones **210** were reacted with hydrazines R^2NHNH_2 to obtain triazolobenzotriazocines **211** (R = H, lower alk; R^1 = hal, NO_2, CF_3, R, OR; R^2 = H, alk, aralk) (74JAP74-26673) (for medicinal uses of these compounds, see Section IV,C).

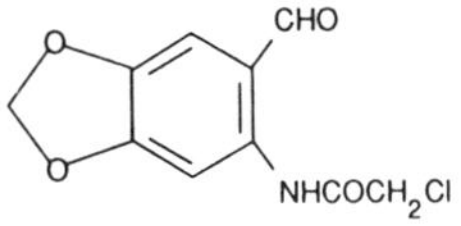

208

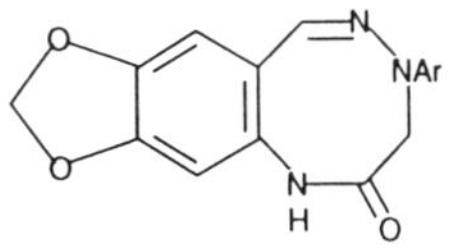

209

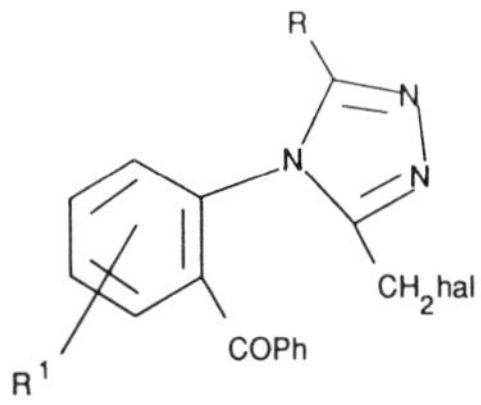

210

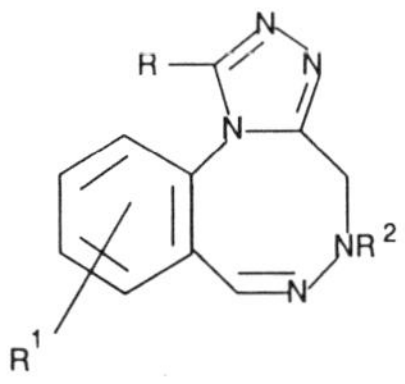

211

Reductive cyclization of the *o*-nitrophenylhydrazones of *o*-nitro-benzaldehyde (26MI1) and *o*-chlorobenzaldehyde (28JIC439) with tin and hydrochloric acid afforded triazocine **212**. The amidodiazonium salt derived from *N*-methyl-*N*-(*m*-diethylaminophenyl)-anthranilamide underwent intramolecular diazo coupling to produce triazocinone **213** (64USP3133086). When either the isoindolopyrazolidocoline **214** or its anhydro derivative **215** is refluxed with aqueous sulfuric acid, ring-expanded triazocine **216** results (Scheme 12) (35JCS1796).

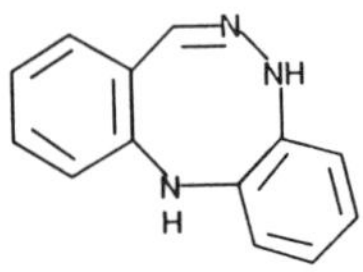

212

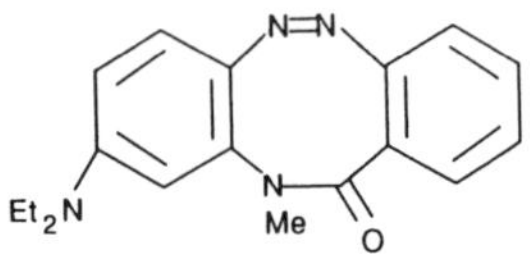

213

3. *1,3,5-Triazocines*

Chapman found that nitration of bistrimethylenedinitramine **217** (R = CH$_2$-morpholinyl) with a mixture of nitric acid, acetic anhydride, and ammmonium nitrate afforded 1,3,5-triazocine **218** (R = NO$_2$) along with 4-nitromorpholine (49JCS1631). In a related reaction, Mannich condensation of **217** (R = H) with formaldehyde in the presence of primary amines yielded **218** [R = Me (49JCS1638), var. alkoxyalk, alk (66JCS(C)870)]. Use

217

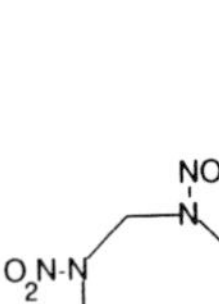

218

219

220

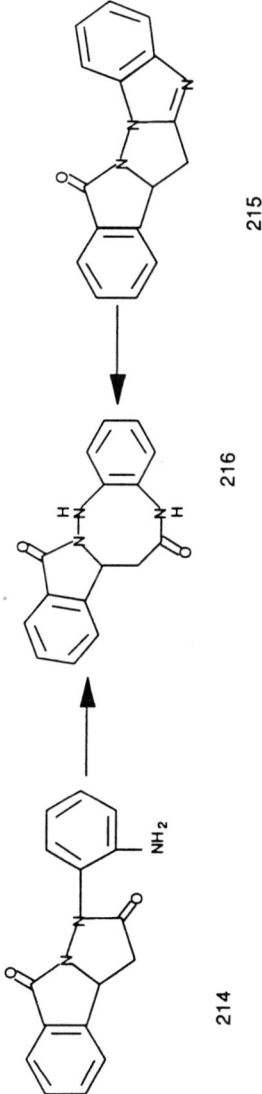

215

216

214

SCHEME 12

of ammonia instead of primary amines resulted in formation of **219** ($n = 1$) [66JCS(C)870]. Similarly, reaction of **217** (R = H) with formaldehyde, followed by ethylenediamine or tetramethylenediamine afforded **219** ($n = 2$) or **219** ($n = 4$), respectively (67IZV1839). Interestingly, compound **217** (R = H) reacted with 1,3,5,7-tetrazocine **220** (Section V,A,1,a) to give **218** (R = CH_2OAc) [66JCS(C)867].

Arylbiguanides **221** reacted with benzoylacetone containing acid to yield triazocine **222**. Use of **221** (Ar = $pMeOC_6H_4$) afforded the corresponding **222** plus tautomer **223**, both compounds being interconverted on heating (72CPB927). Binder *et al.* found that, although the major product of methoxide-catalyzed cyclization of the glycine ester **224** was an imidazo-

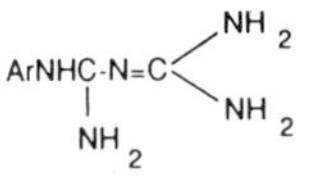

221

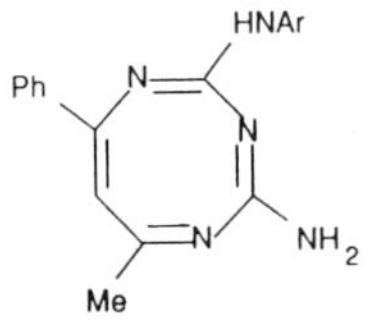

222

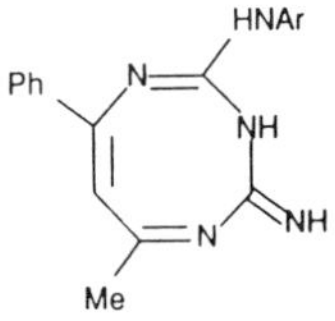

223

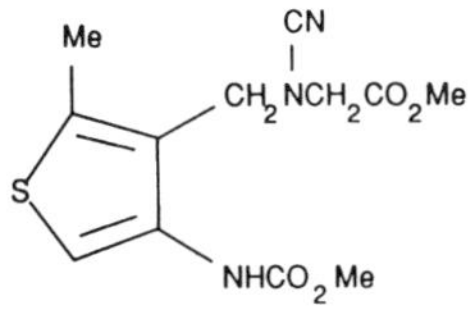

224

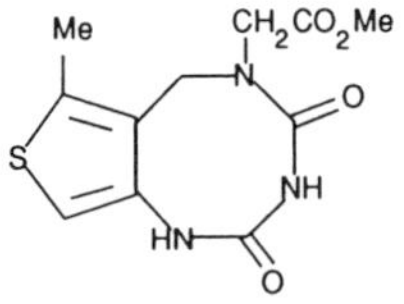

225

thienopyrimidone, workup of the acetone and DMF extracts from the reaction afforded a small amount of triazocine **225** (81AP557). Plescia and co-workers condensed *o*-aminobenzamides **226** [R = Ph, R' = H; RR' = $(CH_2)_4$, $(CH_2)_5$] with triethyl orthoformate to obtain the coresponding triazocinones **227** (75JHC199). Diazotization of **226**, followed by cyclization, afforded tetrazocines (see Section V,B,1,a). In a similar reaction, treatment of diamine **228** with *N,N'*-carbonyldiimidazole afforded compound **229** (82MI3). Finally, when **230** was reacted with polyphosphoric acid, cyclized product **231** was isolated (80IZV1886).

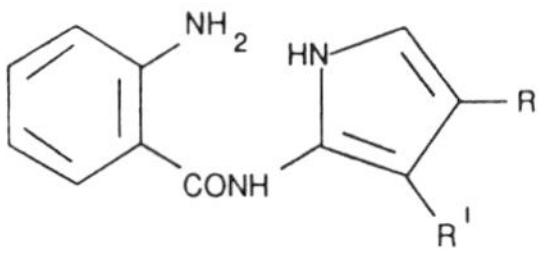

226 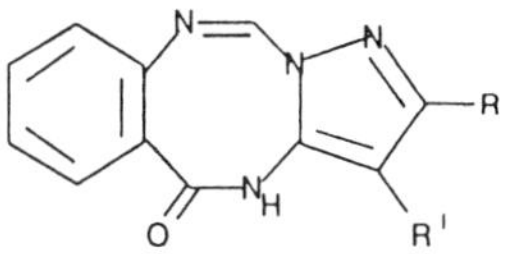   227 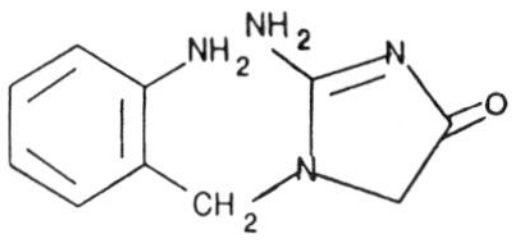   228

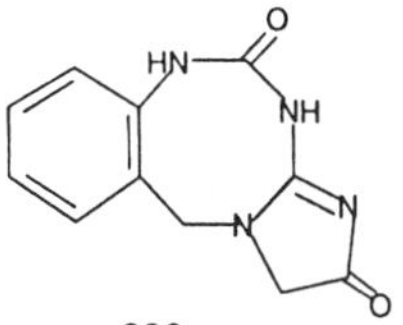

229

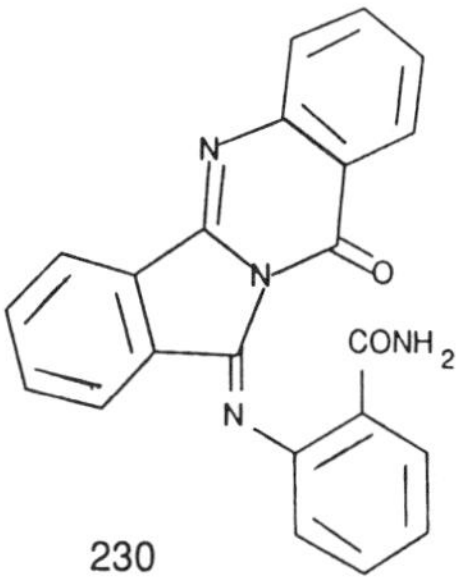

230 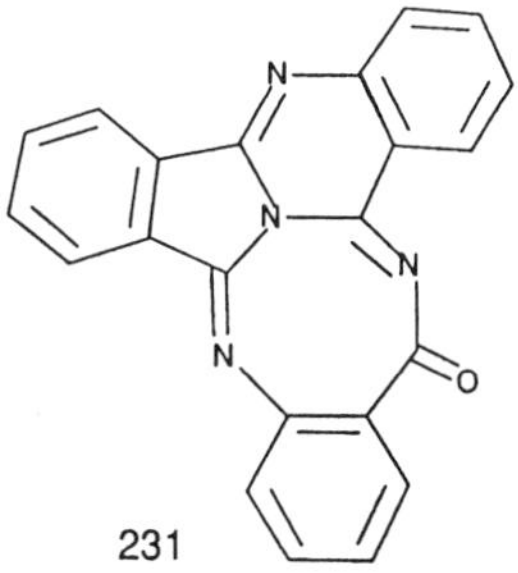   231

4. *1,3,6-Triazocines*

The initial report of the reaction of an imidazoline with thionyl chloride to give a triazocine was found to be in error, the correct structure being an imidazolidone (56JA6144). However, condensation of tristosyltriamine **232** with dibromomethane did afford triazocine **233** (83MI3); 1,3-diazocines were also prepared by this method (see Section III,B,1,c).

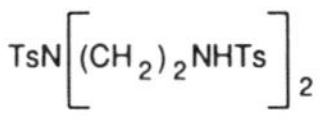

232 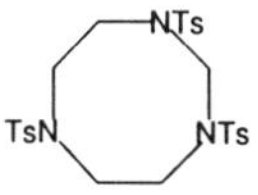   233

The 1,3,6-benzotriazocine ring system was first synthesized by Hornyak, Lempert, and co-workers, who found that *N*-tosyl-*o*-nitroanilines could be haloalkylated and then aminated to yield compounds **234** (R = H, Me; R′ = H, Me, OMe). Hydrogenation of **234**, followed by reaction with orthoesters, $R^2C(OAlk)_3$, produced compounds **235** (R = R^2 = H, Me; R^1 = H, Me, OMe). If cyanogen bromide was used instead of orthoesters, triazocines **235** (R^2 = NH_2) resulted. Reduction of **234** (R = R′ = Me), followed by condensation with diethyl carbonate, gave urea **234** (R = R′ = Me, X = O). A similar reaction starting with **234**

(R = H, R′ = OMe) and using carbonyl and thiocarbonyldiimidazoles yielded urea **236** (R = H, R′ = OMe, X = O) and thiourea **236** (R = H, R′ = OMe, X = S), respectively. Condensation of the latter compound with methyl iodide in the presence of sodium methoxide afforded **235** (R = H, R^1 = OMe, R^2 = SMe) (83T479, 83T1199). (For similar preparations of 1,3-diazocines, see Section II,A,1,c.)

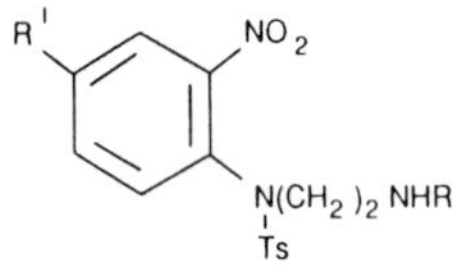

234

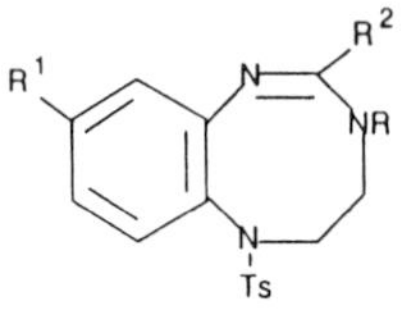

235

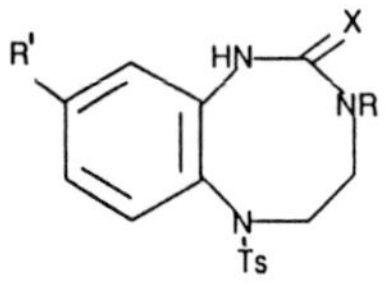

236

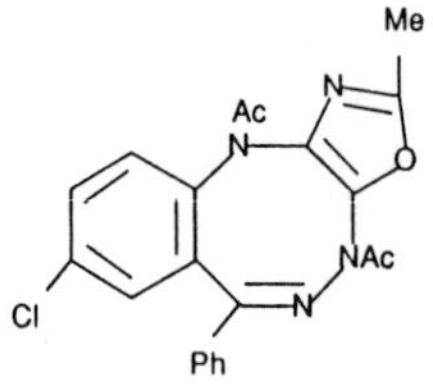

237

B. REACTIONS

1. *Eight-Membered Ring Preserved*

Kamiya *et al.* found that **207** (R = H, R′ = OH) underwent annelation rearrangement in acetyl chloride/DMF to **237**. The structure of this latter compound, as well as derivatives of **207** [*i.e.,* R = R′ = H; R = Ac, R′ = OAc; R = Ac, R′ = OH], were confirmed by X-ray diffraction analysis and were thus incompatible with the earlier benzodiazepine structures assigned to these compounds (73CPB1520) (see Section IV,A,2).

Other reactions include *N*-acylations (83CHE337, 83KGS411), including a reaction with oxalyl chloride to produce an oxalyl bis-(triazocine) (83CHE337), *N*-sulfonations and salt formation (83CHE337), *N*-alkylations (72JAP72-34384, 72JAP72-34386; 73JAP73-00491; 74JAP74-26292, 74JAP74-95987, 74JAP74-95988; 78JHC1309; 81JAP81-01312, 81JAP81-01313; 83CHE337, 83KGS411; 85MI3), including Mannich reactions [i.e., conversion of **199** (R = R′ = H) to **199** (R = H, R′ = CH₂-morpholinyl, CH₂-piperidinyl (85MI3)], *N*-nitrations [66JCS(C)870], *N*-nitrosations [66JCS(C)870], amination of haloacetyl sidechains (72GEP2216837; 74JAP74-51292; 81JAP81-01314), *N*-deacylations (72-GEP2216837; 74JAP74-95990; 81JAP81-00433, 81JAP81-03350), *N*-desulfonations (85MI3), C=N bond reduction (83CHE337), and lactam-to-lactim ether conversion (83CHE337).

2. Ring Contractions

The product mixture of 1,2,4-triazocine **197** and phthalazine derivative (see Section IV,A,1) was converted, on prolonged standing, into tria-zolophthalazine **238** (81JHC1625). 1,2,5-Triazocines **202** and **203** (Section IV,A,2), upon treatment with base, underwent ring contraction to the pharmacologically useful benzodiazepines **239** (73JAP73-33760, 73JAP73-35276; 79CPB2084)

Fujimura *et al.* found that if benzotriazocines **203** (R¹ = Me, R² = H) were allowed to stand in chloroform solution, indazoles **240** were formed. This transformation was rationalized by the reaction sequence shown in Scheme 13 (84CPB3252).

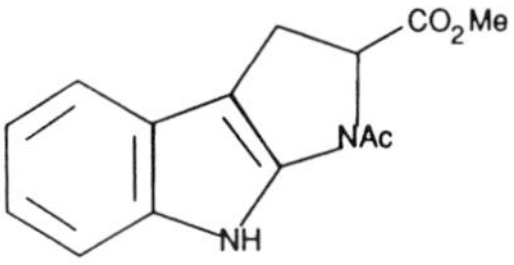

119 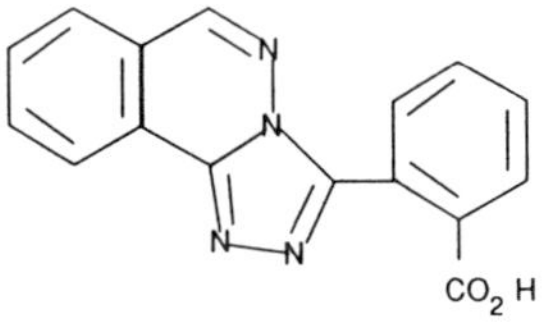   238 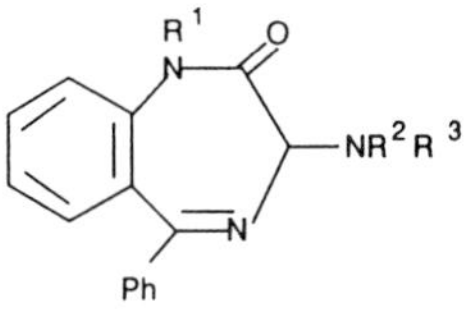   239

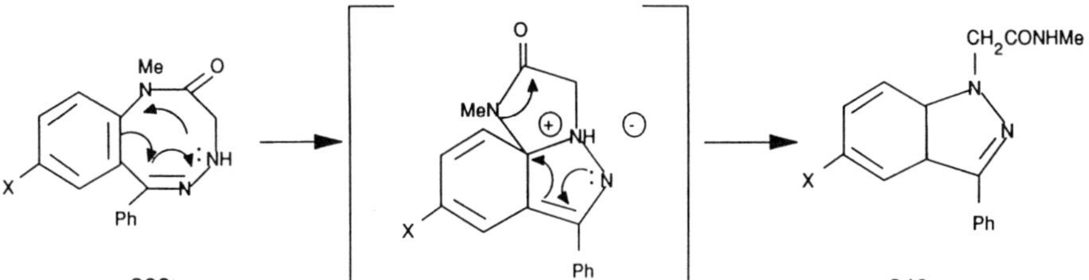

203   240

SCHEME 13

3. Ring Openings

Compound **203** ($R^1 = R^2 = H$, $X = Br$) underwent hydrazone exchange with hydrazine hydrate to give 2-amino-5-bromobenzophenone hydrazone. The same benzotriazocine was ring-opened with lithium aluminum hydride to afford 2-benzyl-4-bromoaniline, while hydrolysis of the compound yielded 2-amino-5-bromobenzophenone (83CHE337).

C. Applications

Benzotriazocines **203** ($R^1 = H$, Alk; $R^2 = COCH_2hal$) were found to act on the CNS (72JAP72-34384, 72JAP72-34386; 73JAP73-00491; 74JAP74-26292), and compounds **203** ($R^1 = $ Alk, $R^2 = COCH_2hal$) were active as psychotropics (81JAP81-01312–01314). These compounds were also found to have sedative and analgesic activity (74JAP74-95989; 81JAP81-03349, 81JAP81-05745). Benzotriazocinium salts **202** were useful as hypoglycemics and as intermediates in the preparation of CNS depressants (73JAP73-33760; 79CPB2084).

Dibenzotriazocine **213** is useful as a dye for wool, polyesters, or polyacrylonitriles (64USP3133086). The syntheses of **218** (*vide supra*) and other nitramines were investigated in order to gain insight into the chemistry of nitramine explosives (49JCS1631, 49JCS1638) (see Section V,A,1).

V. Tetrazocines

A. 1,3,5,7,-Tetrazocines

1. HMX and Related Compounds

a. *Introduction.* This special section is centered around the important explosive perhydro-1,3,5,7-tetetranitro-1,3,5,7-tetrazocine (**241**, $R^1 = R^4 = NO_2$), alternatively called "cyclotetramethylenetetramine," "octogen," or "HMX." This last designation will be used throughout this section. Also discussed will be the preparation and chemistry of those 1,3,5,7-tetra-*N*-substituted-1,3,5,7-tetrazocines and other compounds related to HMX and/or molecules that are part of the HMX synthesis scheme. This is shown in Scheme 14.

HMX has usually been associated with the better known explosive RDX (**242**, $R = NO_2$), with which it has usually been admixed. RDX has been regarded as a somewhat more powerful explosive (67MI1), but HMX is

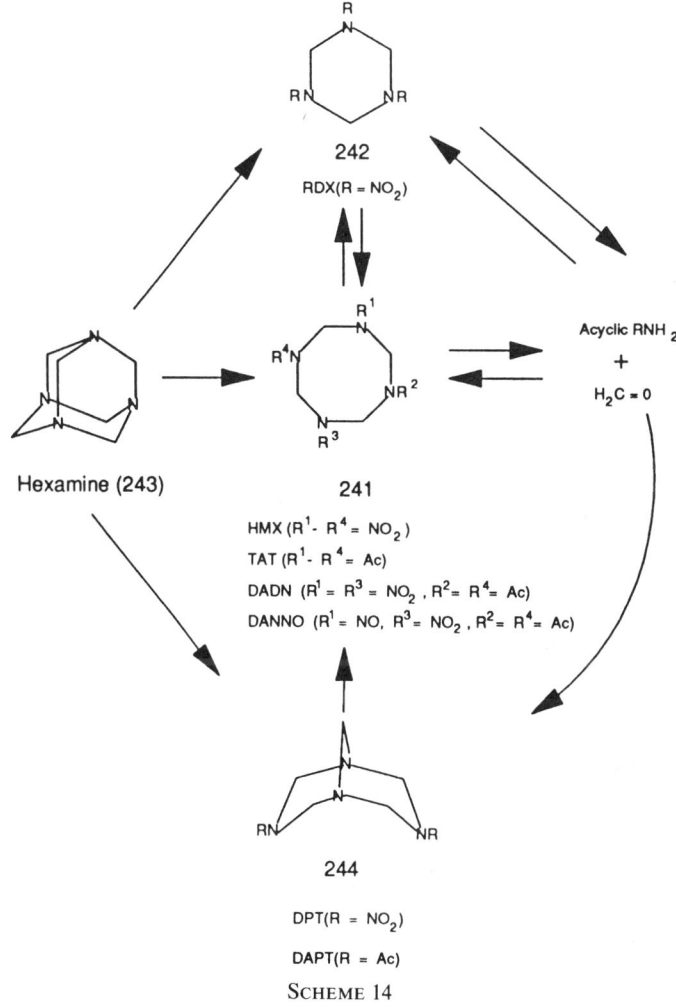

RDX(R = NO₂)

Hexamine (243)

242

241

HMX (R¹- R⁴= NO₂)
TAT (R¹- R⁴= Ac)
DADN (R¹ = R³ = NO₂ , R²= R⁴= Ac)
DANNO (R¹ = NO, R³= NO₂ , R²= R⁴= Ac)

Acyclic RNH₂
+
H₂C = 0

244

DPT(R = NO₂)

DAPT(R = Ac)

SCHEME 14

now considered the most powerful military explosive (81MI7), being 1.6 times more powerful than TNT. We will not discuss the explosive and propellent properties of HMX, rather, the reader is referred to literature such as reviews by Urbanski (67MI1) and Fifer (84MI3). (The latter is a review of the ignition and combustion chemistry of HMX as well as decomposition catalysis mechanisms.) HMX finds many applications in explosive (81MI2; 82MI4, 82USP4331080; 84MI4) and propellant (81USP4288262; 82MI6; 84MI3, 84USP4462848) preparations. Only one nonexplosive application was reported for these types of compounds (**241,**

$R^1 - R^4 = COEt$) which was useful as a dispersant and a suspending and blending agent (76USP3978047).

b. *Preparation and Chemistry.* Siele *et al.* are responsible for an excellent review of the synthetic methodology for HMX (81MI7). One reason that HMX was used as an explosive mixture with RDX was that the materials were synthetic coproducts (see Scheme 14). In fact, Bachmann and Sheehan first prepared HMX (and RDX) from hexamine (**243**) while exploring new procedures for preparing the known RDX (49JA1842). At about the same time, Wright *et al.* (49MI1, 49MI2) and Epstein and Winkler (52CJC734) identified 3,7-dinitro-1,3,5,7-tetraazabicyclo[3.3.1]-nonane (**244,** R = NO₂) (DPT) as a precursor in a Bachmann-type synthesis. The interconnection of all of these molecules is shown in Scheme 14 and will be discussed as outlined here:

(i) Direction degradation of hexamine (**243**) to tetrazocines (**241**).
(ii) Sequential degradation of **243** to bicyclononanes **244;** conversion of **244** to tetrazocines **241.**
(iii) Mannich-type condensation of formaldehyde with acyclic amines to afford **241** and **244,** and the role of these reactions in triazine–tetrazocine (**242**)–(**241**) interconversion.

i. *Direct degradation of hexamine* (**243**) *to 1,3,5,7-tetrazocines* (**241**). The initial method for synthesizing HMX (**241,** $R^1 - R^4 = NO_2$) entailed nitration of hexamine (**243**) with nitric acid, leading to the dinitrate of **243.** Nitrolysis of this material was accomplished by ammonium nitrate in a solvent of nitric acid and acetic anhydride (49JA1842). Accompanying HMX was RDX (**242,** R = NO₂); RDX and other triazines **242** were usually formed in the degradations of hexamine cited here. The low yield of HMX obtained by Bachmann and Sheehan was improved upon by Fischer who modified reagent ratios (49CB192). Subsequent attempts to improve the reaction involved varying the mole ratios of reactants (52CJC734; 82MI2), the number and order of nitration and acetylation steps (67FRP1476116; 68BEP719042; 78USP4086228), and the acetylating reagents (51JA2769; 71FRP2053804). Also, other additives were introduced (60USP2941994; 61USP2983725; 82MI2; 84CZP212600; 86CZP227911).

Several groups reported hexamine (**243**) could be converted directly to perhydro-1,3,5,7-tetraacetyl-1,3,5,7-tetrazocine (**241,** $R^1 = R^4 =$ Ac) (TAT) without the intermediacy of 3,7-diacetyl-1,3,5,7-tetraaza-bicyclo[3.3.1]nonane (**244,** R = Ac) (DAPT) (*vide infra*) [81MI5; 82AKZ315; 86JCS(P2)835] [in addition, **242** (R = Ac) and DAPT were formed (82AKZ315; 86JCS(P2)835)]. TAT can in turn be nitrated to produce HMX (76USP3939148; 83MI2, 83USP399948). Hexamine (**243**) reacted with propionic anhydride to yield **241** ($R^1 = R^4 = $ COEt) (72JOC320; 76USP3978047) [use of acetic instead of propionic anhydride afforded only DAPT (**244**, R = Ac); *vide infra*].

ii. *Sequential degradations* **243–244** *and/or* **244–241**. As mentioned previously, the bicyclic molecule DPT (**244**, R = NO_2) was identified as an HMX precursor at about the same time that HMX was synthesized from hexamine. The intermediacy of DPT in the nitrolysis of hexamine was established by Castorina and co-workers using ^{14}C-tracer studies with paraformaldehyde (60JA1617).

Procedures for preparing DPT from the nitrolysis liquors of hexamine have been described by Wright and Chute (54USP2678927), Bachmann *et al.* (51JA2769), and others (49BRP615419, 49BRP615793; 61MI1; 82MI2). Nitrolysis of DPT afforded HMX (49MI1; 82MI2; 86CZP227907); similar treatment of **244** (R = NO) also gave HMX (82CZP195038). Nitrolysis and acetolysis of DPT yielded **241** ($R^1 - R^3 = NO_2$, $R^4 = CH_2OAc$) (identical to **220**, Section IV,A,3) (51JA2773; 53CJC602); reaction of this latter material with nitric acid/acetic anhydride gave HMX [51JA2773; 69JCS(C)1556]. In an interesting synthesis of an azido analog of HMX, a potential explosive, Frankel and Woolery condensed **241** ($R^1 - R^3 = NO_2$, $R^4 = CH_2OAc$) with acetyl bromide to afford **241** ($R^1 - R^3 = NO_2$, $R^4 = CH_2Br$). This latter compound was reacted with a new "azidation" agent, acetyl azide, to produce **241** ($R^1 - R^3 = NO_2$, $R^4 = CH_2N_3$) (83JOC611; 85USP4534895).

Two groups reported that reaction of hexamine (**243**) with propionic anhydride yielded **244** (R = COEt) along with triazine **242** (R = COEt). However, no tetrazocine (**241**) was found [82AKZ315; 86JCS(P2)835], although previous groups *did* isolate **241** from this reaction (*vide supra*). Similarly, a number of groups reacted hexamine with acetic anhydride to form DAPT (**244**, R = Ac) (72JOC320; 74JHC237; 76USP3978047).

DAPT could be converted into a number of 1,3,5,7-tetrazocines, including HMX, by the use of nitric acid, phosphorus pentoxide, and urea (86CZP227913). Acetolysis of DAPT afford TAT (*vide supra*). (76USP 3979379; 86MI3). Siele *et al.* postulated **241** ($R^1 - R^3 = $ Ac, $R^4 = CH_2Cl$) and **241** ($R^1 - R^3 = $ Ac, $R^4 = CH_2OAc$) to be the intermediates in this reaction (81MI7). Ju and Wang, based on an NMR study, predicted the latter compound as the product of the reaction of DAPT with anhydrous acetic anhydride and predicted **241** ($R^1 - R^3 = $ Ac, $R^4 = $ H) as the product if the acetic anhydride was wet (84MI5). DAPT was condensed with the appropriate acid chlorides to afford compounds **241** [$R^1 = R^3 = $ Ac; $R^2 = R^4 = Cl(CH_2)_2CO$), phthalimido-CH_2CO]; similarly, **244** (R = Ts) reacted with the appropriate acid chlorides to yield **241** ($R^1 = R^3 = $ Ts; $R^2 = R^4 = $ Ac, phthalimido-CH_2CO) (84AKZ185). Wright *et al.* first treated DAPT with nitric acid/dinitrogen tetroxide to produce perhydro-1-

nitroso-3-nitro-2,4-diacetyl-1,3,5,7-tetrazocine (**241**, R^1 = NO, R^2 = NO_2, $R^2 = R^4$ = Ac) (DANNO), which was then oxidized to perhydro-1,3,di-nitro-2,4-diacetyl-1,3,5,7-tetrazocine (**241**, $R^1 = R^3 = NO_2$, $R^2 = R^4$ = Ac) (DADN) (49MI3). DAPT could be converted directly to DADN using a number of reagents: nitric/sulfuric acids (73JHC725); nitric acid/urea (75USP3926953); and potassium nitrate/dinitrogen tetroxide (76USP-3987034).

DADN can be transformed into **241** ($R^1 - R^3 = NO_2$, R^4 = Ac) us-ing trifluoroacetic anhydride/nitric acid (81MI3, 81MI4; 85MI1). DADN can be nitrolyzed one step further to HMX using a variety of chemi-cals: nitric/polyphosphoric acids (81MI6); dinitrogen pentoxide (83MI2, 83USP399948); and nitric acid/phosphorus pentoxide in the presence of ammonium nitrate (85CZP218703). Tetrazocines **241** were the product of reaction of **244** (R = $ArSO_2$) with dinitrogen tetroxide (73JHC279).

iii. *Mannich-type condensations of formaldehyde with acyclic amines.*   Although HMX itself has not been prepared by condensative, as opposed to degradative, methods, a number of related compounds **241** have been made from acyclic precursors using a Mannich-type reaction involving formaldehyde. DPT (**244**, R = NO_2) has been prepared by con-densation of dimethylolnitramide [$O_2NN(CH_2OH)_2$], methylenediamine and formalin. The DPT was then nitrolyzed to HMX (*vide supra*) (49MI1). Dinitrodimethyltetrazocine **241** ($R^1 = R^2 = NO_2$, $R^3 = R^4$ = Me) was formed from methylamine and methylnitramine (MeNHNO$_2$) in formalin (49MI3). Coon obtained TAT (**241**, $R^1 - R^4$ = Ac) through the reaction of paraformaldehyde with methylenebis(acetamide) [$CH_2(NHCOCH_3)_2$] (75MI6; 76USP3978046).

A number of studies have shown production of both triazines **242** and tetrazocines **241** from condensations of acyclics, with perhaps intercon-version between **241** and **242**. In an early first report of a 1,3,5,7-tetrazocine, Giua and Racciu claimed triazine **242** (R = CO_2Alk), produced by condensation of urethanes H_2NCO_2Alk with formaldehyde, underwent conversion to tetrazocine **241** ($R^1 - R^4 = CO_2Alk$) (29MI1). Thyagarajan and Majumdar found that depending on the experimental conditions, these condensations yielded either triazines **242** (R = CO_2Alk) or tetrazocines **241** [R = CO_2Alk (Alk = Me, Et, *n*-Pr, *n*-Bu)]. In the presence of a trace of acid, compounds **241** ring-contracted to the corresponding compound **242**. That these reactions probably involved total degradation of **241** back into urethanes and formaldehyde was demonstrated by the observation that **242** underwent acid-catalyzed transformation into **241** (74JHC937). A simi-lar breakdown may be involved in the interesting reaction of **241** ($R^1 - R^3 = NO_2$, $R^4 = CH_2OAc$) (identical to **220**) with trimethylene dini-tramine **217** to give triazocine **218** (R = CH_2OAc) [66JCS(C)867] (see

Section IV,A,3). Borowski and Haas reported thioamines CF_2X–SNH_2 (X = F,Cl) reacted with formaldehyde to afford, in addition to triazines **242** and 10-membered pentazecine, the tetrazocines **241** ($R^1 - R^4 = SCF_3$, SCF_2Cl) (82CB523).

c. *Theory and Structure.* Most of the investigations described here were on HMX (**241,** $R^1 - R^4 = NO_2$) itself. There have been reports on TAT (**241,** $R^1 - R^4 = Ac$). X-ray diffraction analysis showed it to have a boat-shaped conformation [73AX(B)651]. The protonation equilibria and decomposition reactions of TAT (as well as hexamine **243,** the other acylation products of **243**) were determined using UV and NMR spectroscopy. The pK_a of TAT was -2.5, showing it to be a very feeble base. In 6.6 M acid complete breakdown occurred to yield formaldehyde and acetic acid [86JCS(P2)835]. Studies of HMX (**241,** $R^1 - R^4 = NO_2$) are outlined in the following paragraphs.

i. *Theoretical.* Molecular-orbital calculations on HMX allowed determination of properties such as orbital energies, UV spectra, charge distribution, bond orders and energies, dipole moments, molecular ionization potentials, crystal packing, and bond strengths [69AJC2505, 69AJC2515, 69TFS904; 72JCS(F2)1659; 82MI5].

ii. *Polymorphism of HMX.* HMX exists in four crystalline forms designated α, β, γ, and δ (50AC1225). The β polymorph is the least impact-sensitive (66FRP1463470) and is the isomorph of choice in explosive and propellant preparations. Wright (62CJC2278) and others (66FRP1463470) have outlined methods for isolating β-HMX.

X-ray diffraction studies have revealed that the α and δ forms have a basket-like conformation with the four carbon atoms that are coplanar. The crystal structures possessed a twofold symmetry axis [63AX617]; 74AX(B)1918; 79JPC340]. Additional X-ray data on crystalline HMX showed the nitramine groups of the α, β and γ forms are also planar [55MI1; 75AX(B)2805]. Neutron diffraction analysis indicated the ring in the β-polymorph assumes a chair conformation [70AX(B)1235]. NMR and Raman spectroscopic data suggested a chair–chair shape for γ-HMX analogous to α and δ-HMX (79JPC340; 85MI2; 86MI2). Stefaniak *et al.* predicted tetrazocines **241** may display rapid ring inversion accompanied by isomerization at the nitrogen atoms while in the thermodynamically preferred crown conformation (69RC1687). However, Mikhailov and Shlyapochnikov judged the crown or double-chair forms to be energetically less favorable than the double-chair conformation, based on mechanical models of free, noncrystalline HMX (82ZSK172).

Wright noted IR spectroscopy may be used to differentiate between the solid-state polymorphs (64T159). The IR spectrum of β-HMX was found to

differ markedly from those of the α and γ isomers (73MI1). Raman and IR
data for HMX crystals confirmed the noncentrosymmetry of each crystal's
unit cell in addition to revealing some of the components of its molecular
conformation [71CR(B)658; 74JCP221].

iii. *Other studies.* Also the subject of numerous studies have been
polymorph stabilities (62MI2; 65MI1; 74MI1; 78JPC1912; 80JPC1376); in
conjunction with this have been studies of phase transitions (62MI2;
65MI1; 67MI2; 75MI9; 80JPC3573; 82MI1; 83MI1) and thermodynamics of
solid–liquid, solid–gas, and solid–solid transitions of HMX [69MI2;
71MI1; 74TFS556; 76JCS(F1)723; 78JCS(F1)1339; 80JPC3573]. There
have been several reports of solubility, adsorption, and surface properties
of HMX (63MI1; 67MI4; 68MI1; 76MI1), as well as dipole moment measurement [79JCS(P2)869].

Spectroscopic studies include electronic adsorption and luminescence
(69MI3; 70MI2; 71TFS1739; 75MI5; 76MI2; 77MI2); ^{1}H NMR [68JCS(B)6;
73MI2; 75T1521]; ^{13}C NMR (75T1521; 81OMR52); ^{14}N and ^{15}N NMR
(73CC602); ^{14}N NQR (81JPC2618); IR and Raman [57MI1; 63T159;
71CR(B)658; 73MI1; 74JCP221], and mass spectrometric data (70OMS13;
71TFS1768; 82OMS321). Intermolecular hydrogen-bonding in solid HMX
has also been investigated (58MI1; 63JCS1105; 83ACR146). There have
been reports of HMX complexes and adducts with other molecules
(63MI2; 69MI1; 79MI3; 82MI8).

The decomposition of HMX results in the rapid evolution of NO_2
(75MI3) followed by a slower production of N_2O and formaldehyde
(69MI3). There have been other reports of the decomposition characteris-
tics of HMX (62MI1; 67MI3; 71TFS556; 72MI1; 82MI7; 83JMR143;
84MI1; 85JPC3118). In a structural and Fourier transform infrared de-
tection (FTIR) characterization of the thermal decomposition of **241**
($R^1 - R^3 = NO_2$, $R^4 = CH_2N_3$) (see Section V,A,1,b), X-ray crystallographic analysis showed this compound to have a chair–chair conformation with CH_2—N_3 bonding being very long, which explains the
initial liberation of HN_3. As was the case with HMX (*vide supra*), N_2O and
formaldehyde began evolving as time progresses, overshadowing the initial HN_3 loss (84JPC4138).

2. *1,3,5,7-Tetrazocines not Related to HMX*

a. *1,3,5,7-Tetra-N-Substituted Compounds.* In 1936, Kadowaki
claimed that the reaction of N,N'-dimethylurea with formaldehyde
in the presence of acid or base afforded **245** (R = R′ = Me) (36BCJ248).
In the 1970s, this compound was reported as being formed

from N,N'-dimethyl-N-methoxyurea in the presence of acid, by condensation of bisurea **246** with acidic formaldehyde and by reacting N,N'-dimethylurea with oxadiazine **247** (X = O). Interestingly, **245** (R = R' = Me) underwent acid-catalyzed hydrolytic ring-contraction to **247** (X = NH) (73S243). [For a similar **241–242** reaction, see Section V,A,1, c,iii.] Maas and co-workers found that **245** (R = R' = Me) reacted intramolecularly in the presence of trifluoroacetic anhydride to yield the dicationic salt **248**. This was in contrast to the behavior of the related tetraazabicyclo[3.3.0]octadiene, which gave bimolecular reaction products (85JHC907).

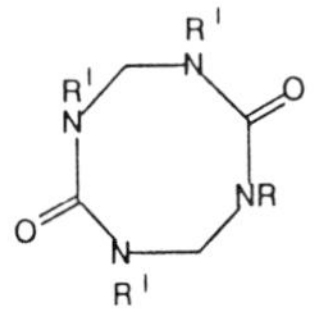

245

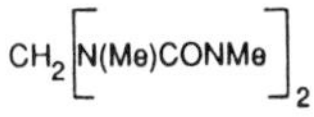

246

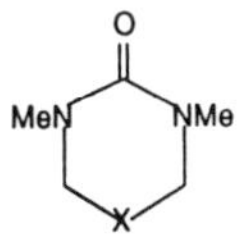

247

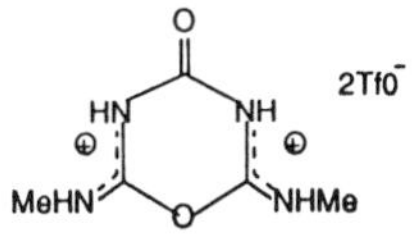

248

b. *Other 1,3,5,7-Tetrazocines.*   Condensation of acetylaldehyde with ammonia afforded tetrazocine **249** (67JAP67-4262). Reaction of 2,2-diaminohexafluoropropane with N-N'-dinitromethylenediamine and formaldehyde resulted in formation of bis(trifluoromethyl)tetrazocine **250** plus a 1,3,5-triazine (71JOC347). Urea–formaldehyde polymerization gave

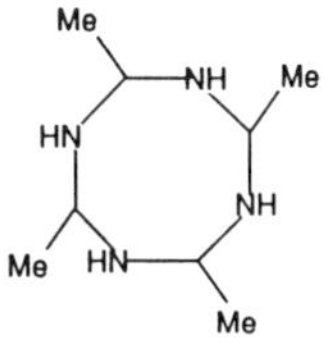

249

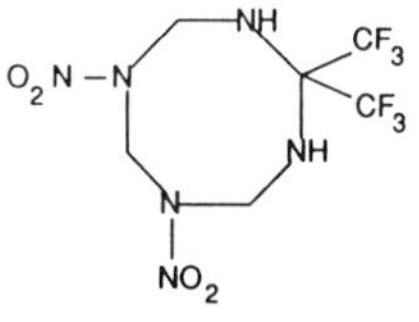

250

the cyclic bisureas **245** as byproducts (R = H, CH$_2$OH; R' = H) (73ZPK1868).

Howard and Michels, repeating the work of Fairfull and Peak, found that the previous assignment (55JCS796) of 2-pyridyl isothiocyanate as the product of the product of the reaction of 2-aminopyridine with carbon disulfide and ammonia or triethylamine (via *N*-2-pyridyldithiocarbamate salts) was incorrect; they assigned dimer **251** as the correct structure (60JOC829). Head-to-tail base-catalyzed condensation of **252** with formaldehyde yielded at 2 : 2 adduct **253** (along with a very small amount of what may have been a 12-ring 3 : 3 adduct (82LA489).

Early attempts to prepare the completely unconjugated unsaturated 1,3,5,7-tetrazocine (**254**) had always led instead to formation of tetraazapentalenes [83BSB781; 87AG(E)1039], but Gompper and co-workers finally synthesized such compounds [**254** R' = OEt; R = Me (83AG(E) 543), Ph (83AG(E)543), RR = var. alkylene bridged (87AG(E)1039)] by treating the appropriate tetraazabicyclooctadienes (**255**) with oxidizing agents, e.g., *t*-butyl hypochlorite/potassium *t*-butoxide.

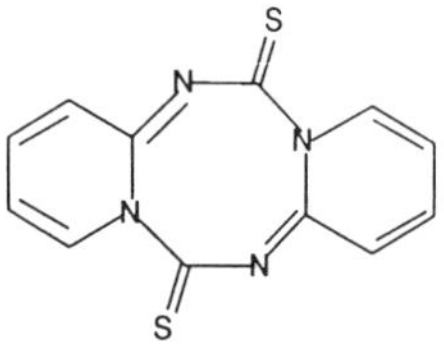

251 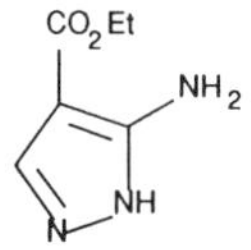   252 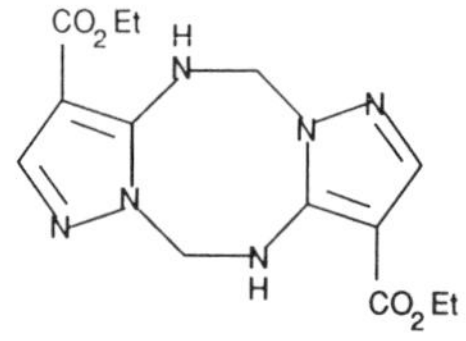   253

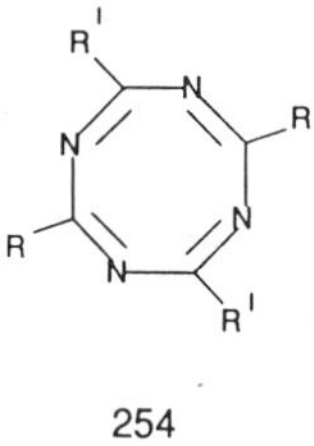

254 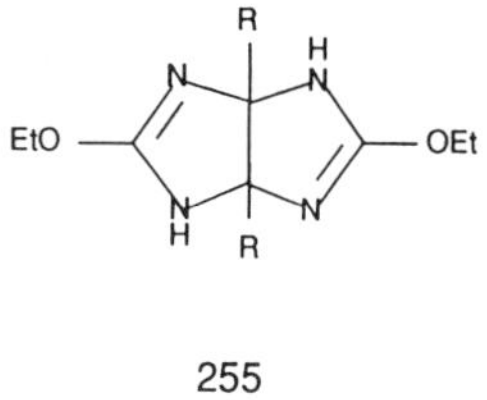   255

X-ray structural analysis of **254** (R = Ph, R' = OEt) showed it to have a boat conformation [83AG(E)543]. Reduction of **254** (R = Me, R' = OEt) with sodium in ammonia was first thought to yield a tetrazocine dianion salt (83BSB781), but later shown to give bicyclic salt **256**, which can be regarded as a bishomoantiaromatic 12π-electron system. Similar salts were prepared from **254** (RR = alkylene bridge; R' = OEt) [87AG(E)1039].

Compound **254** (R = Me, R' = OMe), when treated with silica gel in either solution [83AG(E)543] or with HBr (83BSB781), afforded **257**. When **254** (R = Me, R' = OMe) was heated with diethylamine [83AG(E)543] or dimethylamine (83BSB781), **254** (R = Me, R' = NEt$_2$ or NMe$_2$) resulted. The thermal stability of certain polyacetals was improved by incorporating tetrazocine **258** along with the usual antioxidants (69JAP69-29388).

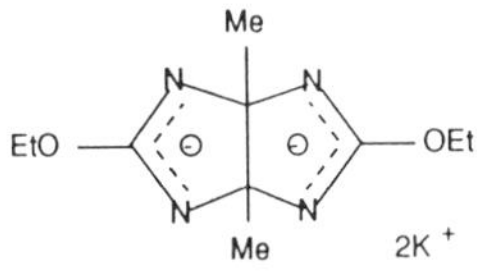

256

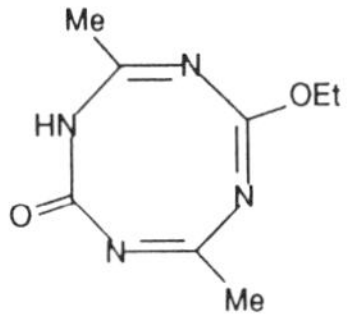

257

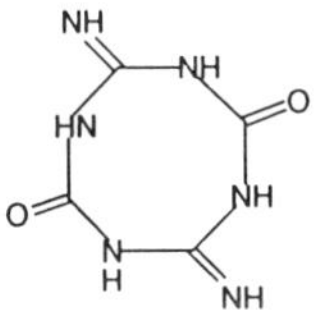

258

B. TETRAZOCINES OTHER THAN 1,3,5,7-TETRAZOCINES

1. Preparation

a. *1,2,3,5-Tetrazocines*. Plescia and co-workers diazotized *N*-pyrazolyl-*o*-aminobenzamides **226** [R = Ph; R' = H; RR' = (CH$_2$)$_4$, (CH$_2$)$_5$] to obtain the corresponding tetrazocines (**259**) (74MI2; 75JHC199). Reaction of **226** with an orthoformate gave a triazocine (see Section IV,A,3).

b. *1,2,4,5-Tetrazocines*. Bis(phenylhydrazone) **260**, when condensed with formaldehyde in the presence of potassium carbonate, afforded tetrazocine **261**. Compound **260** had been prepared by reacting 2 mol of nitroformaldehyde phenylhydrazone with 1 mol of formaldehyde. Compound **261** could also be prepared directly from equimolar amounts of nitroformaldehyde phenylhydrazone and formaldehyde in the presence of potassium carbonate (64RC557). Reaction of triazepines **262** (X = O,

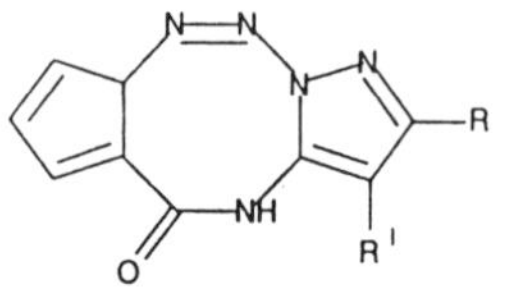

259 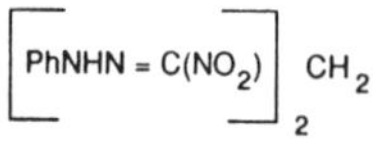   260 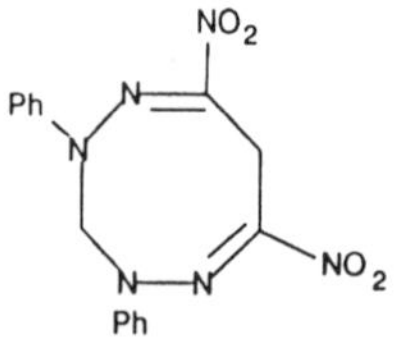   261

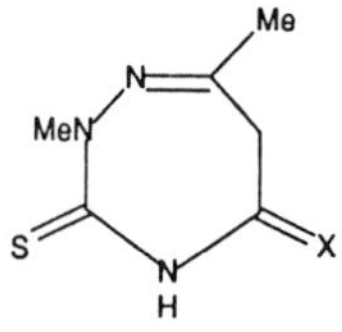

262 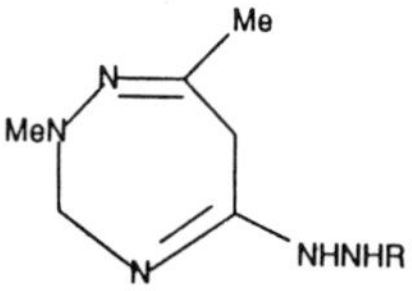   263 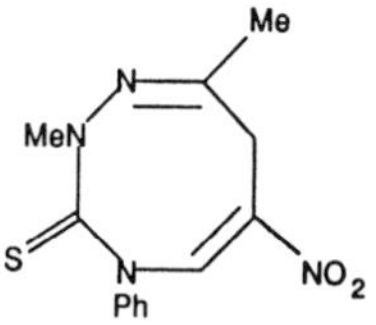   264

S) and **263** (R = H, Ph) with phenylhydrazine yielded, (in addition to pyrazoles, triazepines, and ring-opened products) tetrazocine **264.** Reaction of **262** (X = O, S) or **263** (R = H, Ph) with hydrazine or methylhydrazine failed to produce any tetrazocine (78RTC204).

c. *1,2,4,6-Tetrazocines.* Thiosemicarbazides **265** [R = H (27JIC561; 28JIC439); R = Ph (27JIC561)] underwent hydrochloric acid (27MI2; 28JIC439) or potassium hydroxide catalyzed (27MI2) ring-closure to give benzotetrazocine **266.**

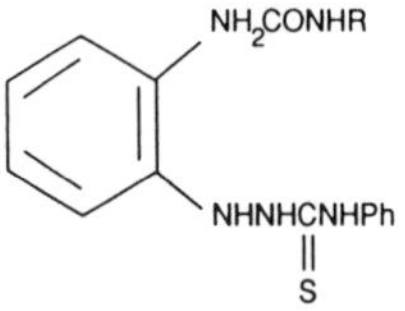

265 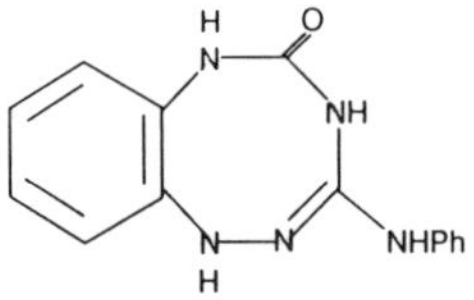   266 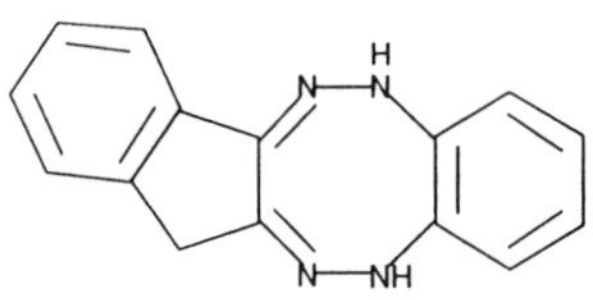   267

d. *1,2,5,6-Tetrazocines.* In a report of a 1,2,5,6-tetrazocine synthesis, Guha and De condensed isatin and phenanthraquinone monoximes with *o*-aminophenylhydrazine to obtain **267** and **268,** respectively (26MI1). The compound initially thought to be tetrazocine **269** (59USP2504544) was found to be the mesoionic heteroaromatic tetraazapentalene **270.** This structure revision was supported by NMR spectroscopy, by independent

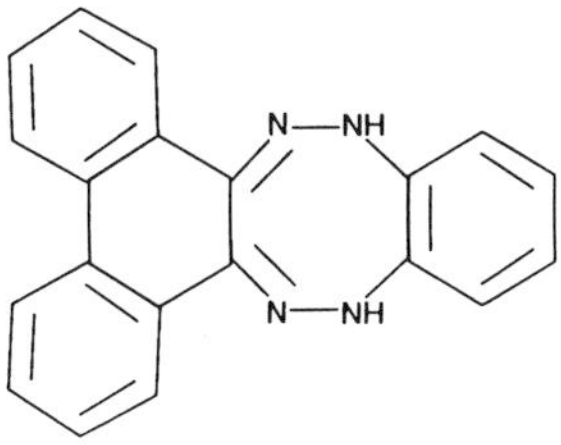

268 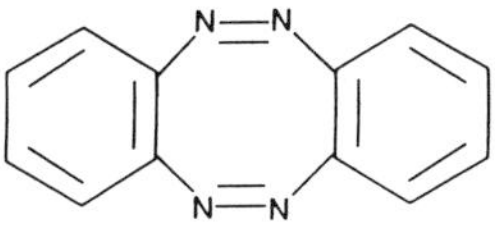   269 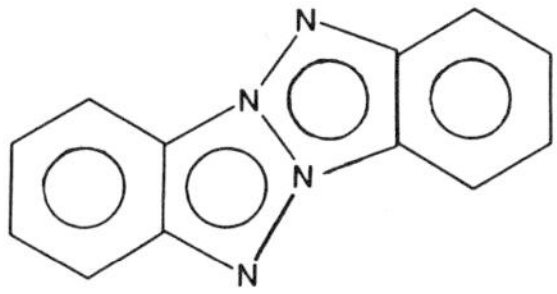   270

synthesis of a compound analogous to **270,** and was conclusively established by X-ray diffraction analysis (62JA2453; 67JA2618). Similar earlier revisions of reported tetrazocine structures were cited (67JA2618). This preference for a bicyclic (i.e., **270**) structure over a monocyclic one (i.e., **269)** is similar to that found for the (1,4)-diazocine analog of **269** [89AHC(45)185] and is supported by molecular orbital calculations on nonfused-ring analogs of **269** and **270** (67JA2618; 67JA2640) (see Section V,B,2).

Ogden condensed azine **46** with oxalyl fluoride in the presence of cesium fluoride to isolate tetrakis(trifluoromethyl)tetrazocine **271** [71JCS(C)2920]. Gol'din and Tsiomo reacted dihydrazine **272** with biacetyl to make polyhy-

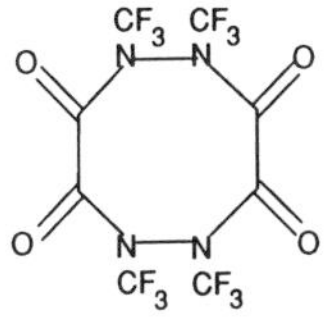

271 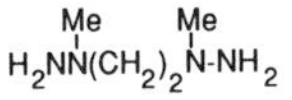   272 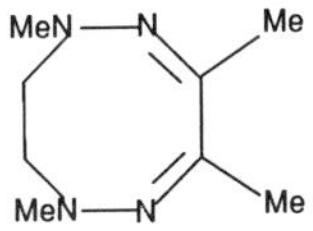   273

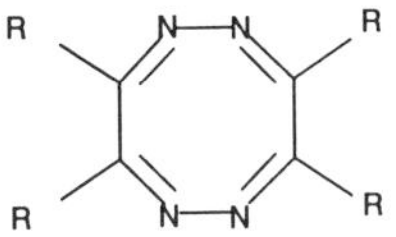

274 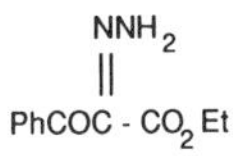   275

drazones, but also obtained the low molecular weight tetrazocine **273** (75MI10). Aydin and Feuer found that when benzil monohydrazone was condensed with ethyl nitroacetate, they obtained, in addition, to deoxybenzoin, benzil azine, and a pyridazinone, small amounts of tetrazocine **274** (R = Ph) (79MI2). Similarly, when hydrazone **275** was kept at room temperature it converted into **274** (R = CO$_2$Et) [87JCS(D)501]. However,

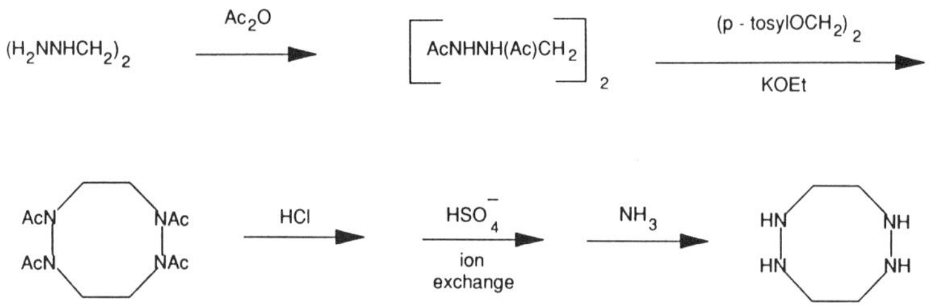

$(H_2NNHCH_2)_2$ $\xrightarrow{Ac_2O}$ $\left[AcNHNH(Ac)CH_2 \right]_2$ $\xrightarrow[\text{KOEt}]{(p\text{-}tosylOCH_2)_2}$

SCHEME 15

the previous erroneous reports of tetrazocine formation (e.g., **269**) (62JA2453; 67JA2618) make structure **274** suspect.

Neilsen succeeded in preparing the first unsubstituted tetrazocine (**276**) by the reaction sequence shown in Scheme 15. This paper also reviewed

277

278

previous erroneous reports of 1,2,5,6-tetrazocine synthesis (76JHC101). Taylor and co-workers reported that noncrystalline, 1-substituted diazetidinones **277** (R = H, Ph; R' = Ph) and **277** [RR' = $(CH_2)_5$] underwent

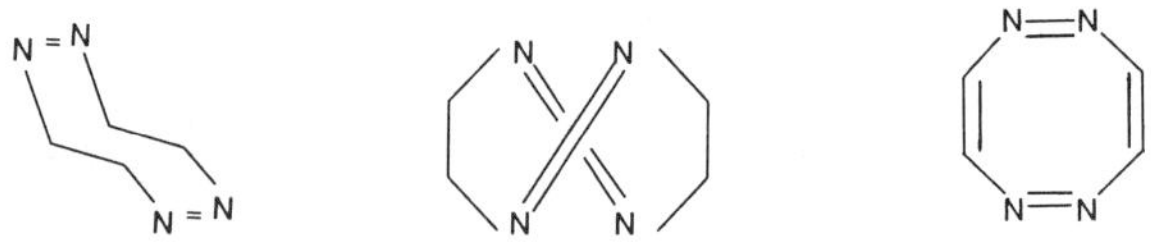

279

280

281

quantitative transformation to tetrazocine dimers **278** [R = H, R' = Ph (81JA7660)] and **278** [R = Me, R' = $CH_2CH(OMe)_2$; RR' = $(CH_2)_5$ (84JOC4415)]. Other crystalline, 1-substituted diazetidinones were indefinitely stable (84JOC4415).

2. Theoretical and Structural Aspects

Molecular mechanics calculations showed that the chair (C_{2h}) conformation of the (Z,Z) cyclic bisazocompound **279** was more stable than the boat (C_{2v}) and twist (D_2) forms by 7-9 kcal/mol, whereas for the (E,E)-isomer the twist (D_2) conformation (**280**) was favored over the chair (C_{2h}) by ~ 5 kcal/mol. Interestingly, these results were the reverse of those for the carbocyclic analogs i.e., (Z,Z)- and (E,E)-1,5-cyclooctadiene (79JA5546). Hückel molecular orbital calculations on the two forms of 1,2,5,6-tetrazocine, the azine form (**274**, R = H) and its double-bond shifted azo tautomer (**281**), showed that both structures have significantly lower delocalization energy that the nonfused-ring parent of tetraazapentalene **270** (67JA2640).

ACKNOWLEDGMENTS

The author wishes to thank Mr. Daniel Tartaglia, who helped obtain and assemble some of the articles used to write this review, and Mr. Brian Lynch, who obtained and assembled articles pertaining to HMX (Section V,A,1) and wrote the preliminary draft of that section of the review.

References

09CR401	H. Duval, *C. R. Hebd. Seances Acad. Sci.* **149**, 401 (1909).
10BSF727	H. Duval, *Bull. Soc. Chim. Fr.* **7**, 727 (1910).
26MI1	P. C. Guha and M. K. De, *Q. J. Indian Chem. Soc.* **3**, 41 (1926).
27MI2	P. C. Guha and T. N. Ghosh, *Q. J. Indian Chem. Soc.* **4**, 561 (1927).
28JIC439	P. C. Guha and T. N. Ghosh, *J. Indian Chem. Soc.* **5**, 439 (1928).
29MI1	M. Giua and G. Racciu, *Atti R. Accad. Sci. Torino, Cl. Sci. Fis., Mat. Nat.* **64**, 300 (1929) [*CA* **24**, 3212 (1930)].
32G20	A. Giacalone, *Gazz. Chim. Ital.* **62**, 20 (1932).
33G764	A. Giacalone, *Gazz. Chim. Ital.* **63**, 764 (1933).
35JCS1796	F. M. Rowe, W. C. Dovey, B. Garforth, E. Levin, J. D. Pask, and A. T. Peters, *J. Chem. Soc.*, 1796 (1935).
36BCJ248	H. Kadowaki, *Bull. Chem. Soc. Jpn.* **11**, 248 (1936).
39JGU606	E. S. Vasserman and G. P. Miklukhin, *J. Gen. Chem. USSR (Engl. Transl.)* **9**, 606 (1939) [*CA* **33**, 7665 (1939)].
40JGU202	E. S. Vasserman and G. P. Miklukhin, *J. Gen. Chem. USSR (Engl. Transl.)* **10**, 202 (1940) [*CA* **33**, 7179 (1940)].
47CR876	J. Bougault, E. Cattelain, and P. Chabrier, *C. R. Hebd. Seances Acad. Sci.* **225**, 876 (1947).
48CR818	P. Chabrier and A. Sekera, *C. R. Hebd. Seances Acad. Sci.* **226**, 818 (1948).
49BRP615419	Honorary Advisory Council for Scientific and Industrial Research, Br. Pat. 615,419 (1949) [*CA* **43**, 9079 (1949)].
49BRP615793	Honorary Advisory Council for Scientific and Industrial Research, Br. Pat. 615,793 (1949) [*CA* **43**, 9079h (1949)].
49CB192	H. Fischer, *Chem. Ber.* **82**, 192 (1949).

49CR1952	J. Bougault and P. Chabrier, *C. R. Hebd. Seances Acad. Sci.* **228**, 1952 (1949).
49JA1842	W. E. Bachmann and J. C. Sheehan, *J. Am. Chem. Soc.* **71**, 1842 (1949).
49JCS1631	F. Chapman, *J. Chem. Soc.*, 1631 (1949).
49JCS1638	F. Chapman, P. G. Owston, and D. Woodcock, *J. Chem. Soc.*, 1638 (1949).
49MI1	W. J. Chute, D. C. Downing, A. F. McKay, G. S. Meyers, and G. F. Wright, *Can. J. Res., Sect. B* **27**, 218 (1949).
49MI2	A. F. McKay, H. H. Richmonds, and G. F. Wright, *Can. J. Res., Sect. B* **27**, 462 (1949).
49MI3	A. Graham, R. H. Meen, G. S. Meyers, and G. F. Wright, *Can. J. Res., Sect. B* **27**, 520 (1949).
50AC225	W. C. McCrone, *Anal. Chem.* **22**, 1225 (1950).
51JA2769	W. E. Bachmann, W. J. Horton, E. L. Jenner, N. W. MacNaughton, and L. B. Scott, *J. Am. Chem. Soc.* **73**, 2769 (1951).
51JA2773	W. E. Bachmann and E. L. Jenner, *J. Am. Chem. Soc.* **73**, 2773 (1951).
52CJC734	S. Epstein and C. A. Winkler, *Can. J. Chem.* **30**, 734 (1952).
53CJC602	R. A. Marcus and C. A. Winkler, *Can. J. Chem.* **31**, 602 (1953).
54USP2678927	G. F. Wright and W. J. Chute, U.S. Pat. 2,678,927 (1954) [*CA* **49**, 7606 (1955)].
55JCS796	A. E. S. Fairfull and D. A. Peak, *J. Chem. Soc.*, 796 (1955).
55MI1	P. E. Eiland and R. Pepinsky, *Z. Kristallogr., Kristallgeom., Kristallphys., Kristallchem.* **106**, 273 (1955) [*CA* **49**, 14416 1955)].
56AG374	H. Beyer and T. Pyl, *Angew. Chem.* **68**, 374 (1956).
56CI(L)1479	R. G. R. Bacon and W. S. Lindsay, *Chem. Ind. (London)*, 1479 (1956).
56JA3671	P. M. Bhargava and C. Heidelberger, *J. Am. Chem. Soc.* **78**, 3671 (1956).
56JA6144	A. F. McKay, W. G. Hatton, and R. O. Brown, *J. Am. Chem. Soc.* **78**, 6144 (1956), and references cited therein.
56JCS3475	M. Hall, J. E. Ladbury, M. S. Lesslie, and E. E. Turner, *J. Chem. Soc.*, 3475 (1956).
56RTC1159	J. M. van der Zanden and G. De Vries, *Recl. Trav. Chim., Pays-Bas* **75**, 1159 (1956).
57JA4358	S. Ozaki, T. Mukaiyama, and K. Uno, *J. Am. Chem. Soc.* **79**, 4358 (1957).
57JCS651	G. H. Beaven and E. A. Johnson, *J. Chem. Soc.*, 651 (1957).
57LA50	H. Beyer and T. Pyl, *Justus Liebigs Ann. Chem.* **605**, 50 (1957).
57LA67	H. Behringer and H. Meier, *Justus Liebigs Ann. Chem.* **607**, 67 (1957).
57MI1	A. Werbin, *U.S. At. Energy Comm.* **UCRL-5078**, 8 (1957) [*CA* **53**, 1920 (1959)].
57NKZ1416	Y. Iwakura, K. Uno, and K. Hamatani, *Nippon Kagaku Zasshi* **78**, 1416 (1957) [*CA* **54**, 1539h (1960)].
57RTC519	J. M. van der Zanden and G. De Vries, *Recl. Trav. Chim., Pays-Bas* **76**, 519 (1957).
58JCS1375	R. G. R. Bacon and W. S. Lindsay, *J. Chem. Soc.*, 1375 (1958).
58MI1	R. Pepinsky, *Rev. Mod. Phys.* **30**, 100 (1958) [*CA* **52**, 12504 (1958)].

59JA217   C. G. Overberger and I. Tashlick, *J. Am. Chem. Soc.* **81**, 217 (1959).
59USP2504544 R. A. Carboni, U.S. Pat. 2,504,544 (1959) [*CA* **54**, 11062d (1960)].
60HCA1890   F. Gerson, E. Heilbronner, A. van Veer, and B. M. Wepster, *Helv. Chim. Acta* **43**, 1890 (1960).
60JA1617   T. C. Castorina, F. S. Holohan, R. J. Graybush, J. V. R. Kaufman, and S. Helf, *J. Am. Chem. Soc.* **82**, 1617 (1960).
60JOC829   J. C. Howard and J. G. Michels, *J. Org. Chem.* **25**, 829 (1960), and references cited therein.
60JOC1509   N. L. Allinger and G. A. Youngdale, *J. Org. Chem.* **25**, 1509 (1960).
60LA150   H. Beyer, T. Pyl, and C. E. Völker, *Justus Liebigs Ann. Chem.* **638**, 150 (1960), and references cited therein.
60USP2941994 L. B. Silberman, U.S. Pat. 2,941,994 (1960) [*CA* **54**, 20211 (1960)].
61CB2148   G. Wittig and J. E. Grolig, *Chem. Ber.* **94**, 2148 (1961).
61MI1   D. B. Parihar, K. K. Shukla, P. K. Saxena, and K. R. D. Chandiramani, *Chem. Process Des. Symp.* **22**, 1961 (1963) [*CA* **59**, 15113 (1963)].
61USP2983725 J. P. Picard, U.S. Pat. 2,983,725 (1961) [*CA* **55**, 20436 (1961)].
61USP2988478 R. N. Gordon, U.S. Pat. 2,988,478 (1961) [*CA* **57**, 6363i (1962)].
62CJC2278   M. Bedard, H. Huber, J. L. Meyers, and G. F. Wright, *Can. J. Chem.* **40**, 2278 (1962).
62JA2453   R. A. Carboni and J. E. Castle, *J. Am. Chem. Soc.* **84**, 2453 (1962).
62MI1   M. J. Urizar, E. D. Loughran, and L. C. Smith, *Explosivstoffe* **10**, (N3), 55 (1962) [*CA* **57**, 6196 (1962)].
62MI2   W. C. McCrone, *Microchem. J., Symp. Ser.* **2**, 243 (1962) [*CA* **58**, 6267 (1962)].
63AP510   H. J. Roth and G. Dvorak, *Arch. Pharm. (Weinheim, Ger.)* **296**, 510 (1963).
63AX617   H. H. Cady, A. C. Larson, and D. T. Cromer, *Acta Crystallogr.* **16** (7), 617 (1963) [*CA* **59**, 12260 (1963)].
63BRP940165   J. R. Geigy, A-G, Br. Pat. 940,165 (1963) [*CA* **61**, 1816a (1964)].
63CR929   J. Kossanyi, *C. R. Hebd. Seances Acad. Sci.* **257**, 929 (1963).
63JA2144   L. A. Carpino, *J. Am. Chem. Soc.* **85**, 2144 (1963).
63JA2752   C. G. Overberger, J.-P. Anselme, and J. R. Hall, *J. Am. Chem. Soc.* **85**, 2752 (1963), and references cited therein.
63JCS1105   D. J. Sutor, *J. Chem. Soc.*, 1105 (1963).
63LA160   H. Lettre and H. Schelling, *Justus Liebigs Ann. Chem.* **669**, 160 (1963).
63MI1   J. Haberman and T. C. Castorina, *NASA, Doc.* **N63-23115** (1963) [*CA* **61**, 6849 (1964)].
63MI2   A. H. Castelli, D. J. Cragle, and W. E. Fredericks, *NASA, Doc.* **N63-19914** (1963) [*CA* **60**, 8659 (1946)].
64RC557   W. E. Hahn and H. Zawadzka, *Rocz. Chem.* **38**, 557 (1964) [*CA* **61**, 10685d (1964)].
64T159   G. F. Wright, *Tetrahedron* **20**, Suppl. 1, 159 (1964).
64USP3133086 W. Bossard, H. Bosshard, and H. E. Wegmueller, U.S. Pat. 3,133,086 (1964) [*CA* **61**, 16200e (1964)].
65MI1   A. S. Teetsov and W. C. McCrone, *Microsc. Cryst. Front* **15** (1), 13 (1965) [*CA* **67**, 76965 (1967)].
65USP3170929 H. S. Lowrie, U.S. Pat. 3,170,929 (1965) [*CA* **62**, 14706 (1965)].
65USP3219522 P. N. Gordon, U.S. Pat. 3,219,522 (1965) [*CA* **64**, 667lf (1966)].

66FRP1463470   Eastman Kodak Co., Fr. Pat. 1,463,470 (1966) [*CA* **67,** 75044 (1967)].

66JCS(C)867   J. A. Bell and I. Dunstan, *J. Chem. Soc. C,* 867 (1966).

66JCS(C)870   J. A. Bell and I. Dunstan, *J. Chem. Soc. C,* 870 (1966).

66JIC723   A. S. Wasfi, *J. Indian Chem. Soc.* **43,** 723 (1966) [*CA* **66,** 55193 (1967)].

66LA109   G. Wittig, P. Borzel, F. Neumann, and G. Klar, *Justus Liebigs Ann Chem.* **691,** 109 (1966).

67CB2569   J. Burkhardt and K. Hamann, *Chem. Ber.* **100,** 2569 (1967).

67FRP1476116   Eastman Kodak Co., Fr. Pat. 1,476,116 (1967) [*CA* **68,** 39192 (1968)].

67IZV1839   S. S. Novikov, A. A. Dudinskoya, N. V. Makarov, and L. I. Khmel'nitskii, *Izv. Akad. Nauk SSSR, Ser. Khim.* 1839 (1967) [*CA* **68,** 29689 (1968)].

67JA2618   R. A., Carboni, J. C. Kauer, J. E. Castle, and H. E. Simmons, *J. Am. Chem. Soc.* **89,** 2618 (1967).

67JA2640   Y. T. Chia and H. E. Simmons, *J. Am. Chem. Soc.* **89,** 2640 (1967).

67JAP67-4262   H. Miyama, Jpn. Pat. 67-4262 (1967) [*CA* **68,** 59628 (1968)].

67MI1   T. Urbanski, "Chemistry and Technology of Explosives," Vol. 3, Chapter IV, Pergamon, Oxford, 1967.

67MI2   B. Suryanarayana, J. R. Actera, and R. J. Graybush, *Mol. Cryst.* **2** (4), 373 (1967) [*CA* **68,** 73104 (1968)].

67MI3   V. L. Zbarskii, Y. Y. Maksimov, and E. Y. Orlova, *Tr. Mosk. Khim.-Tekhnol. Inst. im. D.I. Mendeleeva* **53,** 84 (1967) [*CA* **68** 41745 (1968)].

67MI4   T. C. Castorina and J. Haberman *U.S. C. F. S. T. I. AD Rep.* **653027** (1967) [*CA* **67,** 92461 (1967)].

68BEP719042   Poudreries Réunies de Belgique, A. A., Belg. Pat. 719,042 (1968) [*CA* **71,** 50009 (1969)].

68JCS(B)6   A. H. Lamberton, I. O. Sutherland, J. E. Thorpe, and H. M. Yusuf, *J. Chem. Soc. B,* 6 (1968).

68JCS(C)1103   M. E. Derieg, R. I. Fryer, and L. H. Sternbach, *J. Chem. Soc. C,* 1103 (1968).

68MI1   T. C. Castorina, J. Haberman, and A. F. Smetana, *Int. J. Appl. Radiat. Isot.* **19,** 495 (1968) [*CA* **69,** 32007 (1968)].

68SAP6706503   H. Staele and K. Zeile, South African Pat. 6,706,503 (1968) [*CA* **70,** 47453 (1969)].

68TL789   B. Loev and M. M. Goodman, *Tetrahedron Lett.,* 789 (1968).

69AJC2505   J. Stals, *Aust. J. Chem.,* 2505 (1969) [*CA* **72,** 35977 (1970)].

69AJC2515   J. Stals, *Aust. J. Chem.,* 2515 (1969) [*CA* **72,** 42575 (1970)].

69JA3226   C. G. Overberger, J. W. Stoddard, C. Yaroslavsky, H. Katz, and J. P. Anselme, *J. Am. Chem. Soc.* **91,** 3226 (1969), and references cited therein.

69JAP69-29388   K. Tamura, S. Kataoka, and A. Osakada, Jpn. Pat. 69-29388 (1969) [*CA* **72,** 101416 (1970)].

69JCS(C)1556   J. A. Bell and I. Dunstan, *J. Chem. Soc. C,* 1556 (1969).

69JOC3237   W. W. Paudler and A. G. Zeiler, *J. Org. Chem.* **34,** 3237 (1969).

69MI1   W. Selig, *Explosivstoffe,* 73 (1969) [*CA* **71,** 91449 (1969)].

69MI2   J. N. Maycock, V. R. P. Verneker, and L. Rouch, Jr., *Phys. Status Solidi* **35,** 843 (1969) [*CA* **71,** 117516 (1969)].

69MI3   J. N. Maycock V. R. P. Vernecker, and W. Lochte, *Phys. Status Solidi* **35**, 849 (1969) [*CA* **71**, 107161 (1969)].
69RC1687   L. Stefaniak, T. Urbanski, M. Witanowski, and H. Januszewski, *Rocz. Chem.* **9**, 1687 (1969).
69TFS904   J. Stals, C. G. Barraclough, and A. S. Buchanan, *Trans. Faraday Soc.*, 904 (1969) [*CA* **70**,101359 (1969)].
70AJC1913   N. V. Riggs and S. M. Verma, *Aust. J. Chem.* **23**, 1913 (1970).
70AX(B)1235 C. S. Choi and H. P. Boutin, *Acta Crystallogr., Sec. B* **B26**, 1235 (1970) [*CA* **73**, 134889 (1970)].
70JA4922   C. G. Overberger and J. W. Stoddard, *J. Am. Chem. Soc.* **92**, 4922 (1970).
70MI1   D. Danchev, K. Khristova, and D. Sidzhakova, *Farmatsiya (Sofia)* **20**, (1970) [*CA* **74**, 31740 (1971)].
70MI2   K. Suryanarayanan and S. Bulusu, *U.S. Gov. Res. Dev. Rep.* **70**, 74 (1970) [*CA* **75**, 12946 (1971)].
70OMS13   S. Bulusu, T. Axenrod, and G. W. A. Milne, *Org. Mass Spectrom* **3** 13 (1970) [*CA* **72**, 110451 (1970)].
71BSF3245   A. LeBerre, C. Renault, and P. Giraudeau, *Bull. Soc. Chim. Fr.*, 3245 (1971).
71CR (B)658 R. Cavagnat, M. T. Forel, and M. Rey-Lafon, *C. R. Hebd. Seances Acad. Sci., Ser. B* **273**, 658 (1971).
71FRP2053804 R. Cartalas, J. P. Konrat, and L. Molard, Fr. Pat. 2,053,804 (1971) [*CA* **76**, 85850 (1972)].
71JA5573   B. M. Trost and R. M. Cory, *J. Am. Chem. Soc.* **93**, 5573 (1971).
71JCS(C)2920 P. H. Ogden, *J. Chem. Soc. C,* 2920 (1971).
71JOC347   J. A. Young, J. J. Schmidt-Collerus, and J. A. Krimmel, *J. Org. Chem.* **36**, 347 (1971).
71TFS556   P. G. Hall, *Trans. Faraday Soc.* **67**, 556 (1971) [*CA* **74**, 103832 (1971)].
71TFS1739   J. Stals, *Trans. Faraday Soc.* **67**, 1739 (1971) [*CA* **75**, 43034 (1971)].
71TFS1768   J. Stals, *Trans. Faraday Soc.* **67**, 1768 (1971) [*CA* **75**, 43032 (1971)].
72AAC541   M. Misiek, T. A. Pursiano, L. B. Crast, F. Leitner, and K. E. Price, *Antimicrob. Agents Chemother.* **1**, 54 (1972) [*CA* **76**, 121458 (1972)].
72CPB927   M. Furukawa, K. Mitsuru, Y. Kajima, and S. Hayashi, *Chem. Pharm. Bull.* **20** 927 (1972) [*CA* **77**, 101550 (1972)].
72GEP2216837 M. Shindo, M. Kakimoto, and H. Nagano, Ger. Pat. 2,216,837 (1972) [*CA* **78**, 43542 (1973)].
72JAP72-34384 M. Shindo, M. Kakimoto, and H. Nagano, Jpn. Pat. 72-34384 (1972) [*CA* **78**, 58483 (1973)].
72JAP72-34386 M. Shindo, M. Kakimoto, and H. Nagano, Jpn. Pat. 72-34386 (1972) [*CA* **78**, 111391 (1973)].
72JCS(F2)1659 J. Stals and M. G. Pitt, *J. C. S., Faraday Trans. 2* **68**, 1659 (1972).
72JOC320   E. B. Hodge, *J. Org. Chem.* **37**, 320 (1972).
72JOC1851   L. A. Carpino and J. P. Masaracchia, *J. Org. Chem.* **37**, 1851 (1972).
72LA112   M. Bruggar and F. Korte, *Justus Liebigs Ann. Chem.* **764**, 112 (1972).
72MI1   C. Darnez and J. Paviot, *Int. J. Radiat. Phys. Chem.* **4**, 11 (1972) [*CA* **77**, 18922 (1972)].

72TL4565   C. G. Overberger, M. S. Chi, and J. A. Barry, *Tetrahedron Lett.*, 4565 (1972).

73AC(P)329   A. Yanagida and C. Gansser, *Ann. Chim. (Paris)* **8**, 329 (1973) [*CA* **80**, 132937 (1974)].

73AX(B)651   C. S. Choi, J. E. Abel, B. Dickens, and J. M. Stewart, *Acta Crystallogr., Sect. B* **29**, 651 (1973).

73BSF2029   G. Leclerc, *Bull. Soc. Chim. Fr.*, 2029 (1973).

73CC602   S. Bulusu, J. R. Autera, and T. Axenrod, *J. C. S., Chem. Commun.*, 602 (1973).

73CPB1520   K. Kamiya, Y. Wada, and M. Nishikawa, *Chem. Pharm. Bull.* **21**, 1520 (1973), and references cited therein.

73CPB2375   K. Meguro and Kuwada, *Chem. Pharm. Bull.* **21**, 2375 (1973).

73GEP2308064   M. Shindo, M. Kakimoto, H. Nagano, and Y. Fujimura, Ger. Pat. 2,308,064 (1973) [*CA* **79**, 126533 (1973)].

73HCA705   E. Hasselbach, A. Henriksson, A. Schmelzer, and H. Berthou, *Helv. Chim. Acta* **56**, 705 (1973).

73JA2738   T. J. Katz and N. Acton, *J. Am. Chem. Soc.* **95**, 2738 (1973).

73JAP73-00491   M. Shindo, M. Kakimoto, and H. Nagano, Jpn. Pat. 73-00491 (1973) [*CA* **78**, 111391 (1973)].

73JAP73-33760   H. Natsukari, K. Meguro, and Y. Kuwada, Jpn. Pat. 73-33760 (1973) [*CA* **80**, 121025 (1974)].

73JAP73-35276   H. Natsukari, K. Meguro, and B. Kuwata, Jpn. Pat. 73-35276 (1973) [*CA* **81**, 13567 (1974)].

73JHC43   J. J. Porter, J. L. Murray, and K. B. Takvorian, *J. Heterocycl. Chem.* **10**, 43 (1973).

73JHC279   H. Yoshida, G. Sen, and B. S. Thyagarajan, *J. Heterocycl. Chem.* **10**, 279 (1973).

73JHC423   W. W. Paudler, A. G. Zeiler, and M. M. Goodman, *J. Heterocycl. Chem.* **10**, 423 (1973).

73JHC725   H. Yoshida and B. S. Thyagarajan, *J. Heterocycl. Chem.* **10**, 725 (1973).

73MI1   V. Kucera and E. Hromadkova, *Chem. Prum.* **23**, 184 (1973) [*CA* **79**, 10719 (1973)].

73MI2   I. J. Solomon, R. K. Momii, F. H. Jarke, A. I. Kacmarek, J. K. Raney, and P. C. Adlaf, *J. Chem. Eng. Data* **18**, 335 (1973).

73S243   H. Peterson, *Synthesis*, 243 (1973).

73SC249   T. Sasaki, H. Kanematsu, Y. Yukimoto, and E. Kato, *Synth. Commun.* **3**, 249 (1973).

73ZPK1868   B. M. Bulygin, L. A. Aleksandrova, and N. I. Borodkina, *Zh. Prikl. Khim. (Leningrad)* **46**, 1868 (1973); *J. Appl. Chem. USSR (Engl. Transl.)* **46**, 1991 (1973) [*CA* **80**, 60432 (1974)].

74AP673   D. Danchev, K. Khristova, V. Mutaftschieva, and D. Daleva, *Arch. Pharm. (Weinheim, Ger.)* **307**, 673 (1974) [*CA* **81**, 169521 (1974)].

74AX(B)1918   R. E. Cobbledick and R. W. H. Small, *Acta Crystallogr., Sect. B*, **30**, 1918 (1974) [*CA* **81**, 112335 (1974)].

74GEP2323554   C. M. Roldan, M. F. Brana, and J. M. C. Berlanga, Ger. Pat. 2,323,554 (1974) [*CA* **81**, 91226 (1974)].

74GEP2365309   E. R. Squibb and Sons, Inc., Ger. Pat. 2,365,309 (1974) [*CA* **81**, 63697 (1974)].

74JAP74-26292   M. Shindo, M. Kakimoto, H. Nagano, and Y. Fujimura, Jpn. Pat.
                 74-26292 (1974) [CA **81,** 120720 (1974)].
74JAP74-26673   H. Tawada, K. Meguro, and Y. Kuwada, Jpn. Pat. 74-26673 (1974)
                 [CA **82,** 140218 (1975)].
74JAP74-51292   H. Nagano, M. Shindo, and M. Kakimoto, Jpn. Pat. 74-51292
                 (1974) [CA **81,** 136198 (1974)].
74JAP74-95987   M. Shindo, M. Kakimoto, H. Nagano, and Y. Fujimura, Jpn. Pat.
                 74-95987 (1974) [CA **82,** 156404 (1975)].
74JAP74-95988   M. Shindo, H. Nagano, and Y. Fujimura, Jpn. Pat. 74-95988 (1974)
                 [CA **82,** 156403 (1974)].
74JAP74-95989   M. Shindo, M. Kakimoto, H. Nagano, and Y. Fujimura, Jpn. Pat.
                 74-95989 (1974) [CA **82,** 156402 (1975)].
74JAP74-95990   M. Shindo, M. Kakimoto, H. Nagano, and Y. Fujimura, Jpn. Pat.
                 74-95990 (1974) [CA **82,** 156405 (1975)].
74JCP221   Z. Iqbal, S. Bulusu, and J. R. Autera, *J. Chem. Phys.* **60,** 221
                 (1974).
74JHC237   V. I. Siele, M. Warman, and E. E. Gilbert, *J.Heterocycl. Chem.*
                 11, 237 (1974).
74JHC937   B. S. Thyagarajan and K. C. Majumdar, *J. Heterocycl. Chem.* **11,**
                 937 (1974), and references cited therein.
74JOC1710   R. Sarges and J. R. Tretter, *J. Org. Chem.* **39,** 1710 (1974).
74KGS233   M. A. Mikhaleva, N. L. Il'chenko, and V. P. Mamaev, *Khim.
                 Geterosikl. Soedin.,* 233 (1974) [CA **80,** 133404 (1974)].
74MI1   J. Rylance, R. W. H. Small, and D. Stubley, *J. Chem. Thermodyn.*
                 6, 1103 (1974) [CA **82,** 57338 (1975)].
74MI2   S. Plescia, E. Ajello, and V. Sprio, *Atti Accad. Sci., Lett. Arti
                 Palermo, Parte 1,* **33,** 301 (1974) [CA **83,** 97241 (1975)].
75AX(B)2805   R. E. Cobbledick and R. W. H. Small, *Acta Crystallogr., Sect. B*
                 31, 2805 (1975).
75JCS(P1)45   B. M. Adger, C. W. Rees, and R. C. Storr, *J. C. S., Perkin Trans.
                 1,* 45 (1975).
75JHC199   S. Plescia, E. Ajello, and V. Sprio *J. Heterocycl. Chem.* **12,** 199
                 (1975).
75LA1357   H. Stetter and H. Koch, *Liebigs Ann. Chem.* 1357 (1975).
75MI1   C. M. Roldan, M. F. Brana, J. M. Castellano, and F. Rabadan,
                 Arch. Farmacol. Toxicol. 1, 37 (1975) [CA **83,** 164147 (1975)].
75MI2   J. M. Tedder, A. Nechvatal, and A. H. Jubb, "Basic Organic
                 Chemistry, Part 5. Industrial Products," Wiley, New York,
                 1975.
75MI3   S. Bulusu, *Compat. Propellants, Explos. Pyrotechnics Plast.
                 Addit. Conf.* II-C, 1p. *1974 [CA* **86,** 142410 (1975)].
75MI4   C. Gansser, *Eur. J. Med. Chem. Chim. Ther.* **10,** 273 (1975), and
                 references cited therein [CA **84,** 31027 (1976)].
75MI5   P. L. Marinkas, *Gov. Rep. Announce. Index (U.S.)* **75,** 56 (1975)
                 [CA **84,** 97376 (1976)].
75MI6   C. L. Coon and J. M. Guilmont, *Gov. Rep. Announce. Index (U.S.)*
                 75, 71 (1975) [CA **83,** 118022 (1975)].
75MI7   G. Kempter, *Mekh. Deistviya Gerbits. Sint. Regul. Rosta Rast.
                 Ikh Sud'ba Biosfere, Mater. Mezhdunar. Simp. Stran-Chlenov.
                 SEV, 10th, 1975,* Vol. 1, p. 147 (1975) [CA **89,** 6313 (1978)].

75MI8　　　　　E. A. Jauregui, F. Ferretti, and C. A. Ponce, *Rev. Latinoam. Quim.* **6**, 13 (1975) [*CA* **83**, 27346 (1975)].

75MI9　　　　　J. Rylance and D. Stubley, *Thermochim. Acta* **13**, 253 (1975) [*CA* **84**, 22899 (1976)].

75MI10　　　　G. S. Gol'din, and S. N. Tsiomo, *Vysokomol. Soedin., Ser. B.* **17**, 463 (1975) [*CA* **83**, 147771 (1975)].

75T295　　　　B. A. Hess, Jr., L. J. Schaad, and C. W. Holyoke, Jr., *Tetrahedron* **31**, 295 (1975).

75T1521　　　 A. R. Farminer and G. A. Webb, *Tetrahedron* **31**, 1521 (1975).

75USP3926953　M. D. Coburn and T. M. Benziger, U.S. Pat. 3,926,953 (1975) [*CA* **84**, 90192 (1976)].

75USP3926994　A. C. White and R. M. Black, U.S. Pat. 3,926, 994 (1975) [*CA* **85**, 21505 (1976)].

76AX(B)2314　 R. L. Harlow and S. H. Simonsen, *Acta Crystallogr. Sect. B* **B32**, 2314 (1976).

76BSF1962　　 D. Cartier, J. Levy, and J. Le Men, *Bull Soc. Chim. Fr.* 1962 (1976).

76CB518　　　 G. Vitt, E. Haedicke, and G. Quinkert, *Chem. Ber.* **109**, 518 (1976).

76H53　　　　　I. Saito, Y. Takahashi, M. Imuta, S. Matsugo, H. Kaguchi, and T. Matsuura, *Heterocycles* **5**, 53 (1976).

76IJC(B)773　 V. Arya, V. Honkan, and S. J. Shenoy, *Indian J. Chem., Sect. B* **14B** 773 (1976) [*CA* **87**, 5932 (1977)].

76JA4320　　　 N. Turro, C. A. Renner, W. H. Waddell, and T. J. Katz, *J. Am. Chem. Soc.* **98**, 4320 (1976).

76JAP76-38425　O. Wakabayashi, K. Matsuya, H. Ohta, T. Jikihara, and H. Watanabe, Jpn. Pat. 76-38425 (1976) [*CA* **85**, 73445 (1976)].

76JAP76-65757　O. Wakabayashi, K. Matsuya, H. Ota, T. Jikihara, and H. Watanabe, Jpn. Pat. 76-65757 (1976) [*CA* **86**, 29859 (1977)].

76JAP76-86489　O. Wakabayashi, K. Matsuya, T. Jikihara, and S. Suzuki, Jpn. Pat. 76-86489 (1976) [*CA* **86**, 72656 (1977)].

76JCS(Fl)723　J. W. Taylor and R. J. Crookes, *J. C. S., Faraday Trans. l*, 723 (1976).

76JCS(Pl)2281　N. Dennis, A. R. Katritzky, and R. Ramaiah, *J. C. S., Perkin Trans. l*, 2281 (1976).

76JHC101　　　 A. T. Nielsen. *J. Heterocycl. Chem.* **13**, 101 (1976), and references cited therein.

76JOC3229　　 H. W. Heine, L. M. Baclawski, S. M. Bonser, and G. D. Wachab, *J. Org. Chem.* **41**, 3229 (1976).

76MI1　　　　　B. Singh, L. K. Chaturvedi, C. M. P. Kujur, P. S. Bhatia, and P. N. Gadhikar, *Indian J. Technol.* **14**, 675 (1976) [*CA* **87**, 29775 (1977)].

76MI2　　　　　P. L. Marinkas, J. E. Mapes, D. S. Downs, P. J. Kemmey, and A. C. Forsyth, *Mol. Cryst. Liq. Cryst.* **35**, 15 (1976) [*CA* **86**, 54873 (1977)].

76TL1569　　　 N. Dennis, A. R. Katritzky, E. Lunt, M. Ramaiah, R. L. Harlow, and S. H. Simonsen, *Tetrahedron Lett.*, 1569 (1976).

76USP3939148　V. I. Siele and E. E. Gilbert, U.S. Pat. 3,939,148 (1976) [*CA* **84**, 180317 (1976)].

76USP3978046　C. L. Coon, U.S. Pat. 3,978,046 (1976) [*CA* **86**, 16714 (1977).

76USP3978047　E. B. Hodge, U.S. Pat. 3,978,047 (1976) [*CA* **86**, 6152 (1977)].

76USP3979379　V. I. Siele, U.S. Pat. 3,979,379 (1976) [*CA* **86**, 72719 (1977)].

76USP3987034   B. S. Thyagarajan. U.S. Pat. 3,987,034 (1976) [*CA* **86**, 121385 (1977)].

77CPB1559   K. Murato, T. Shioiri, and S.-I. Yamada, *Chem. Pharm. Bull.* **25**, 1559 (1977).

77CPB1911   Y. Nagai, H. Uno, and S. Umemoto, *Chem. Pharm. Bull.* **25**, 1911 (1977).

77IJC(B)867   E. Jena, V. G. Nikade, J. P. Sheth, and S. C. Bhattacharyya, Indian J. Chem. *Sect. B* **15B**, 867 (1977).

77JAP77-83552   O. Wakabayashi, M. Osama, H. Ohta, T. Jikihara, and H. Watanabe, Jpn. pat. 77-83552 (1977) [*CA* **88**, 50904 (1978)].

77JCS(P2)2068   I. Perillo, B. Fernandez, and S. Lamdan, *J. C. S., Perkin Trans. 2*, 2068 (1977).

77MI1   I. Janev, L. Fukara-Jovevska, M. Jancevska, and I. Janculey, *God. Zb.—Prir.—Mat. Fak. Univ. Kiril Metodij—Skopje, Sek. A* **25–26**, 205 (1976) [*CA* **87**, 117825 (1977)].

77MI2   P. L. Marinkas, *J. Lumin.* **15**, 57 (1977) [*CA* **86**, 173768 (1977)].

77MI3   I. Janev, M. Jancevska, and I. Janculev, *Prilozi—Makedon. Akad. Nauk. Umet., Odd. Prir.—Mat. Nauki* **9**, 59 (1977) [*CA* **92**, 22487 (1980)].

77MI4   J. Michl. *Spectrosc. Lett.* **10**, 509 (1977) [*CA* **87**, 167297 (1977)].

77MI5   G. Kempter, H.-J. Ziegner, G. Moser, and W. Natho, *Wiss. Z. Paedagog Hochsch. "Karl Liebknecht" Potsdam* **21**, 5 (1977) [*CA* **89**, 163510 (1978)].

77USP4001233   H. L. Yake and R. Petigara, U.S. Pat. 4,001,233 (1977) [*CA* **86**, 171493 (1977)].

78AF1435   G. H. Douglas, J. Diamond, W. L. Studt, G. N. Mir, R. I. Alioto, K. Auyang, B. J. Burns, J. Cias, P. R. Darkes, S. A. Dodson, S. O'Connor, N. J. Santora, C. T. Tsuei, J. J. Zalipski, and H. K. Zimmerman, *Arzneim.-Forsch.* **28** (II), 1435 (1978).

78CB596   W. Duisman and C. Rüchardt, *Chem. Ber.* **111**, 596 (1978).

78JCS(F1)1339   R. B. Cundall, T. F. Palmer, and C C. Wood, *J. C. S., Faraday Trans. 1,* 1339 (1978).

78JHC649   M. Abbasi, M. Nasr, H. H. Zoorob, and J. M. Michael, *J. Heterocycl. Chem.* **15**, 649 (1978).

78JHC1309   J. K. Daniel and N. P. Peet, *J. Heterocycl. Chem.* **15**, 1309 (1978).

78JOC1544   H. Ulrich, B. Tucker, and R. Richter, *J. Org. Chem.* **43**, 1544 (1978).

78JOC4150   R. Richter, B. Tucker, and H. Ulrich, *J. Org. Chem.* **43**, 4150 (1978).

78JPC1152   F. A. Neugebauer and H. Weger, *J. Phys. Chem.* **82**, 1152 (1978).

78JP1912   F. Goetz, T. B. Brill, and J. R. Ferraro, *J.Phys. Chem.* **82**, 1912 (1978).

78RTC204   A. Hasnaoui, J.-P. Lavergne, and P. Viallefont, *Recl. Trav. Chim., Pays-Bas* **97**, 204 (1978).

78USP4086228   I. J. Solomon and L. B. Silberman, U.S. Pat. 4,086,228 (1978) [*CA* **89**, 62015 (1978)].

79CPB2084   H. Natsukari, K. Meguro, and Y. Kuwada, *Chem. Pharm. Bull.* **27**, 2084 (1979).

79JA5546   J. Kao and T. N. Huang, *J. Am. Chem. Soc.* **101**, 5546 (1979).

79JCS(P2)869   K. E. Calderbank and R. K. Pierens, *J. C. S., Perkin Trans. 2,* 869 (1979).

79JOC1264   B. M. Trost, P. H. Scudder, R. M. Cory, N. J. Turro, V. Ramamurthy, and T. J. Katz, *J. Org. Chem.* **44**, 1264 (1979).

79JPC340   F. Goetz and T. B. Brill, *J. Phys. Chem.* **83**, 340 (1979) [*CA* **90**, 94653 (1979)].

79MI1   K. Hokama, S. Yogi, and M. Higa, *Bull. Sci. Eng. Div. Univ. Ryukus, Math. Nat. Sci.* **27**, 77 (1979) [CA **94**, 120410 (1981)].

79MI2   A. Aydin and H. Feuer, *Chim. Acta Turc.* **7**, 122 (1979) [*CA* **93**, 220680 (1980)].

79MI3   W. Selig, *Energy Res. Abstr.* **4**, 47115 (1979) [*CA* **92**, 198375 (1979)].

79RTC173   N. J. Turro and V. Ramamurthy, *Rec. Trav. Chim., Pays-Bas* **98**, 173 (1979).

80IZV1886   F. L. Ponomarev, N. I. Vasyukova, S. A. Siling, B. V. Lokshin, S. V. Vinogradova, and V. V. Korshak, *Izv. Akad. Nauk SSSR Ser. Khim.,* 1886 (1980) [*CA* **94**, 103320 (1981)].

80JHC1409   G. Daidone, S. Plescia, and J. Fabra, *J. Heterocycl. Chem.* **17**, 1409 (1980).

80JOC4597   M.A. Smith, B. Weinstein, and F. D. Greene, *J. Org. Chem.* **45**, 4597 (1980).

80JO5216   S. Wawzonek and J. M. Shradel, *J. Org. Chem.***45**, 5216 (1980).

80JPC1376   T. B. Brill and C. O. Reese, *J. Phys. Chem.* **84**, 1376 (1980). [*CA* **93**, 94587 (1980)].

80JPC3573   A. G. Landers and T. B. Brill, *J. Phys. Chem.* **84**, 3573 (1980) [*CA* **94**, 138982 (1980)].

80MI1   A. A. Dvorkin, A. A. Mazurov, Yu. A. Simonov, S. A. Andronati, A. V. Bogatskii, and A. S. Yavorskii, *Dopov. Akad. Nauk Ukr. RSR, Ser. B; Geol. Khim Biol. Nauki,* 36 (1980) [*CA* **94** 39781 (1981)].

80MI2   W. J. Rodewald and K. Olejniczak, *Pol. J. Chem.* **54**, 1055 (1980).

81AP557   D. Binder, C. R. Noe, and M. Zahora, *Arch. Pharm. (Weinheim, Ger.)* **314**, 557 (1981) [*CA* **95**, 115446 (1981)].

81IJC(B)672   K. Nagarajan, V. P. Arya, T. N. Parthasarathy, S. J. Shenoy, R. K. Shah, and Y. S. Kulkarni, *Indian J. Chem., Sect. B* **20B**, 672 (1981) [*CA* **105**, 226-266 (1986)].

81JA7660   E. C. Taylor, H. M. L. Davies, R. J. Clemens, H. Yanagisawa, and N. F. Haley, *J. Am. Chem. Soc.* **103**, 7660 (1981).

81JAP81-00433   Chugai Pharmaceutical Co., Ltd., Jpn. Pat. 81-00433 (1981) [*CA* **95**, 7359 (1981)].

81JAP81-01312   Chugai Pharmaceutical Co., Ltd., Jpn. Pat. 81-01312 (1981) [*CA* **95**, 7362 (1981)].

81JAP81-01313   Chugai Pharmaceutical Co., Ltd., Jpn. Pat. 81-01313 (1981) [*CA* **95**, 7363 (1981)].

81JAP81-01314   Chugai Pharmaceutical Co. Ltd., Jpn. Pat. 81-01314 (1981) [*CA* **95**, 7364 (1981)].

81JAP81-03349   Chugai Pharmaceutical Co. Ltd., Jpn. Pat. 81-03349 (1981) [*CA* **95**, 115622 (1981)].

81JAP81-03350   Chugai Pharmaceutical Co., Ltd., Jpn. Pat. 81-03350 (1981) [*CA* **95**, 115623 (1981)].

81JAP81-05745   Chugai Pharmaceutical Co. Ltd., Jpn. Pat. 81-05745 (1981) [*CA* **95**, 81040 (1981)].

81JHC1625   A. Amen and H. Zimmer. *J. Heterocycl. Chem.***18**, 1625 (1981).

81JMC662   P. S. Liu, V. E. Marquez, J. S. Driscoll, R. W. Fuller, and J. J. McCormack, *J. Med. Chem.* **24**, 662 (1981).

81JOC303   C. G. Overberger and M. S. Chi, *J. Org. Chem.* **46**, 303 (1981), and references cited therein.

81JOC3306   J. W. Waluk, E. Vogel, and J. Michl, *J. Org. Chem.*, **46**, 3306 (1981).

81JPC2618   A. G. Landers, T. B. Brill, and R. A. Marino, *J. Phys. Chem.* **85**, 2618 (1981).

81MI1   H. Ulrich, *ACS Symp. Ser* **172** (Urethane Chem. Appl.), 519 (1981).

81MI2   J. M. Short, F. H. Helm, M. Finger, and M. J. Kamlet, *Combust. Flame* **43**, 99 (1981) [*CA* **96**, 37867 (1982)].

81MI3   C. D. Bedford, S. J. Staats, M. A. Geigel, and D. L. Ross, *Gov. Rep. Announce Index (U.S.)*, 280 (1981) [*CA* **95**, 7242 (1981)].

81MI4   C. D. Bedford, S. J. Staats, M. A. Geigel, and C. W. Marynowski, *Gov. Rep. Announce. Index (U.S.)*, 4393 (1981) [*CA* **96**, 68958 (1981)].

81MI5   K. Nagarajan, V. P. Aryo, C. L. Kaul, J. David, and R. S. Grewal, *Indian J. Exp. Biol.* **19**, 1150 (1981). [*CA* **96**, 115505 (1982)].

81MI6   S. Wang and F. Li, *Int. Jahrestag—Fraunhofer-Inst. Treib—Explosivst.*, 589 (1981) [*CA* **97**, 130024 (1982)].

81MI7   V. I. Siele, M. Warman, J. Leccacorvi, R. W. Hutchinson, R. Motto, E. E. Gilbert, T. M. Benziger, M. D. Coburn, R. K. Rohwer, and R. K. Davey, *Propellants Explos.* **6**, 67 (1981).

81OMR52   S. Bulusu, T. Axenrod, and J. R. Autera, *Org. Magn. Reson.* **16**, 52 (1981) [*CA* **95**, 96408 (1981)].

81USP4283334 W. B. Wright, Jr. and J. W. Marsico, Jr., U.S. Pat. 4,283,334 (1981) [*CA* **95**, 204013 (1981)].

81USP4288262 J. E. Flanagan and J. C. Gray, U.S. Pat. 4,288,262 (1981) [*CA* **96**, 8939 (1982)].

82AHC115   H. D. Perlmutter and R. B. Trattner, *Adv. Heterocycl. Chem.* **31**, 115 (1982).

82AKZ315   T. E. Agadzhanyan and G. G. Minasyan, *Arm. Khim. Zh.* **35**, 315 (1982) [*CA* **97**, 182382 (1982)].

82CB523   H. E. Borowski and A. Haas, *Chem. Ber.* **115**, 523 (1982).

82CL1579   S. Yogi, K. Hokama, and O. Tsuge, *Chem. Lett.*, 1579 (1982).

82CZ95038   M. Dimun, A. Oswald, and A. Mikestik, Czech, Pat. 195,038 (1982) [*CA* **97**, 6336 (1982)].

82H1595   Y. Tamura, M. Tsunekawa, H. Ikeda, and M. Ikeda, *Heterocycles* **19**, 1595 (1982).

82JIC1170   S. N. Dehuri and A. Nayak, *J. Indian Chem. Soc.* **59**, 1170 (1982) [*CA* **98**, 179345 (1983)].

82LA489   H. Wamhoff, G. Hendrikx, and M. Ertas, *Liebigs Ann. Chem.*, 489 (1982).

82MI1   R. J. Karpowicz and T. B. Brill, *AIAAJ.* **20**, 1586 (1982) [*CA* **98**, 37064 (1983)].

82MI2   B. Singh, P. S. Bhatia, and H. Singh, *Def. Sci. J.* **32**, 297 (1982) [*CA* **98**, 163352 (1983)].

82MI3   F. Kienzle, A. Kaiser, and M. S. Chodnekar, *Eur. J. Med. Chem.—Chim. Ther.* **17**, 547 (1982) [*CA* **98**, 143363 (1983)].

82MI4   P. Nikowitsch and M. Held, *Int. Jahrestag.—Fraunhofer-Inst.*

Treib- Exposivst., 427 (1982) [*CA* **99**, 73155 (1983)].

82MI5　　　　P. C. Harihan, W. S Koski, J. J. Kaufman, R. S. Miller, and A. H. Lowrey, *Int. J. Quantum Chem., Quantum Chem. Symp* **16**, 363 (1982).

82MI6　　　　C. W. Fong. *J. Ballist* **6**, 1534 (1982) [*CA* **98**, 218164 (1983)].

82MI7　　　　Y. B. Burov and G. M. Nazin, *Kinet. Katal.* **23**, 12 (1982) [*CA* **96**, 180549 (1982)].

82MI8　　　　W. Selig, *Propellants, Explos., Pyrotech.* **7**, 70 (1982) [*CA* **97**, 130025 (1982)].

82OMS321　　J. Yinon, D. J. Harvan, and J. R. Hass, *Org. Mass Spectrom.* **17**, 321 (1982) [*CA* **97**, 181537 (1982)].

82TL4207　　V. V. Kaminski, J. S. Swenton, and C. E. Cottrell, *Tetrahedron Lett.*, **23**, 4207 (1982), and references cited therein.

82USP4331080　M. M. West and P. D. Zavitsanos, U.S. Pat. 4,331,080 (1982) [*CA* **97**, 75013 (1982)].

82USP4340734　A. S. Tomcufcik, W. B. Wright, Jr., and J. W. Marsico, Jr., U.S. Pat. 4,340,734 (1982) [*CA* **97**, 216213 (1982)].

82USP4347248　W. B. Wright, Jr., A. S. Tomcufcik, and J. W. Marsico, Jr., U.S. Pat. 4,347,248 (1982) [*CA* **98**, 53925 (1983)].

82ZSK172　　V. K. Mikhailov and V. A. Shlyapochnikov, *Zh. Strukt. Khim.* **23**, 172 (1982).

83ACR146　　F. H. Allen, O. Kennard, and R. Taylor, *Acc. Chem. Res.* **16**, 146 (1983).

83AG(E)543　R. Gompper and M.-L. Schwarzensteiner, *Angew. Chem., Int. Ed. Engl.* **22**, 543 (1983), and references cited therein.

83BSB781　　R. Gompper, *Bull. Soc. Chim. Belg.* **92**, 781 (1983), and references cited therein.

83CHE337　　A. Mazurov, V. V. Antonenko, and S. A. Andronati, *Chem. Heterocycl. Compd. (Engl. Transl.)* 337 (1983) [*CA* **99**, 22440 (1983)].

83GEP3220118　J. Ippen, E. Perzborn, W. Puls, K. Schaller, and F. Seuter, Ger. Pat. DE3,220,118 (1983) [*CA* **100**,103330 (1984)].

83JIC970　　S. N. Dehuri and A. Nayak, *J. Indian Chem. Soc.* **59**, 970 (1982) [*CA* **98**, 179345 (1983)].

83JMR143　　M. D. Pace and B. S. Holmes, *J. Magn. Reson.* **52**, 143 (1983).

83JOC611　　M. B. Frankel and D. O. Woolery, *J. Org. Chem.* **48**, 611 (1983).

83JOC1694　　R. Richter, B. Tucker, and H. Ulrich, *J. Org. Chem.* **48**, 1694 (1983).

83JOC2337　　V. V. Kaminski, R. N. Comber, A. J. Wexler, and J. S. Swenton, *J. Org. Chem.* **48**, 2337 (1983), and references cited therein.

83KGS411　　A. A. Mazurov, V. V. Antonenko, and S. A. Andronati, *Khim. Geterotsikl. Soedin*, 411 (1983) [*CA* **99**, 22440 (1983)].

83LA1623　　C. Klein, G. Schulz, and W. Steglich, *Liebigs Ann. Chem.,* 1623 (1983).

83M1231　　W. O. Lin and E. De S. Coutinho, *Monatsh. Chem.* **114**, 1231 (1983) [*CA* **100**, 191847 (1984)].

83MI1　　　R. J. Karpowicz and T. B. Brill, *Appl. Spectrosc.* **37**, 79 (1983) [*CA* **98**, 62462 (1983)].

83MI2　　　J. E. Harrar and R. K. Pearson, *J. Electrochem. Soc.* **130**, 108 (1983).

93MI3   K. Karkowiak, P. Idowski, and B. Kotelko, *Pol. J. Chem.* **57**, 1381
               (1983) [*CA* **103**, 71297 (1985)].
83MI4   M.N. Glukhovtsev and B. Ya. Simkin, *Zh. Strukt. Khim.* **24**, 31
               (1983); *J. Struct. Chem. USSR (Engl. Transl.)* **24**, 356 (1983)
               [*CA* **99**, 157447 (1983)].
83T479   G. Hornyak and K. Lempert, *Tetrahedron* **39**, 470 (1983).
83T1199   F. Bertha, G. Hornyak, K. Zauer, and K. Lempert, *Tetrahedron*
               39, 1199 (1983).
83T3971   H. Bieraugel, R. Plemp, H. C. Hiemstra, and U. K. Pandit, *Tetra-
               hedron* **39**, 3971 (1983).
83USP4391751   C.-Y. Lin, U.S. Pat 4,391,751 (1983) [*CA* **99**, 105883 (1983)].
83USP399948   R. R. McGuire, C. L. Coon, J. E. Harrar, and R. K. Pearson, U.S.
               Pat Appl. 399,948 (1983) [*CA* **100**, 123574 (1984)].
84AF9   A. Galabov, E. Velichkova, A. Karparov, D. Sidzhakova, D. Dan-
               chev, and N. Chakova, *Arzneim.-Forsch.* **34**, 9 (1984) [*CA* **100**,
               99743 (1984)].
84AKZ185   T. E. Agadzhanyan and G. G. Minasyan, *Arm. Khim. Zh.* **37**, 185
               (1984) [*CA* **101**, 171226 (1984)].
84CHE451   S. A. Andronati, V. S. Yur'eva, A. A. Mazurov, and L. Ni-
               vorzhkin, *Chem. Heterocycl. Compd. (Engl. Transl.)*, 451 (1984)
               [*CA* **101**, 89982 (1984)].
84CPB3252   Y. Fujimura, H. Nagano, I. Matsunaga, and M. Shindo, *Chem.
               Pharm. Bull.* **32**, 3252 (1984) [*CA* **101**, 211121 (1984).
84CZP212600   M. Soucek and L. Pavlickova, Czech. Pat. 212, 600 (1984) [*CA* **102**,
               203699 (1985)].
84CZP214311   S. Zeman and M. Dimun, Czech. Pa. 214, 311 (1984) [*CA* **101**,
               191980 (1984)].
84EUP113017   A. Avram and K. Shi, Eur. Pat. EP113,017 (1984) [*CA* **102**, 123205
               (1985)].
84JOC4415   E. C. Taylor and H. M. L. Davies, *J. Org. Chem.* **49**, 4415 (1984).
84JPC4138   T. B. Brill, R. J. Karpowitz, T. M. Haller, and A.L. Rheingold,
               J. Phys. Chem. **88**, 4138 (1984).
84KGS552   S. A. Andronati, V. S. Yur'eva, A. A. Mazurov, and L. Ni-
               vorzhkin, *Khim. Geterotsikl. Soedin.*, 552 (1984).
84MI1   R. J. Karpowicz and T. B. Brill, *Combust. Flame* **56**, 317 (1984)
               [*CA* **101**, 57281 (1984)].
84MI2   K. Krakowiak, P. Idowski, and B. Kotelko, *Pol. J. Chem.* **58**, 251
               (1984) [*CA* **103**, 37462 (1985)].
84MI3   R. A. Fifer, *Prog. Astronaut. Aeronaut.* **90**, 177 (1984) [*CA* **102**,
               64397 (1984)].
84MI4   V. K. Mohan and T. B. Tong, *Propellants, Explos., Pyrotech* **9**, 30
               (1984) [*CA* **100**, 158995 (1984)].
84MI5   C. Ju and S. F Wang. *Propellants, Explos., Pyrotech* **9**, 58 (1984)
               [*CA* **101**, 57298 (1984)].
84TL2577   M. J. Haddadin, B. J. Agha, and M. S. Salka, *Tetrahedron Lett.* **25**,
               2577 (1984).
84TL4223   Z. Yoshida, M. Shibata, and T. Sugimoto, *Tetrahedron Lett.* **25**,
               4223 (1984).
84USP4454133   P. A. Berke and W. E. Rosen, U.S. Pat. 4,454,133 (1984) [*CA* **101**,
               130692 (1984)].

84USP4462848   D. E. Elrick, U.S. Pat. 4,462,848 (1984) [*CA* **101**, 154312 (1984)].
84USP4465679   F. C. Huang, H. Jones, and W. K. Chan, U.S. Pat. 4,465,679 (1984)
                    [*CA* **101**, 211193 (1984)].
85CCC33   J. Hlavaty, *Collect. Czech. Chem. Commun.* **50**, 33 (1985) [*CA* **103**,
                    13430 (1985)].
85CJC1829   W. Adam and T. Oppenländer, *Can J. Chem.* **63**, 1829 (1985).
85CZ218703   S. Zeman and M. Dimun, Czech. Pat 218,703 (1985) [*CA* **105**,
                    229374 (1986)].
85IZV1612   N. E. Agafonov and G. Y. Kondrat'eva, *Izv. Akad. Nauk SSSR,
                    Ser. Khim.*, 1612 (1985) [*CA* **104**, 5860 (1986)].
85JHC357   W. T. Brady and C. H. Shieh, *J. Heterocycl. Chem.* **22**, 357 (1985).
85JHC907   G. Maas, Brückmann, and B. Feith, *J. Heterocycl. Chem.* **22**, 907
                    (1985).
85JOC270   A. T. M. Marcelis, H. C. van der Plas, and S. Harkema, *J. Org.
                    Chem.* **50**, 270 (1985).
85JPC3118   S. A. Shackelford, M. B. Coolidge, B. B. Goshgarian, B. A. Lov-
                    ing, R. N. Rogers, J. L. Janney, and M. H. Ebinger, *J. Phys.
                    Chem.* **89**, 3118 (1985).
85MI1   C. D. Bedford, *Gov. Rep. Announce. Index (U.S.),* **88** (1985) [*CA*
                    103, 6322 (1985)].
85MI2   A. G. Landers, T. M. Apple, C. Dybowski, and T. B. Brill, *Magn.
                    Reson. Chem.* **23**, 158 (1985).
85MI3   J. Soloducho, *Pol. J. Chem.* **59**, 1115 (1985) [*CA* **107**, 59010 (1987)].
85USP4532301   B. D. Dean U.S. Pat. 4,532,301 (1985) [*CA* **103**, 179127 (1985)].
85USP4534895   M. B. Frankel and D. O. Woolery, U.S. Pat. 4,534,895 (1985) [*CA*
                    104, 36417 (1986)].
85USP4556512   B. D. Dean, U.S. Pat. 4,556,512 (1985) [*CA* **104**, 169045 (1986)].
86BCJ1087   S. Yogi, K. Hokama, K. Ueno, and O. Tsuge, *Bull. Chem. Soc.
                    Jpn.* **59**, 1087 (1986) [*CA* **106**, 84579 (1987)].
86BSB141   L. K. A. Rahman, *Bull. Soc. Chim. Belg.* **95**, 141 (1986) [*CA* **106**,
                    156441 (1987)].
86CZ227907   S. Zeman and M. Dimun, Czech. Pat. 227,907 (1986) [*CA* **106**,
                    35608 (1987)].
86CZ27911   S. Zeman and M. Dimun, Czech. Pat. 227,911 (1986) [*CA* **106**,
                    35607 (1987)].
86CZ227913   D. Zeman, Czech Pat. 227,913 (1986) [*CA* **106**, 52780 (1987)].
86EGP235452   G. Kempter, H. J. Mengs, J. Spindler, and S. Kuehne, Ger. (East)
                    Pat. DD235,452 (1986) [*CA* **107**, 236677 (1987)].
86JAP6133158   K. Kondo, H. Kosai, and H. Shidara, Jpn. Pat 6133158 (1986) [*CA*
                    105, 42382 (1986)].
86JCS(P2)835   A. P. Cooney, M. R. Crampton, and P. Golding, *J. C. S. Perkins
                    Trans 2*, 835 (1986), and references cited therein.
86JHC975   J. Hénin and J. Gardnet, *J. Heterocycl. Chem.* **23**, 975 (1986).
86JHC1435   L. K. A. Rahman, *J. Heterocycl. Chem.* **23**, 1435 (1986).
86JOC67   A. T. M. Marcelis and H. C. Van der Plas, *J. Org. Chem.* **51**, 67
                    (1986).
86JOC417   R. Richter and E. A. Barsa, *J. Org. Chem.* **51**, 417 (1986).
86MI1   A. Karparov, A. Galabov, D. Sidzhakova, and D. Danchev, *Dokl.
                    Bolg. Akad. Nauk* **39**, 149 (1986) [*CA* **106**, 27366 (1987)].

86MI2   S. Bulusu, *Sagamore Army Mater. Res. Conf. Proc., 1984,* 479
               (1986) [*CA* **105,** 81790 (1989)].
86MI3   C. R. Surapaneni and N. S. Gelber, *Statutory Invent. Regist. U.S.*
               Cl. 260-23.9BC; C07D257/02 (1986) [*CA* **106,** 158922 (1987)].
86PCl   H. Ulrich, private communication (1986).
87AG(E)1039   R. Gompper, H. Noeth, W. Rattay, M.-L. Schwarzensteiner,
               P. Spes, and H. U. Wagner, *Angew. Chem., Int. Ed. Engl.* **26,**
               1039 (1987).
87BCJ335   S. Yogi, K. Hokama, and O. Tsuge, *Bull. Chem. Soc. Jpn.* **60,** 335
               (1987) [*CA* **107,** 154314 (1987)].
87BCJ343   S. Yogi, K. Hokama, and O. Tsuge, *Bull Chem. Soc. Jpn.* **60,** 343
               (1987) [*CA* **107,** 134291 (1987)].
87BCJ731   S. Yogi, K. Hokama, S. Takayoshi, and O. Tsuge, *Bull. Chem.
               Soc. Jpn.* **60,** 731 (1987) [*CA* **107,** 198287 (1987)].
87CB251   W. Adam and H. Platsch, *Chem. Ber.* **120,** 251 (1987).
87CL157   S. Yogi, K. Hokama, and O. Tsuge, *Chem. Lett.,* 157 (1987) [*CA*
               107, 217609 (1987)].
87JAP6216921   B. R. Jeimuzu, Jpn. Pat. 6216921 (1986) [*CA* **107,** 231462 (1987)].
87JCS(D)501   C. Glidewell, D. Lloyd, and K. W. Lumbard, *J. C. S., Dalton
               Trans.,* 501 (1987).
87JOU306   E. I. Klabunovskii, E. S. Levitina, L. N. Kaigorodova, D. D.
               Gogoladze, L. F. Godunova, E. I. Karpeiskaya, and G. O.
               Chivadze, *J. Org. Chem. USSR (Engl. Transl.)* **23,** 306 (1987)
               [*Ca* **107,** 237234 (1987)].
87ZC36   J. Spindler and G. Kempter, *Z. Chem.* **27,** 36 (1987) [*CA* **107,**
               198289 (1987)].
89AHC(45)185   H. D. Perlmutter, *Adv. Heterocycl. Chem.* **45,** 185 (1989).
89AHC(46)1   H. D. Perlmutter, *Adv. Heterocycl. Chem.* **46,** 1 (1989).

1,3-Thiazines

HERVÉ QUINIOU AND ODETTE GUILLOTON

Faculté des Sciences et des Techniques,
Université de Nantes, Nantes, France

I. Introduction

A. Nomenclature

1,3- Thiazines and dihydro-1,3-thiazines are named as follows

2H-1,3-thiazine   4H-1,3-thiazine

6H-1,3-thiazine

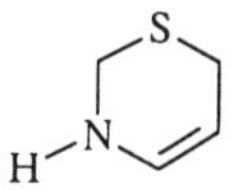

3,6-dihydro-2*H*-1,3-thiazine 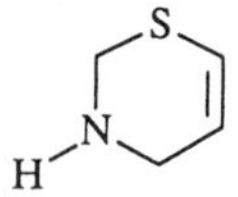   3,4-dihydro-2*H*-1,3-thiazine

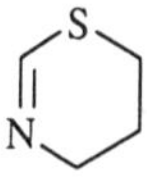

5,6-dihydro-4*H*-1,3-thiazine 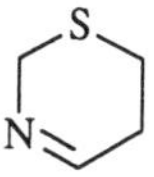   5,6-dihydro-2*H*-1,3-thiaine

Often in the past, 3,6-dihydro-2*H*-1,3-thiazines were incorrectly named 2,3-dihydro-6*H*-1,3-thiazines. It is, however, necessary to apply the IUPAC Rule: the atoms bearing the extra hydrogens have numbers as small as possible.

B. Scope and Limitations

In this chapter, consideration is restricted to monocyclic thiazines and their functionalized derivatives. Benzo-fused systems and condensed-fused systems in general are not included. To date there have been few reviews devoted to 1,3-thiazines, probably because of the difficulty dealing with them without including cephalosporins and their derivatives. (Cephalosporine C is shown here.)

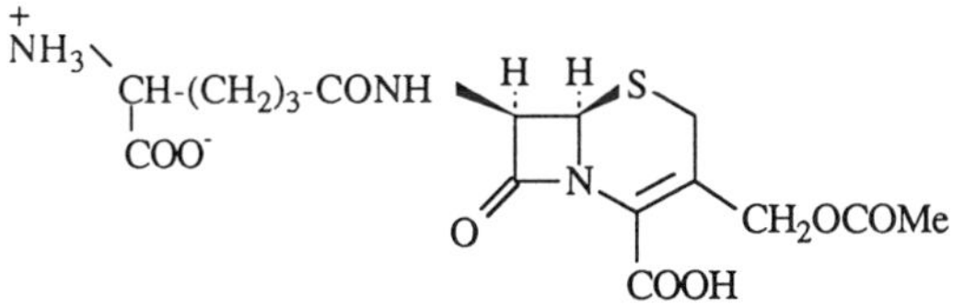

Cephalosporins have a 3,6-dihydro-2*H*-1,3-thiazine nucleus, and some of their total syntheses start from 6*H*-1,3-thiazines. Since the discovery of the cephalosporins by Brotzu in 1943 and the enormous developments occuring in the chemistry of these antibiotics since the 1940s, the 1,3-thiazine nucleus has become one of the most important six-membered heterocycles. Cephalosporins now exceed penicillins in importance and presently account for 31% of the world market of all antibiotics.

Brief mention should be made of the few review articles devoted to thiazines. In 1957, Elderfield and Harris, in a chapter entitled "Thiazines and Benzothiazines,'' devoted eight pages to 1,3-thiazines, which were called at that time "metathiazines" (57MI1). The sp^3-hybridized saturated carbon bearing the extra hydrogen was indicated by a number placed after those for the heteroatoms. Hence, 2H-1,3-thiazine was named 1,3,2-thiazine. In 1960, Ramage et al. described in three pages 1,3-thiazines and their hydroderivatives (60MI1). In 1975 and 1977, the "Specialist Periodical Reports" of Reid and Prota devoted a few pages to 1,3-thiazines (75MI1, 75MI2). In 1978, the second edition of E. H. Rodd (78MI1), dealt in eight pages with the development of chemistry of 1,3-thiazines and 1,3-thiazinium salts between 1960 and 1978. In 1986, Mironova et al. described in thirteen pages and 81 references (86KGS3) the preparation, properties, reactions and tautomerization of oxo- and thio-derivatives of 1,3-thiazines.

II. Synthesis of 1,3-Thiazines

In order to classify the types of reactions that can be used for 1,3-thiazine synthesis, Cain and Warrener (70AJC51) chose a way similar to that proposed for pyrimidines by Kenner and Todd (57M12).

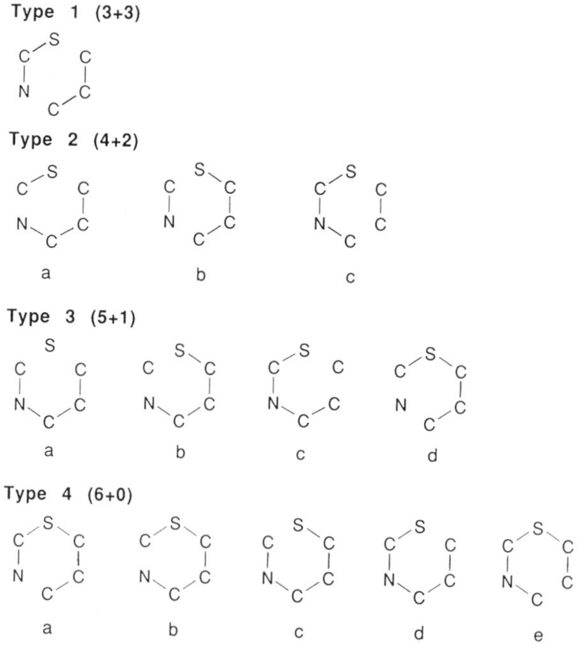

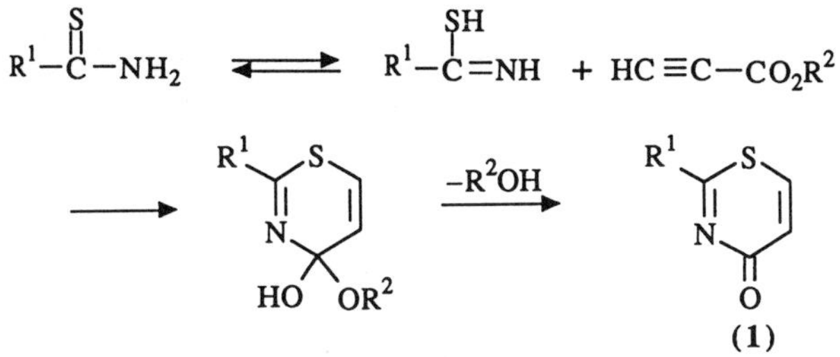

SCHEME 1

A. TYPE 1, (3 + 3) SYNTHESIS

This section will be divided into a discussion of reactions carried out starting from thioamides, thioureas, substituted thioureas, dithiocarbamic acids and their ammonium salts, and miscellaneous syntheses.

1. Syntheses Starting from Thioamides

Thioamides react with acrylates to yield mainly thiazine-4-ones, thiazine-6-ones, or their corresponding thiones. A few reactions proceed by addition of SH to the conjugated triple bond with further cyclization, or by these steps occurring in the reverse order (Scheme 1) (74G849). $4H$-1,3-Thiazine-4-ones (1) with R^1 = aromatic, H_2N—C(Me) = CH—, and also with an aromatic substituent in position 6 have been obtained by such a snythesis (55UKZ732; 70GEP2025339; 70NEP6903457; 85H1225). In a further example (Scheme 2), at least one of the chlorine atoms appears to

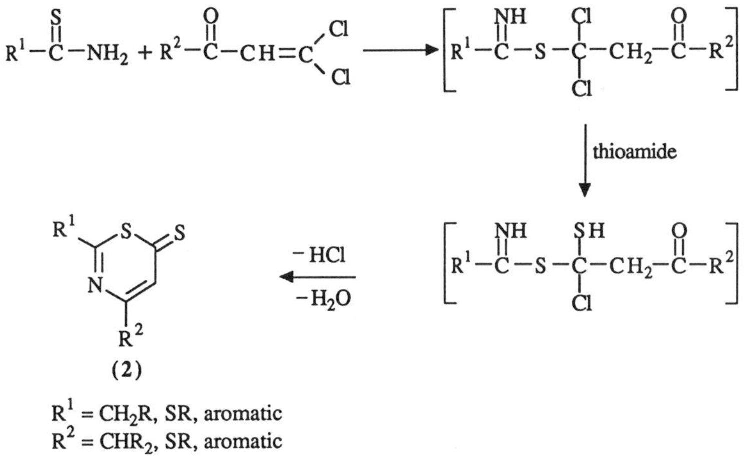

R^1 = CH$_2$R, SR, aromatic
R^2 = CHR$_2$, SR, aromatic

SCHEME 2

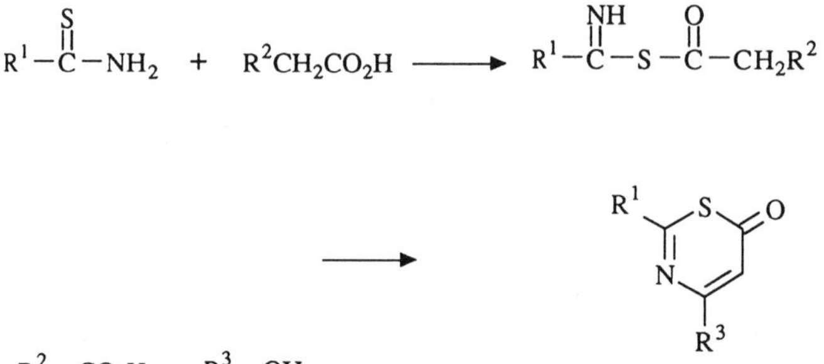

$$R^1-\underset{\underset{S}{\|}}{C}-NH_2 \ + \ R^2CH_2CO_2H \longrightarrow R^1-\underset{\underset{NH}{\|}}{C}-S-\underset{\underset{O}{\|}}{C}-CH_2R^2$$

$$\longrightarrow$$

(3)

R^2 = CO$_2$H   R^3 = OH
R^2 = CN   R^3 = NH$_2$

SCHEME 3

be substituted by SH via the thiouronium salt formed from the thioamide. Other mechanisms have been proposed by Schroth and Dill (74EGP106649-50).

Other compounds with R^2 = H and H, Me, or Ph in position 5 were described later (84ZC435). A third scheme involves esterification of the thiol by either a carboxylic group or an acid chloride followed by cyclization (Scheme 3) (62USP3062816; 64M147; 64M1550; 65M411). When one of the reagents is an acid, or in the case of acid catalysis, we may consider that the protonated form of the thioamide $R^1-C(SH)NH_2^+X^-$ occurs, but that does not fundamentally change the proposed mechanism.

Malonyl dichloride furnishes 2-aryl-5,6-dihydro-4H-1,3-thiazine-4,6-diones (4) (68USP3408348). The condensation of carbon suboxide, C_3O_2, with arenethiocarboxamide also leads to thiazinediones (4) or 4-hydroxy-6H-1,3-thiazine-6-ones (5) (Scheme 4) (62M26; 64M1061).

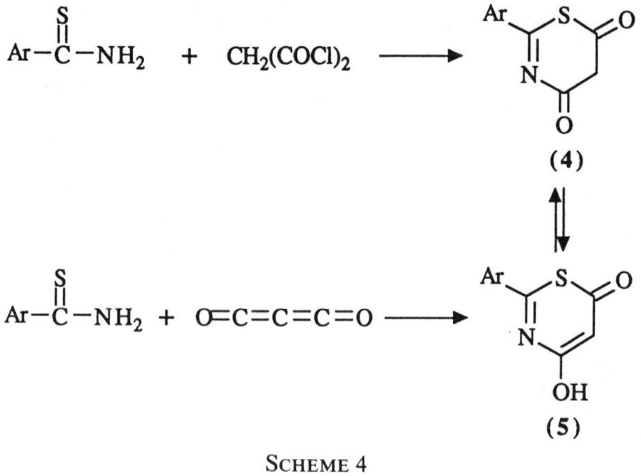

$$Ar-\underset{\underset{S}{\|}}{C}-NH_2 \ + \ CH_2(COCl)_2 \longrightarrow$$

(4)

$$Ar-\underset{\underset{S}{\|}}{C}-NH_2 \ + \ O{=}C{=}C{=}C{=}O \longrightarrow$$

(5)

SCHEME 4

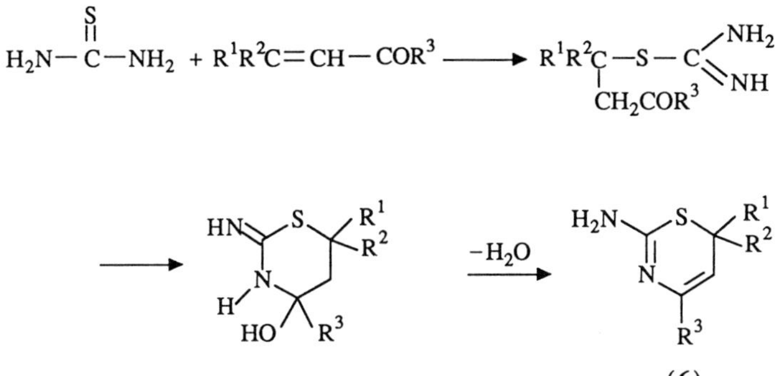

$$H_2N-\overset{\overset{\text{S}}{\|}}{C}-NH_2 + R^1R^2C=CH-COR^3 \longrightarrow R^1R^2\underset{\underset{CH_2COR^3}{|}}{C}-S-C\overset{\nearrow NH_2}{\underset{\searrow NH}{}}$$

(6)

SCHEME 5

2. Syntheses Starting from Thioureas

We will successively consider the action of thioureas and substituted thioureas on conjugated double bonds, allenes, alkynes, and β-chloroketones [62CPB13; 64CPB683; 72MI1; 78AJC2307; 80JCS(P1)1013]
Reaction with α,β-ethylenic ketones leads to thiazines **(6)** (Scheme 5). Reaction of thiourea with β-dichloro-α,β-ethylenic ketones yields thiazine-5-thiones **(7)**. A mechanism analogous to that mentioned previously can be used to explain the result (76ZOR904) (Scheme 6). Reaction of thioureas with α-chloro-esters leads to 5-chloro-4-hydroxythiazines **(8)** (56GEP954417), while esters bearing an ether group cyclize to yield thiazine-5-carboxylic esters **(9)** (Scheme 7) (65JOC2290; 69JAP06820).

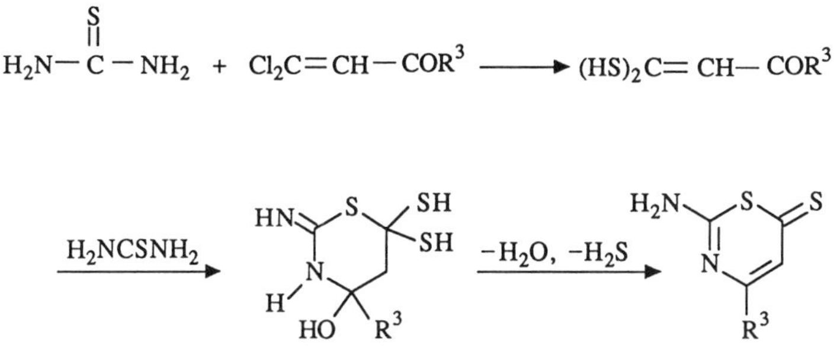

$$H_2N-\overset{\overset{\text{S}}{\|}}{C}-NH_2 + Cl_2C=CH-COR^3 \longrightarrow (HS)_2C=CH-COR^3$$

(7)

SCHEME 6

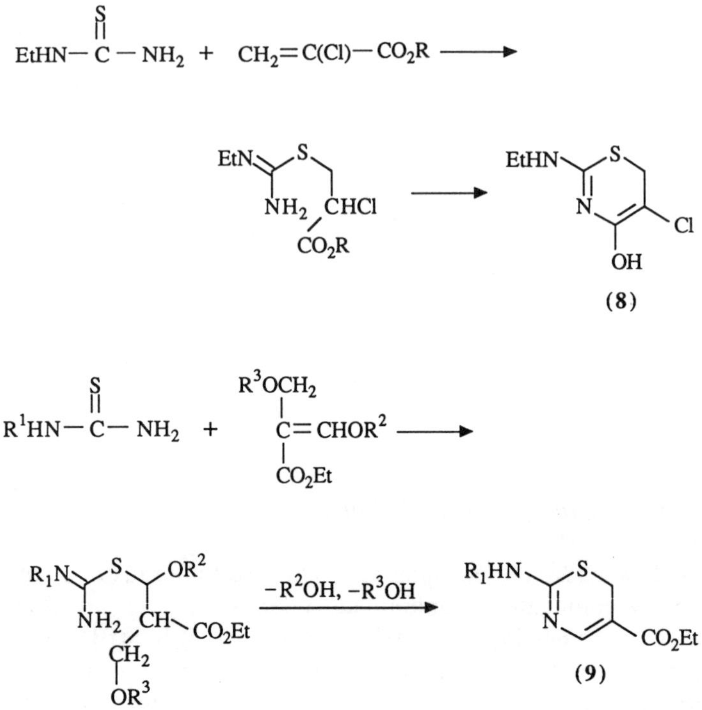

$$EtHN-\overset{\overset{S}{\parallel}}{C}-NH_2 \ + \ CH_2{=}C(Cl){-}CO_2R \longrightarrow$$

(8)

$$R^1HN-\overset{\overset{S}{\parallel}}{C}-NH_2 \ + \ \begin{matrix} R^3OCH_2 \\ | \\ C{=}CHOR^2 \\ | \\ CO_2Et \end{matrix} \longrightarrow$$

$$\xrightarrow{-R^2OH,\ -R^3OH}$$

(9)

SCHEME 7

Allenes react with thiourea to yield thiazine-4-ones, which are further isomerized to 4-oxothiazine-6-ylidene acetic esters (**10**) (Scheme 8) [(82JCS(P1)1905)]. Acetylenic esters [propiolate and (DMAD)] react with thiourea to give thiazine-4-ones **11** and **12**. Substituted thioureas behave like ambident reagents with dimethylacetylene dicarboxylate (DMAD) (Scheme 9) [(69ZOR621; 78JCS(P1)1428)]. Finally, the thiazines

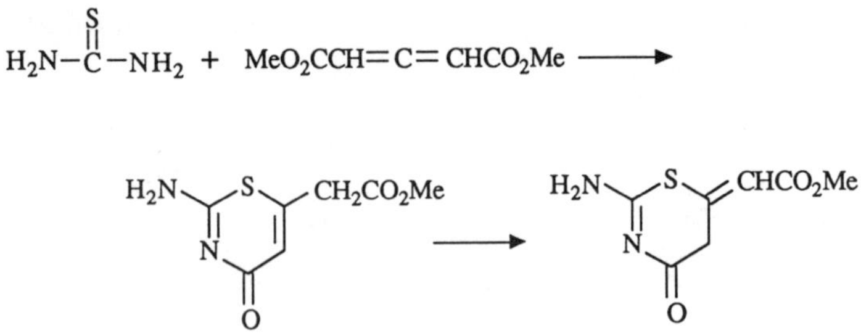

$$H_2N-\overset{\overset{S}{\parallel}}{C}-NH_2 \ + \ MeO_2CCH{=}C{=}CHCO_2Me \longrightarrow$$

(10)

SCHEME 8

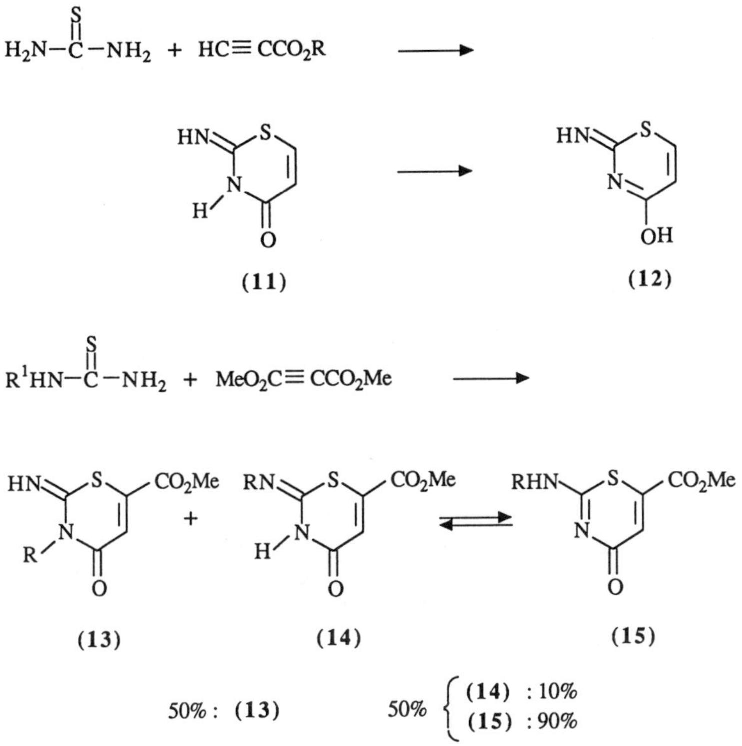

$$H_2N-\overset{\overset{S}{\parallel}}{C}-NH_2 \ + \ HC\equiv CCO_2R \quad \longrightarrow$$

(11)   (12)

$$R^1HN-\overset{\overset{S}{\parallel}}{C}-NH_2 \ + \ MeO_2C\equiv CCO_2Me \quad \longrightarrow$$

(13)   (14)   (15)

50% : (13)   50% $\left\{ \begin{array}{l} (14) : 10\% \\ (15) : 90\% \end{array} \right.$

SCHEME 9

(16) may be obtained starting from substituted thioureas and β-chloroke-
tones (Scheme 10) (75GEP2426653).

3. Syntheses Starting from Dithiocarbamic Acids, Their Ammonium Salts, and Miscellaneous Syntheses

Reaction of dithiocarbamic acids with enones leads to 2-mercaptothia-
zines (17), if the nitrogen is monosubstituted, and to thiazine-2-thiones (18)

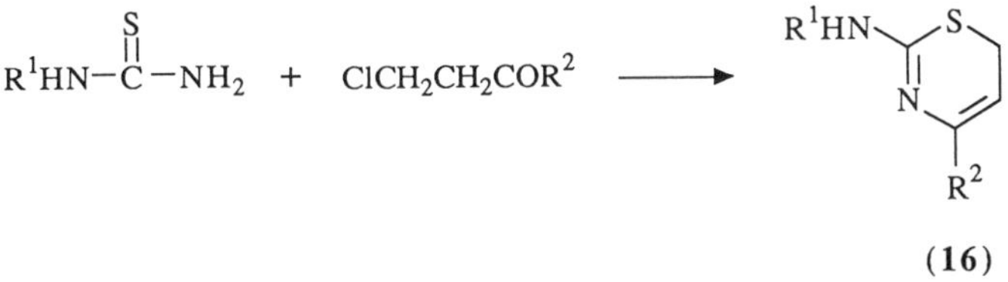

$$R^1HN-\overset{\overset{S}{\parallel}}{C}-NH_2 \ + \ ClCH_2CH_2COR^2 \quad \longrightarrow$$

(16)

SCHEME 10

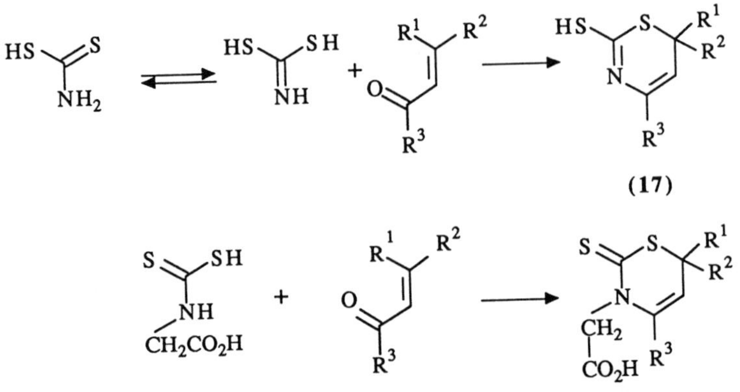

(17)

(18)

SCHEME 11

if the nitrogen is disubstituted (Scheme 11) (48USP2440095; 50JA1879; 51USP2568633; 55JA2866). 4-Hydroxythiazine-6-one is produced by reaction of malonyl dichloride on a thiocarbamic ester (63USP3098071). Reaction of malonyl diesters and dithiocarbamide yields thiazine-4-ones **(20)** together with the bisthiazinyl compounds **(21)** (Scheme 12) (73JHC223).

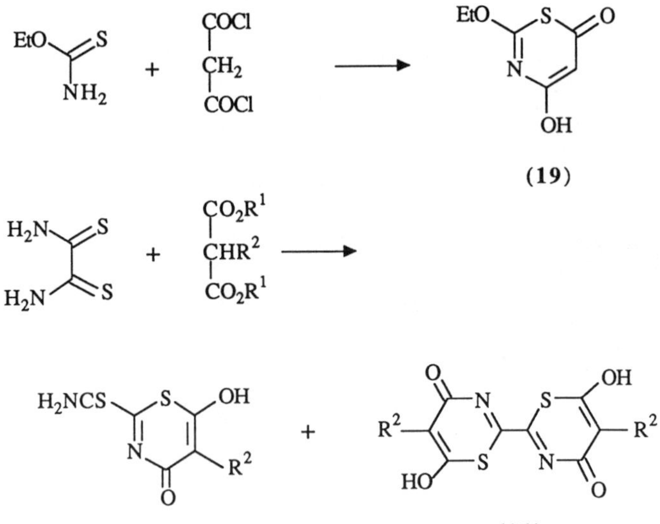

(19)

(20)   **(21)**

SCHEME 12

B. TYPE 2, (4 + 2) SYNTHESIS

The condensation of type 2a is illustrated by only one reaction, and there are few reactions of type 2c.

1. C—S Addition

C-Enamino-imines obtained by photolysis of 1*H*-pyrimidin-2-ones have been treated with carbon disulfide, giving 2*H*-1,3-thiazine-2-thiones (**22**) and/or 1-aryl-1*H*-pyrimidine-2-thiones (**23**) (Scheme 13) [82JCS(P1)2149].

2. C—C Addition

There are many syntheses involving cyclocondensation reactions starting from thiazadienes, S=C—N=CXY. The substituents X and Y may either both be electron-withdrawing (CF₃), or one of them may be an amino group (thioacylformamidine), or CXY may be a part of dithioimidocarbonates as either a carbonyl or a thiocarbonyl (thioacylisocyanate, thioacylisothiocyanate). We will also consider addition reactions that in-

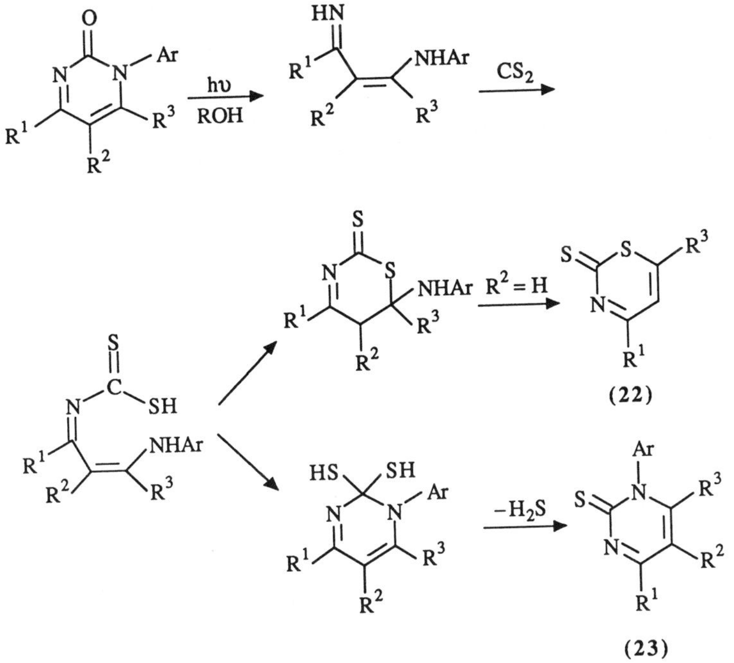

SCHEME 13

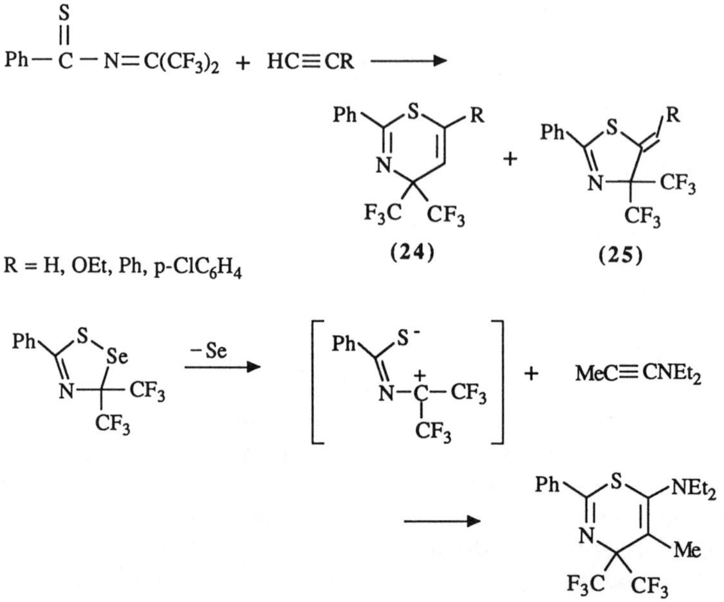

$$Ph-\overset{\overset{\displaystyle S}{\|}}{C}-N=C(CF_3)_2 \ + \ HC\equiv CR \longrightarrow$$

R = H, OEt, Ph, p-ClC$_6$H$_4$

(24)   (25)

(26)

SCHEME 14

volve masked thiazadienes (Dithiazolium salts). Free or masked (3H-thiaselenazoles) 4,4-bistrifluoromethylthiazadienes react with acetylenic compounds to yield the thiazines **24** and **26** (Scheme 14) (80CB2699; 83CC945).

Many reactions of thioacylformamidines have been described, especially those with acrylic compounds (Scheme 15) and ketenes. Acrolein

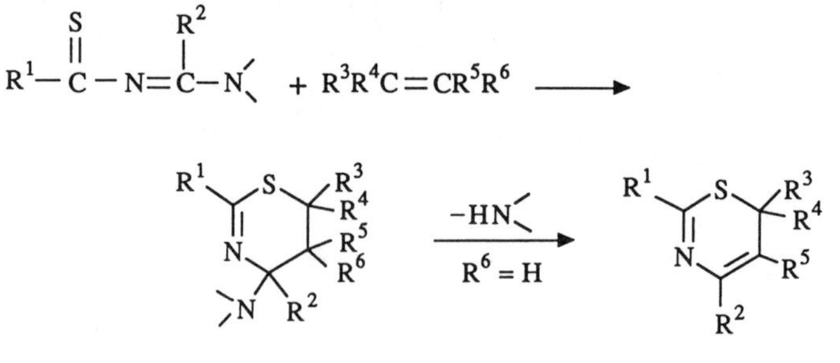

$$R^1-\overset{\overset{\displaystyle S}{\|}}{C}-N=\overset{\overset{\displaystyle R^2}{|}}{C}-N\diagdown \ + \ R^3R^4C=CR^5R^6 \longrightarrow$$

(27)   (28)

$-N\diagdown$ = dimethylamino, piperidino, morpholino etc ...

SCHEME 15

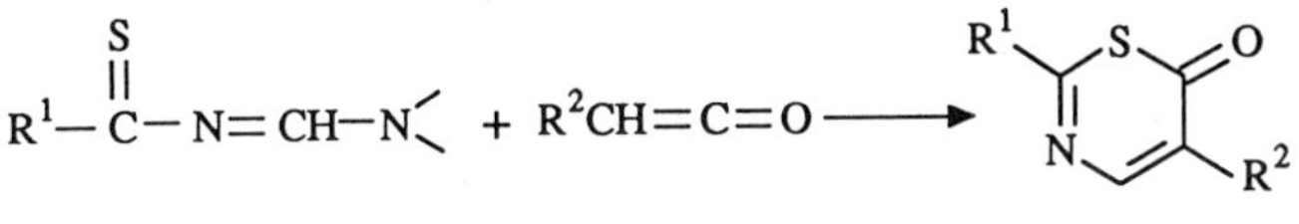

$$R^1-\overset{\overset{\textstyle S}{\|}}{C}-N=CH-N\diagdown \quad + \quad R^2CH=C=O \longrightarrow$$

(structure **29**)

(**29**)

$R^2 = H, Ph$

SCHEME 16

and methyl vinyl ketone (R^3, R^4 = H R^5 = CHO or COMe) yielded the thiazines (**28**) after elimination of the amine [75CR(C)119]. In the case of acrylonitrile and methyl or ethyl acrylate (R^5 = CN, CO_2Me or CO_2Et), the authors originally noted production of 5,6-dihydro-4H-thiazines (**27**) without being able to eliminate the amine (79BSF347). Since then, the elimination has been obtained directly by acid catalysis of the cyclo-condensation ($AlCl_3$, Amberlyst). Acid catalysis is also absolutely neces-sary when R^2 is an electron-withdrawing group (e.g. esters) (87SC1971). In some other cases, pressures up to 12–15 kbars have been used to effect reaction (80S453; 82TL2315; 83CJC1169, 83MI2; 84CC157; 86PS327).

The reaction between thioacylformamidines and ketenes leads to thiazine-6-ones (**29**) (Scheme 16) (75T3055). 4-Amino-4H-thiazines (**30**) are obtained by condensation of thioacylformamidines with acetylenic com-pounds (Scheme 17) (85JOC1545). Thiobenzoyl-1 and N,N-diethylami-nothiocarbonyl isocyanates and isothiocyanates have been reacted with enaminoketones and ynamines. Enaminoketones and ynamines reacted with thiobenzoyl isocyanate give thiazine-4-ones **31** and **32**, respectively (Scheme 18) (76BCJ2828; 85ZC324).

N,N-Diethylaminothiocarbonyl isothiocyanate reacts with an ynamine to give both (4 + 2) and (2 + 2) cycloadditions; the latter is followed by subsequent rearrangement. The mixture consists of **33** (42%) and **34** (29%) (Scheme 19). Thiazine-6-ones have been prepared in 41–92% yield by cyclocondensation of dithiazolium compounds with β-bifunctional com-pounds which have a leaving group R^2 (Scheme 20) (85EGP222310, 85ZC327).

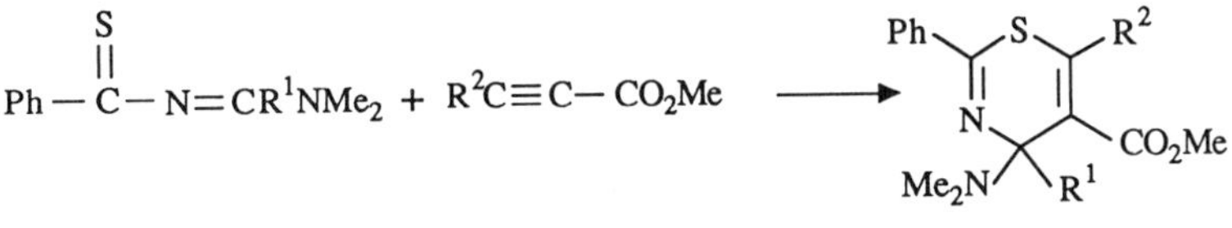

$$Ph-\overset{\overset{\textstyle S}{\|}}{C}-N=CR^1NMe_2 \quad + \quad R^2C\equiv C-CO_2Me \longrightarrow$$

(structure **30**)

(**30**)

$R^1 = H, Me, CO_2Et$

$R^2 = H, CO_2Me$

SCHEME 17

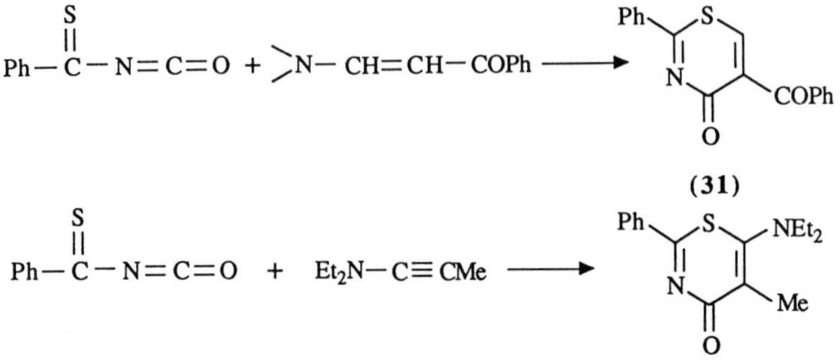

(31)

(32)

SCHEME 18

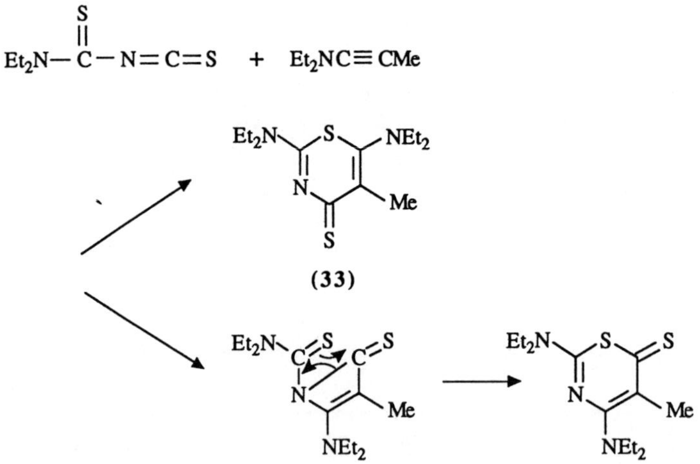

(33)

(34)

SCHEME 19

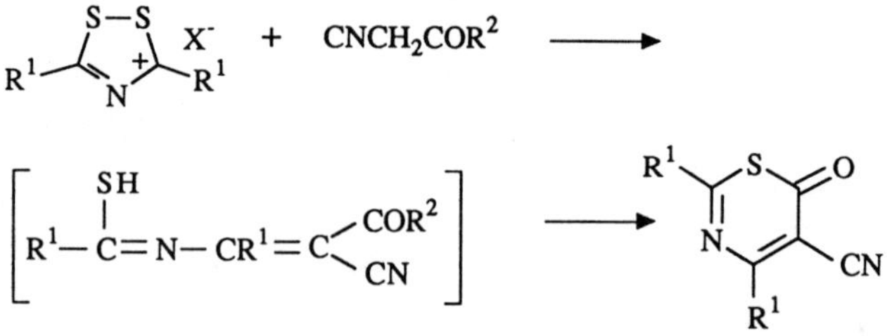

(35)

SCHEME 20

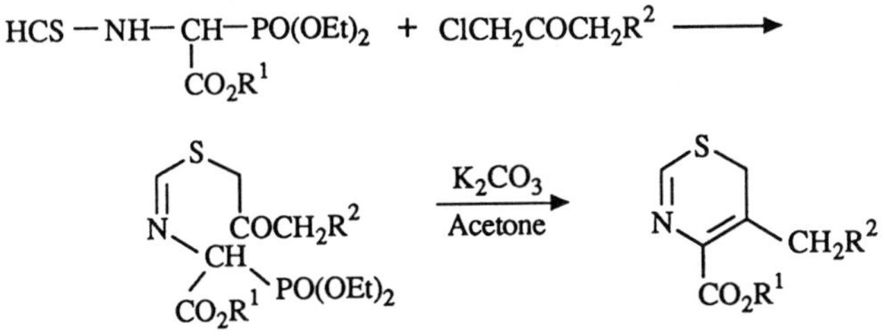

HCS$-$NH$-$CH$-$PO(OEt)$_2$ + ClCH$_2$COCH$_2$R^2 $\longrightarrow$
          |
         CO$_2$R^1

$$\xrightarrow[\text{Acetone}]{K_2CO_3}$$

(36)

SCHEME 21

The formal dihydroderivatives of thiazadienes, α-thioformamido diethylphosphonoacetates, have been cyclized with chloropropanones to give the thiazines (**36**) in the first step of the total synthesis by Merck of cephalosporin (Scheme 21) (73TL4649; 74GEP2356388, 74TL3567). Compound **36** was unstable and was not purified, but was immediately added to azidoacetyl chloride to give cephems.

C. TYPE 3, (5 + 1) SYNTHESIS

All four kinds of condensation 3a–d are met here.

1. S Addition

The sulfur is introduced by way of P_4S_{10}, Lawesson's reagent, or H_2S. The starting materials are saturated γ-amidoketones, γ-amidoesters, γ-oxo-γ-amidoketones, unsaturated γ-amidoesters, or even 1,3-oxazinium salts. P_4S_{10} and ethyl N-benzoyl-3-aminopropionate in benzene gave a thiobenzyl derivative, which was heated with P_4S_{10} to give 2-phenyl-6-ethoxy "metathiazine" (**37**) (28JPJ802). Similar reactions have been achieved (Scheme 22) [78JAP(K)78-111080; 81BSB75]. N-Acetoacetylcarboxamides react with $HClO_4$, giving 4-hydroxyoxazinium salts which, with hydrogen sulfide in the presence of acetic anhydride, lead to thiazine-4-ones (**41**) in satisfactory yields (Scheme 23). On standing at room temperature **41** is isomerized to the 2-ylidenic derivative (**42**, R^1 = H, R^2 = Ph; R^1 = R^2 = Me) (81H851).

2. C-2 Addition

Here the five atoms to add to the carbon C-2 may be furnished by β-mercaptoacrylamides or isothiazoles (Scheme 24). Condensation of 2-cyano-3-mercapto-3-methylthioacrylamide with benzoic acid in the

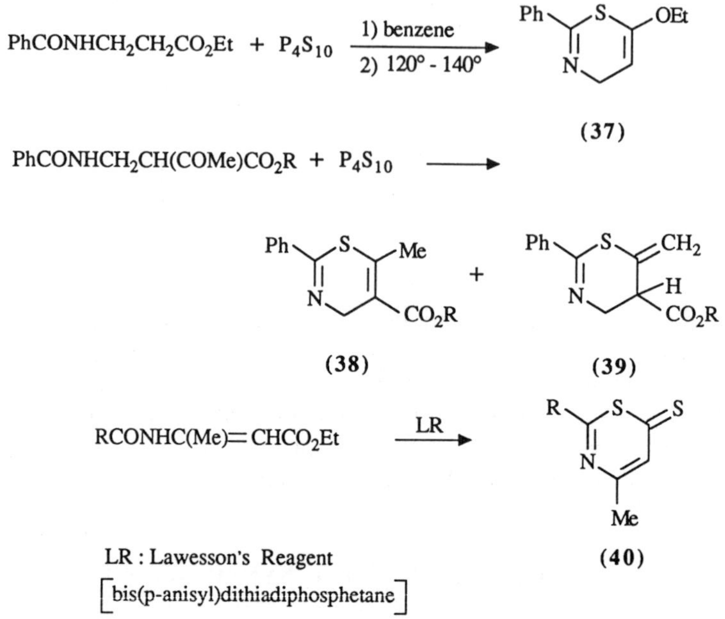

$PhCONHCH_2CH_2CO_2Et$ + P_4S_{10} $\xrightarrow[\text{2) } 120° - 140°]{\text{1) benzene}}$

(37)

$PhCONHCH_2CH(COMe)CO_2R$ + P_4S_{10} $\longrightarrow$

(38)   +   **(39)**

$RCONHC(Me){=}CHCO_2Et$ $\xrightarrow{\text{LR}}$

(40)

LR : Lawesson's Reagent

$\left[\text{bis(p-anisyl)dithiadiphosphetane}\right]$

SCHEME 22

presence of polyphosphoric ester gives 5-carbamoyl-4-methylthio-2-phenyl-6*H*-1,3-oxazine-6-thione which, on treatment with boiling ethanol, is easily converted to 5-carbamoyl-4-methylthio-2-phenyl-6*H*-1,3-thiazine-6-one (**43**). The mechanism, called an S,N double rearrangement, was studied by ^{13}C-labelling and crossover reactions. The results indicated the condensation was a thiaallylic rearrangement of the initial ring-closure product formed from reaction of the cyano group and the mercapto group of the starting compound with benzoic acid (Scheme 24) (73JOC802; 82JOC1090; 84JOC74).

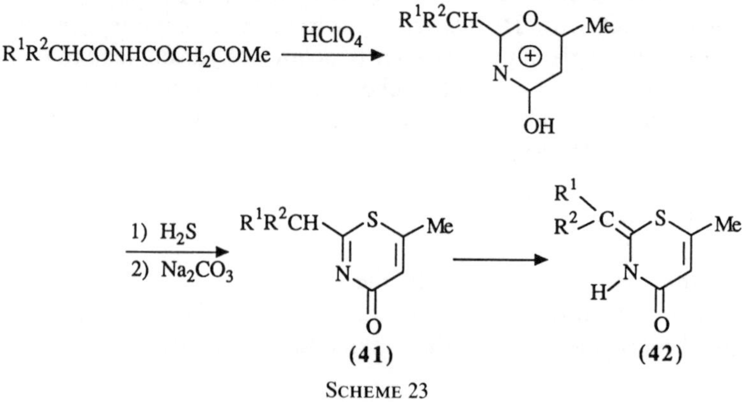

$R^1R^2CHCONHCOCH_2COMe$ $\xrightarrow{HClO_4}$

$\xrightarrow[\text{2) } Na_2CO_3]{\text{1) } H_2S}$

(41) $\longrightarrow$ **(42)**

SCHEME 23

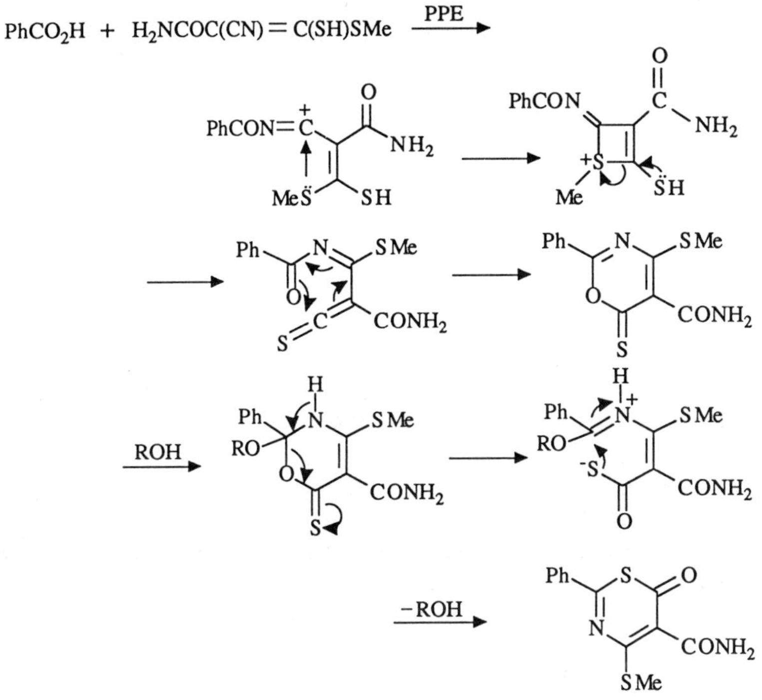

$$PhCO_2H \ + \ H_2NCOC(CN) = C(SH)SMe \xrightarrow{\ PPE\ }$$

(43)

SCHEME 24

2-Phenacylisothiazolium salts were prepared by fusion of isothiazoles with phenacyl bromide (85BSB149). Treatment of the salt with pyridine afforded the 2-benzoyl-2H-1,3-thiazines (**44**) via intramolecular nucleophilic attack on the sulfur atom (Scheme 25).

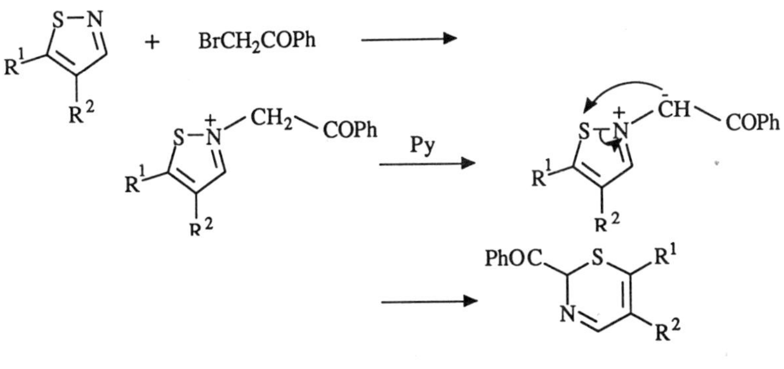

(44)

SCHEME 25

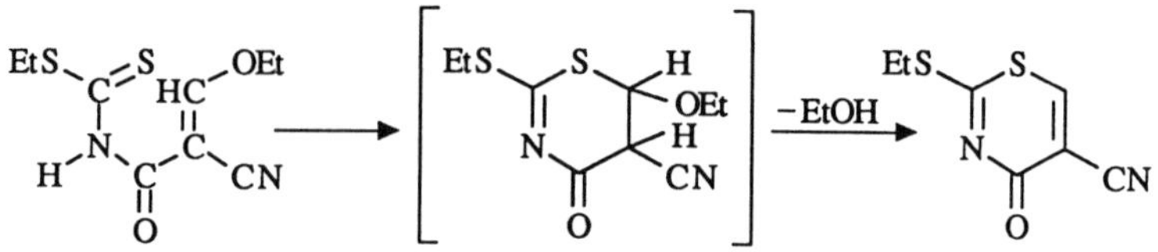

SCHEME 26

3. C-6 Addition

Cyanoacetyl dithiocarbamate reacts with ethyl orthoformate in acetic anhydride giving, after Michael addition, 5-cyano-2-ethylthio-4-oxo-4H-1,3-thiazine (**45**) (Scheme 26) (56JCS3847).

4. N Addition

2-Methyl-3-thiocyanatobuten-2-al, with hydroxylamine, gives thiazine **46,** which is further rearranged to an isothiazole with the help of poly-phosphoric acid (Scheme 27) (81ZC326; 83EGP156908).

D. TYPE 4, (6 + 0) SYNTHESIS

In these syntheses, the linear heteroatomic chain must contain the six atoms of the future thiazine skeleton. The chain may be cyclized between N and C-4 (type 4a), between N and C-2 (type 4b), between S and C-2 (type 4c), between S and C-6 (type 4d), or finally between C-4 and C-5 (type 4e).

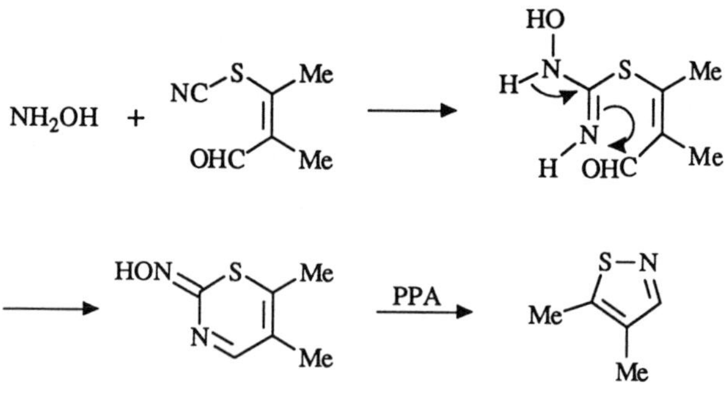

(**46**)

SCHEME 27

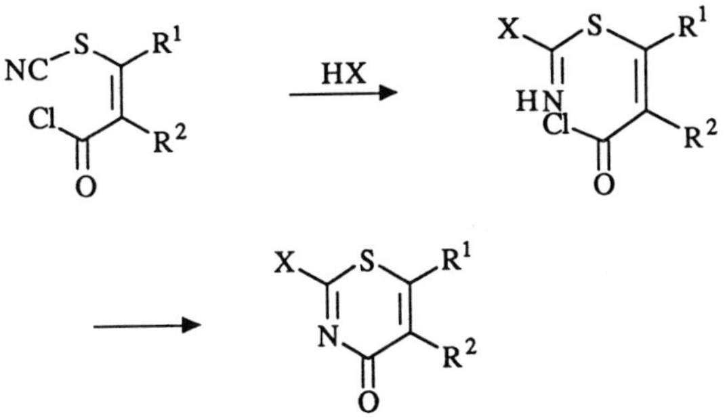

(47)

$$R^1 = R^2 = H \quad X = Cl, I$$
$$R^1R^2 = (CH_2)_3$$

SCHEME 28

1. *N—C-4 Ring Closure*

2-Halo-1,3-thiazine-4-ones **(47)** have been prepared by reaction of α,β-unsaturated thiocyanatocarboxylic acids with PCl_5 and subsequent cyclization by addition of hydrogen halides (Scheme 28) (71GEP2010558; 77LA1249).

2. *N—C-2 Ring Closure*

Thiourea and α-chloropropenoic acid give isothiuronium salts (or their esters) which, when cooled and treated very gradually with aqueous NaOH, lead to the 4-hydroxythiazine **48** or its tautomer **49** (Scheme 29) (51LA140).

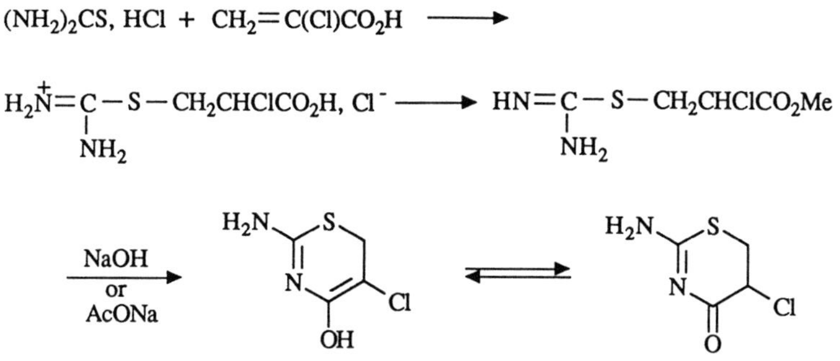

(48)   **(49)**

SCHEME 29

$$Et_2NH + ClCH_2C(Me){=}CH{-}NCS \longrightarrow$$

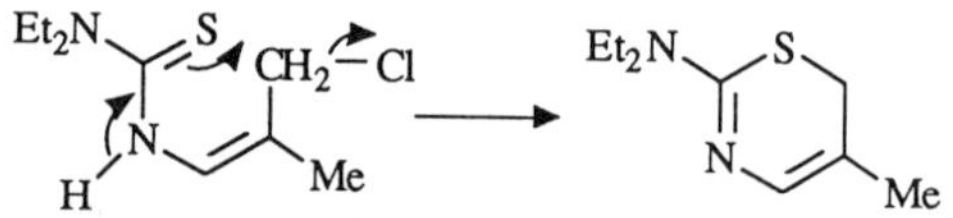

(50)

SCHEME 30

3. S—C-2 Ring Closure

Thiazines such as **50** have been prepared by the reaction of $ClCH_2$-$CR^1{=}CR^2NCS$ (R^1 = alkyl, cycloalkyl, aryl) with a secondary amine in an organic solvent (Scheme 30) (77EGP128124).

4. S—C-6 Ring Closure

Ring closure by linkage of S to C-6 is obtained by adding SH to a carbon–carbon double bond, by adding SH to a carbonyl function, or by variations of these processes such as internal esterification or transesterification of the mercaptan, or nucleophilic attack of the sulfur on a saturated carbon bearing a nucleofugal halogen group. Thiazine **51** may be regarded as a compound resulting from the oxidation of the addition product with hydroxylamine (Scheme 31) (37JA1486). Reaction of CR^2-$Cl{=}CR^1CONCS$ with HNR^3R^4 gave $CR^2Cl{=}CR^1CONHCSNR^3R^4$, which cyclized to give 58–84% of thiazinones **52** or 51% of the thiouracil derivatives **53** (Scheme 32) (77PHA461). Similarly, 3,4-dihydroquinazoline reacts with β-chlorocrotonylisothiocyanate, and the initial reaction is followed by cyclization to the thiazine-4-one **(54)** (Scheme 33) (81EGP149807).

On treatment of N-(2-methyl-4-oxopentyl)dithiocarbamic acid with sulfuric acid or acetic anhydride, 4,4,6-trimethyl-4H-1,3-thiazine-2-thiol **(55)** is obtained (Scheme 34) (55JA5431). If RC_6H_4NMe takes the place of SH

(51)

SCHEME 31

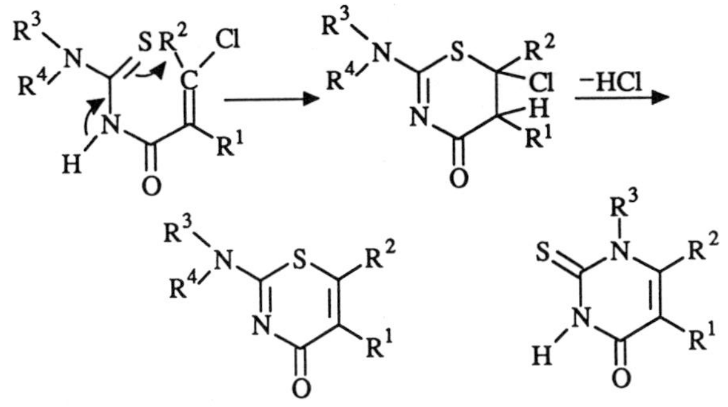

(52) **(53)**

R^1 = H, Cl R^2 = Cl, Me, OH

R^3 = Aromatic R^4 = H, Me, Et

SCHEME 32

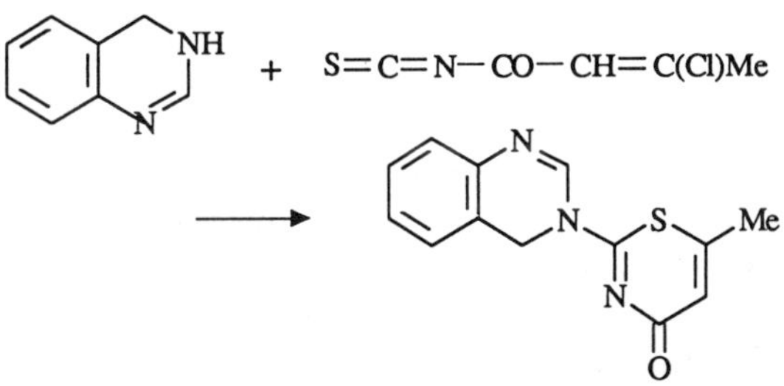

(54)

SCHEME 33

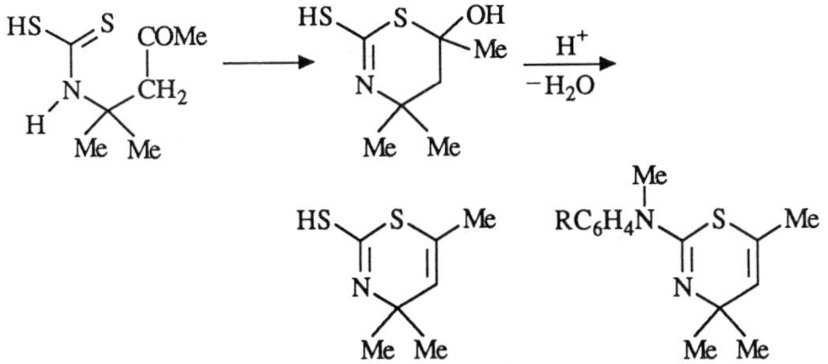

(55) **(56)**

SCHEME 34

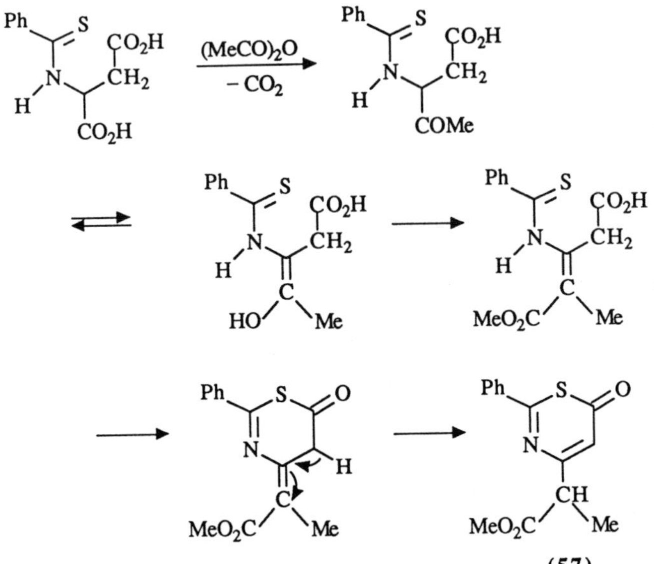

SCHEME 35

in this sequence, compounds 56 (R = H, *p*-Me, *m*-Me, *p*-EtO, *p*-MeO, *p*-Br) are obtained (74MI2). *N*-Thiobenzoylaspartic acid has been converted to 6*H*-1,3-thiazine-6-one (57) by treatment with acetic anhydride and 3-picoline (Scheme 35). This reaction was unexpected (74LA1753).

In contrast to literature data (1886LA1; 12CB1557), base-catalyzed cyclization of 3-ethyl-2-thioureidobutenoate has been shown not to produce thiouracil 58, but to produce 2-amino-4-methyl-6-oxo-6*H*-1,3-thiazine 59 (Scheme 36) (80CCC732). Ethyl 2-isothiocyanato-3-phenyl-2-butenoate

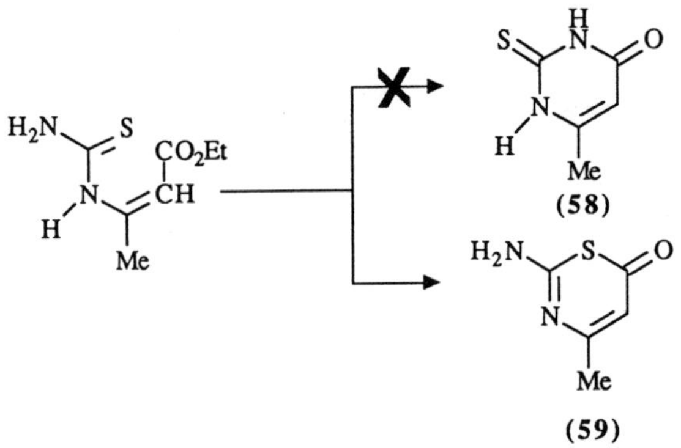

SCHEME 36

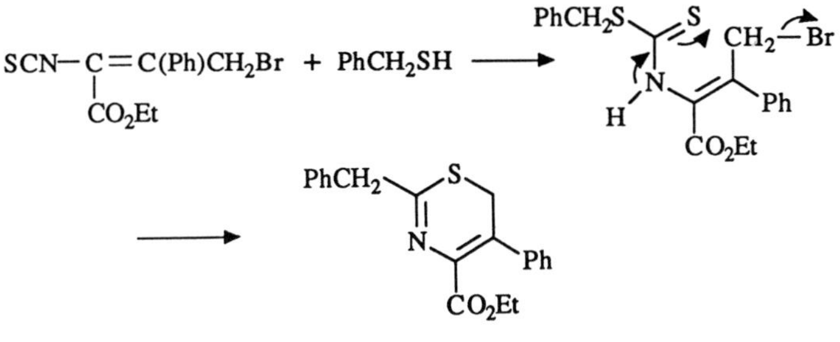

$$SCN-\underset{\underset{CO_2Et}{|}}{C}=C(Ph)CH_2Br \ + \ PhCH_2SH \ \longrightarrow$$

(60)

SCHEME 37

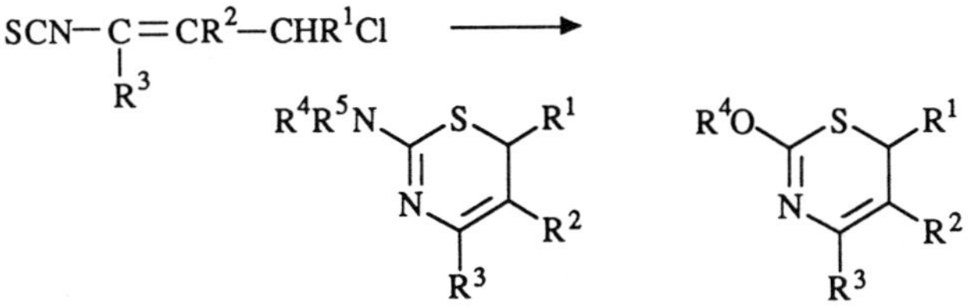

$$SCN-\underset{\underset{R^3}{|}}{C}=CR^2-CHR^1Cl \ \longrightarrow$$

(61)   **(62)**

$$R^1 = H, \ R^4R^5 = (CH_2)_n, \ n = 3, 4, 5, 6$$

SCHEME 38

has been brominated by means of N-bromosuccinimide. When the product is reacted with the nucleophile benzylmercaptan, there is a prior addition to the isocyanato group, followed by cyclization to provide the substituted ethyl 2-benzyl-5-phenyl-6H-1,3-thiazine-4-carboxylate (60) (Scheme 37) (79CB3939). Four years later a similar reaction was observed with chloride as the nucleofuge (Scheme 38) (84JPR101). 2-Alkoxythiazines (62) are prepared in a similar manner.

5. C-4—C-5 Ring Closure

Reaction of R^1NCS with $H_2NCR^2R^3C\equiv CH$ gives benzamidothiazolidines 63, which rearrange to thiazines 64, possibly through the reversibility of the cyclization reaction (Scheme 39) (80MI1). Thiazine-

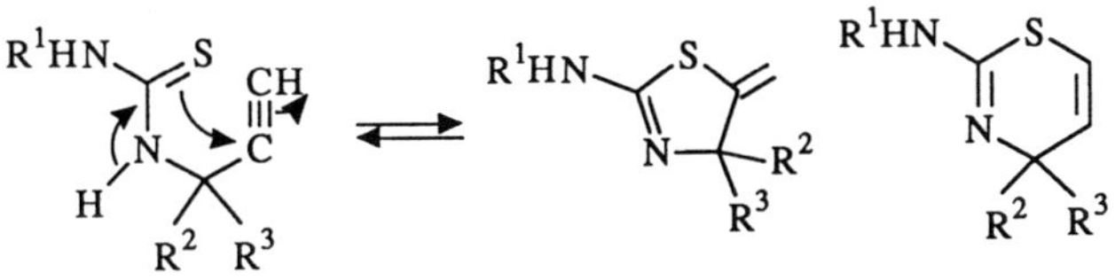

(63)   **(64)**

SCHEME 39

(65)

SCHEME 40

carboxylates **65** (R = Et, trityl) were prepared by cyclocondensation of $HC(S)NHCO_2R$ with $ClCH_2COMe$ via the intermediates $MeCOCH_2$-$SCH=NCH_2CO_2R$ (Scheme 40) (82MI3).

E. MISCELLANEOUS METHODS

The opening of the β-lactam ring of some cephems has led to 4*H*-thiazine-carboxylates (**66**) (Scheme 41) (77ABC65). On heating with 36% hydrochloric or 85% orthophosphoric acids, substituted 4-hydroxyhexa-hydropyrimidine-2-thiones undergo an intramolecular rearrangement involving conversion of the cyclic forms to linear tautomers. This is accompanied by dehydration with formation of substituted 2-alkylamino- or 2-arylamino-4*H*-1,3-thiazines (**67**) (Scheme 42) (69KGS896; 70KGS1690; 72KGS937). A few thiazine-4-ones that are substituted at C-2 by a nitrogen (amine, ergoline, etc.) have been prepared by further miscellaneous methods [68CB1428; 79JAP(K)79-20504; 81EGP147359].

III. Physical Properties

A. ULTRAVIOLET AND VISIBLE SPECTRA

1. *2-Mercapto-1,3-Thiazines: Thione–Enethiol Tautomerism*

It seems very likely that structure **68** predominates in neutral solution (cf. ^{13}C). Otherwise, it has been claimed that in alkaline conditions, the thiolate form **69** occurs (55JA2866; 64JCS4008). Compound **68** exhibited

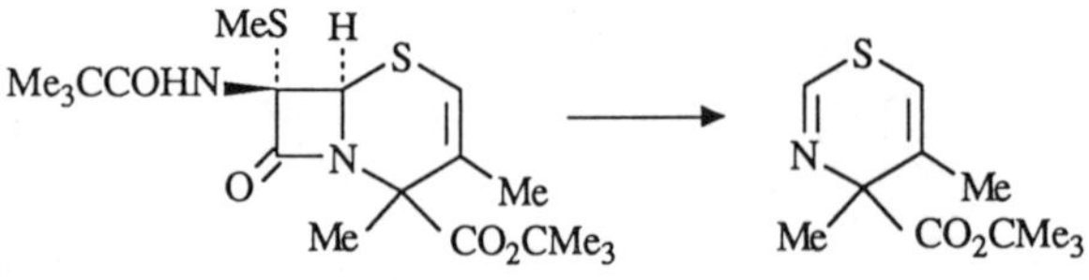

(66)

SCHEME 41

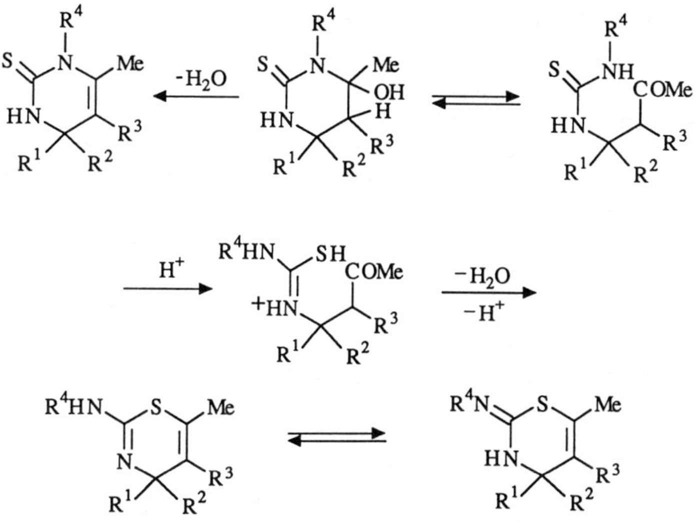

(67)

SCHEME 42

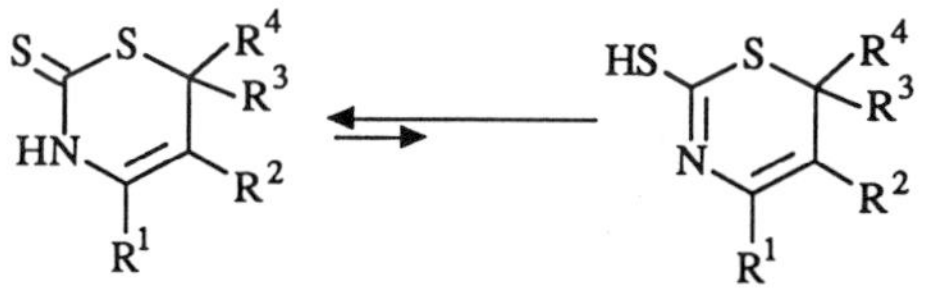

(68)   **(69)**

one band in alcohol at approximately 310–320 nm, while in water, two bands were observable at 320 and 290 nm. A third band of lower intensity has been observed in the region 360–380 nm in some compounds. The spectra remained unchanged in the presence of acid, but alkaline conditions in alcohol resulted in a band at 270–290 nm. With the nonenolisable 2-thioxothiazines **70** and **71,** absorptions appear at the same wavelength (315–321 nm) [82JCS(P1)2149].

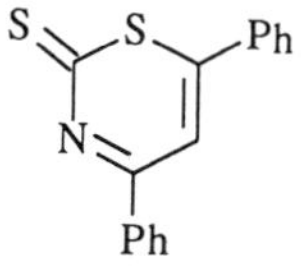

(70) 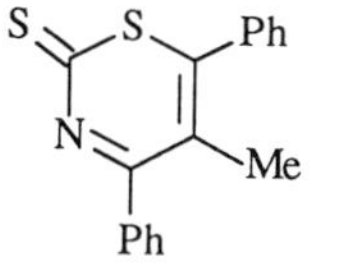   **(71)**

λ max   205, 242, 321, 479   206, 234, 315, 459

2. 2-Aminothiazines: Amino–Imino Tautomerism

N-Methylamino and anilino thiazines **72** and **73**, respectively, where the amino group is conjugated with an ester function at the 5-position, show the indicated absorptions. The authors are of the opinion that these compounds are 2-amino- and not 2-imino-thiazines (65JOC2290). The same observation was made for 4H-thiazines, which were not substituted by a functional group (75KGS1614). The authors have compared the tautomerizable thiazines **74** and **75** to models **76** and **77**, which cannot exhibit tautomerism due to the additional N-methyl groups. However, the N-methyl model compounds that have fixed amine and imine structures have less distinctly expressed absorption maxima than **74,** and the intensities of these maxima are considerably lower.

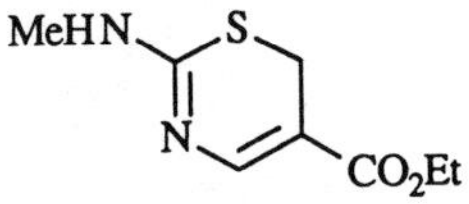

(72) 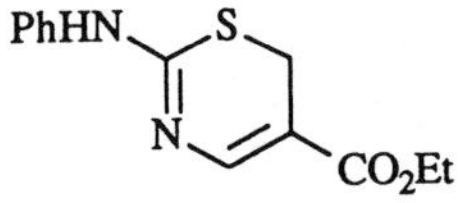   (73)

λ max EtOH	240 266 330
λ max EtOH, HCl	261 297

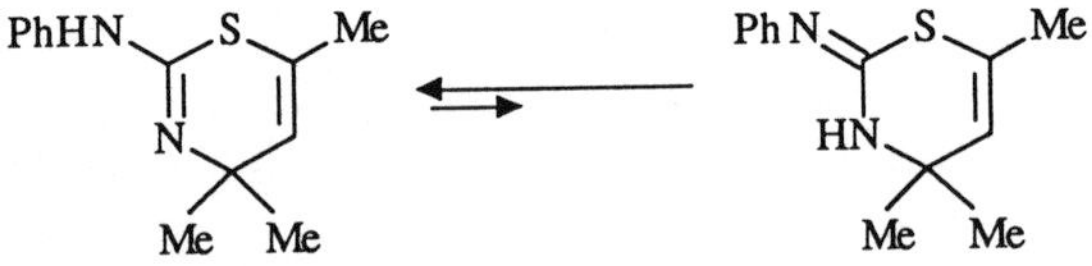

(74)   (75)

λ max   263
dioxanne

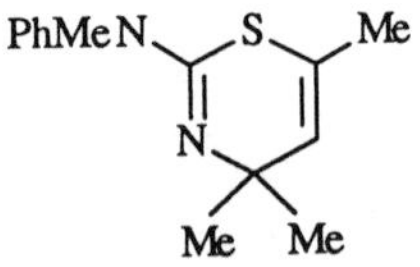

(76)   (77)

λ max   264   236, 280, (sh)

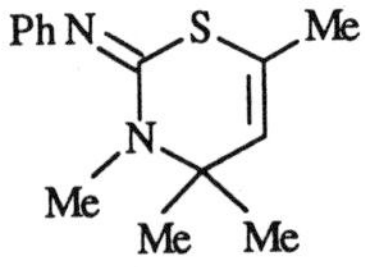

3. *2H-Thiazine-4-ones*

For 4*H*-thiazine-4-ones (**78**), two maxima are generally noted: one between 233 and 258 nm and the other between 278 and 298 nm [66TL3225; 77LA1249; 78JCS(P1)1428]. The differences in the long-wave maxima between amino and imino structures in the UV spectra are not usually substantial, with the exception of the dimethylamino derivative (**79**) (82CCC3268). This compound (λ max = 330 nm) shows a bathochromic shift of 60–80 nm compared to all other thiazine derivatives, probably due to zwitterionic character (**80**).

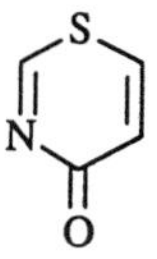

(78) 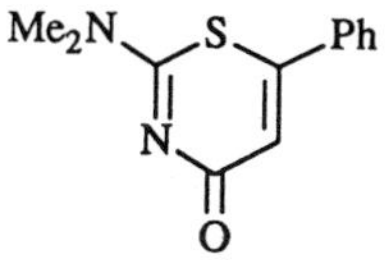   (79) 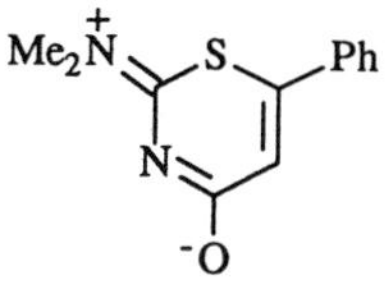   (80)

4. *6H-Thiazine-6-ones*

The UV spectra of compounds **81** in ethanol (λ max = 248 nm) are similar to the spectra of compounds for which an enol structure has been proved (79KGS44). Three absorption maxima at 250, 294, and 430 nm can be linked with the absorption of the two tautomeric forms **82** and **83**.

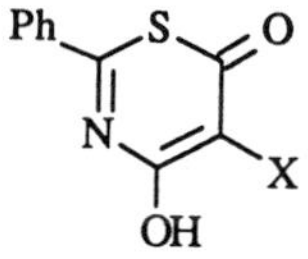

(81)

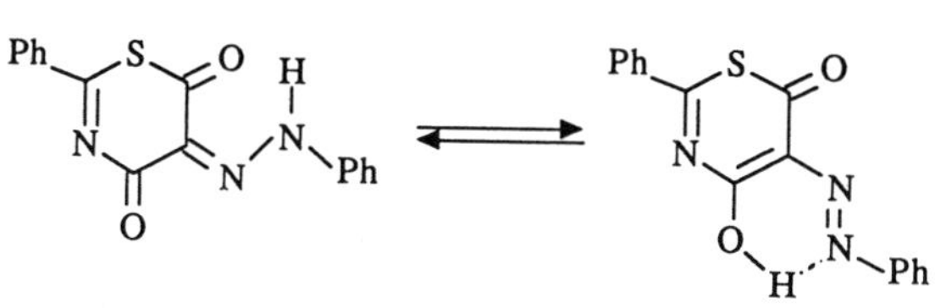

(82)   (83)

B. INFRARED SPECTRA

Randall *et al.* have examined seven examples of thiazoles and found absorptions in the range 1634–1570 cm^{-1}, which they regard as being typical of the thiazole structure. Randall reported it is clearly necessary to regard the system as a simple unit for correlation purposes (48MI1). Theoretical analysis of single 4H-pyrans (or thiopyrans) concluded that a doublet at 1634 and 1686 cm^{-1} could be attributed to coupled antisymmetric and symmetric vibrations of the two double bonds (71MI1).

For 6H-1,3-thiazines, the absorptions belonging to functional groups, which are often carbonyl, are generally found in two absorption ranges: one between 1520 and 1480 cm^{-1}, the other between 1600 and 1520 cm^{-1} (65JOC2290). However, electron-releasing substituents, such as those in 2-anilino-4,6,6-trimethylthiazine (**84**) can raise the high absorption band to 1660 cm^{-1} (78AJC2307). The high absorption is often weak. The absorption in the lower area, 1480–1520 cm^{-1}, is often stronger.

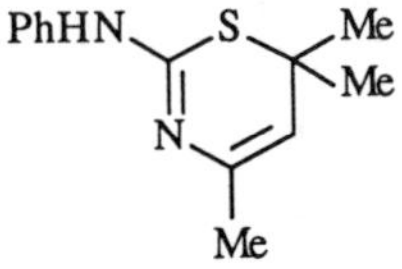

(**84**)

An IR study of 4H-thiazines, undertaken in 1970, relies on two structures **85** and **86** that look well-established, and on a third, **87**, which unfortunately is less safe (71KGS946). If compound **87** does not tautomerize to an imino-thiol, the highest vibration can be assigned to the C=C double bond. By removing the C=C vibration from the IR spectra of **85** and **86,** we can see the C=N vibration. The similarity between the spectra of **85** and **88** has allowed the authors to state that the example of tautomerism shown here (**88, 89**) is strongly shifted in favor of the 2-amino form **88.**

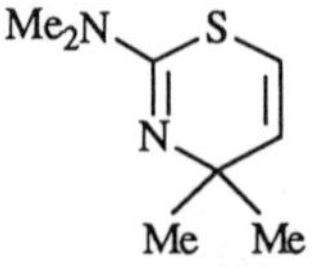

(**85**) 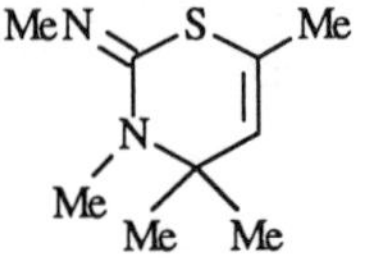   (**86**) 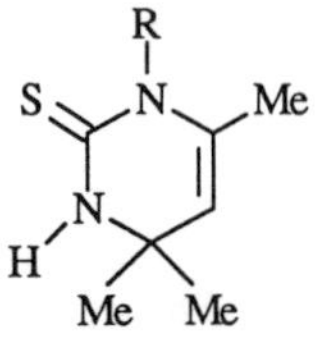   (**87**)

(**88**)   (**89**)

TABLE I
IR DATA OF THIAZINES **90–91**

Compound*	IR, cm⁻¹ (CHCl₃)				IR, cm⁻¹ (KBr)		
	ν (C=N)	ν (C=C)	ν (C=O)	ν (NH)	ν (C=N)	ν (C=C)	ν (C=O)
a	1 523	1 604	1 617	3 435	1 537	1 602	1 621
	1 628[b]		1 696	3 408	1 635[b]		1 693
b	1 515	1 602	1 615	3 426	1 540	1 600	1 617
	1 633[b]		1 661[b]	3 392			
c	1 514	1 601	1 617	3 417	1 522	1 597	1 614
	1 633		1 663	3 377			
d	1 515	1 606	1 618	3 410	1 519	1 603	1 615
	1 635[b]		1 654[b]	3 381			
e	1 627	1 587	1 672	3 363	1 633[b]	1 589	1 661[b]
					1 505	1 601	1 625
f	1 628	1 584	1 670	3 365	1 632[b]	1 594	1 653[b]
					1 577	1 606	1 622
g	1 628	1 585	1 671	3 366	1 636[b]	1 594	1 659[b]
					1 507	1 611	1 619

* a, R_1 = Me; b, C_4H_9; c, CH_2Ph; d, C_6H_{11}; e, Ph; f, MeC_6H_4; g, $MeOC_6H_4$.

These examples are interesting because there is an interruption of conjugation between C=C and C=N which allows more specific behavior. According to this work, the highest absorption, 1655–1680, is assigned to C=C and the lowest band, 1588–1635, is assigned to C=N. However, the high absorptions are probably due to methyl substituents. In thiazine-6-ones the absorption assigned to carbonyl is located between 1635 and 1680 cm⁻¹. The high values agree with structures for which an ester group, for example at the 4-position, restricts the carbonyl conjugation. For thiazine-4-ones, the IR study uses examples of compounds able to give tautomeric equilibria (Table I).

With R^2 = Ph, the IR spectra in solution (CHCl₃) indicate the *N*-alkyl derivatives appear in the imino form (**91**) to some extent. In fact, we find strong absorption bands ν (C=O) at 1615–1618 cm⁻¹ with characteristic shoulders at 1654–1696 cm⁻¹ and two absorption bands ν (NH) at 3377–3408 and 3410–3435 cm⁻¹. The shoulders ν (C=O), as well as the lower

(90)   (91)

R^2 = Ph, CO_2Me

intensity of the absorption band ν (NH) at 3377–3408 cm^{-1} are characteristic of the less conjugated form (91). With the exception of N-methyl derivatives, which show ν (C=O) bands of both tautomeric forms both in chloroform and KBr discs, the equilibria of N-alkyl and N-aryl derivatives are shifted in the solid state in favor of the amino forms (90).

In the case of thiazine-4,6-diones, in contrast to the spectrum of the unsubstituted compound 92 (C=O, 1640 cm^{-1}), there are three absorption bands. One is at 1640 cm^{-1}, corresponding to the enol form, and two are at 1690 and 1730 cm^{-1}, corresponding to the diketo form. In this example, the ^{1}H-NMR spectrum suggests the presence of three tautomeric forms 93, 94 and 95. In conclusion, we record a few absorptions bands for two thiazine-2-thiones (96) [82JCS(P1)2149].

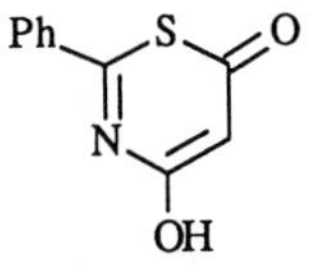

(92)

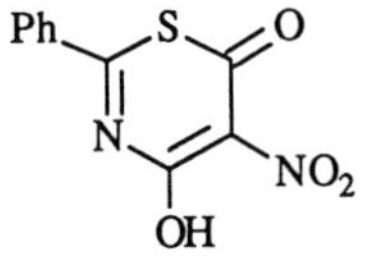

(93) 8%

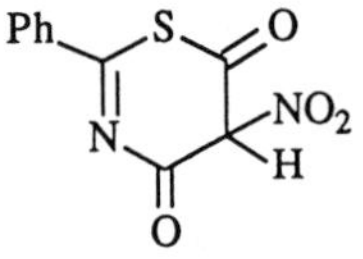

(94) 27%

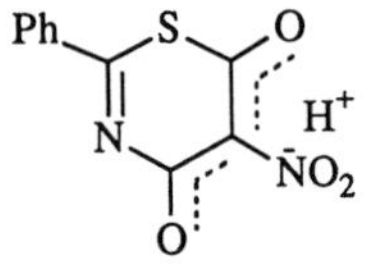

(95) 65%

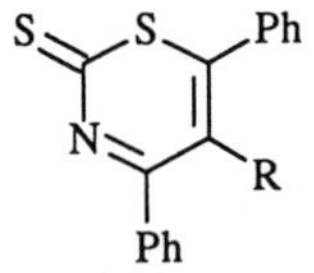

(96)

R = H   1580 1560 1455 1240 1060 760 690
R = Me   1595 1545 1445 1250 1045 755 690

96	Absorption bands						
R = H	1580	1560	1455	1240	1060	760	690
R = CH$_3$	1595	1545	1445	1250	1045	755	690

C. NMR SPECTRA

1. *¹H Spectra*

In the case of 6*H*-thiazines (**97**), the range of values for the chemical shifts for the hydrogens directly bonded to the ring are shown here. The two hydrogens bonded to C-6 are generally considered to be equivalent (73TL4649). With one hydrogen at C-2, the signal for the C-6 protons shows coupling, and when there is one hydrogen at C-4, W-coupling

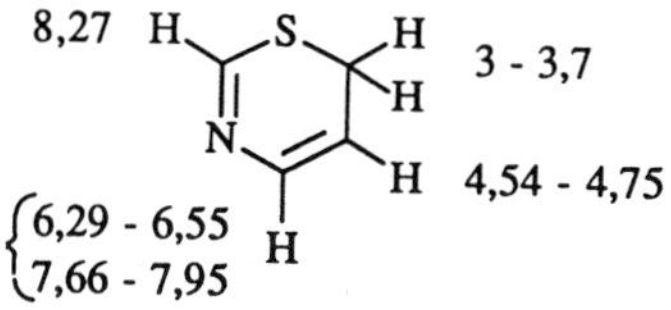

(97)

systems with $J = 1$Hz are noted [79BSF347; 80JCS(P1)1013; 83CJC1169; 84JPR101; 87BSF149]. The hydrogen at the 4-position is situated between δ 6.29 and 6.55 ppm, but when the 5-position is substituted by an electron-withdrawing group, a downfield shift is observed (δ 7.66–7.95 ppm). The chemical shift of the hydrogen at the 5-position in the 4*H*-thiazines (**88**) is δ 5.35–5.4 ppm (71KGS946).

Structure **98** shows the chemical shifts for the hydrogens at the 4 and 5-positions in the rings of thiazine-6-ones [74LA1753; 76BCJ2828; 80CCC732; 86PS327]. Structure **99** gives the same information for thiazine-4-ones: δ 7.25–8.36 ppm for the hydrogen at the 6-position and δ 6.60–6.98 ppm for the hydrogen at the 5-position in inert solvents [66TL3225; 78JCS(P1)1428; 82CCC3268; 82JOC1090; 84JOC74]. A coupling constant of 7.4 Hz appears between both the 4 and 5-position hydrogens in the thiazine-6-thiones (**100**) (65JOC2290). For 4,6-diphenylthiazine-2-thione, Schroth suggests a shift of δ 8.02 ppm for the hydrogen at the 5-position (83S827). Nishio indicates, for the same compound, two groups of signals at δ 8.15–8.30 ppm (2*H*) and δ 7.40–7.75 ppm (9*H*). At least one of the two aromatic hydrogens appears in the same region as H-5 [82JCS(P1)2149].

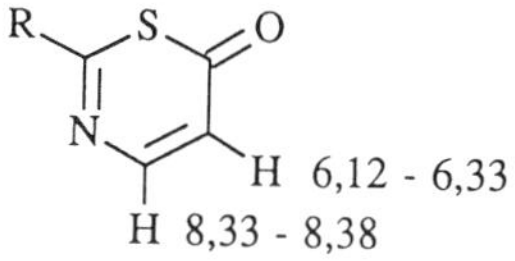

(98)

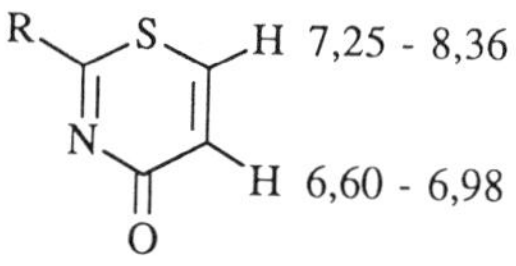

(99)

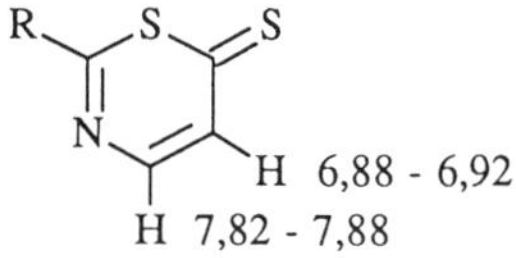

(100)

2. ^{13}C Spectra

The ranges of chemical shifts noted for 6H-1,3-thiazines (**101**) in the ^{13}C-NMR spectra are given (see structure **101**) with the ranges of C—H coupling constants. The compounds concerned often bear an electron-withdrawing function at C-5 (800MR479; 85JOC1545; 86CJC597; 87BSF 149; 87SC1971). Without making assignments, we note the spectrum of 2-anilino-4,6,6-trimethylthiazine: δ19.7, 24.0, 27.3, 31.4, 119.1, 120.7, 123.1, 124.7, 126.7, 130.6, and 130.7 ppm (78AJC2307).

We also note the chemical shifts at C-2 and C-4 for two 2-aryl-4,4-bis(trifluoromethyl)-4H-thiazines (**24**) (83CC945) along with the chemical shifts of four carbons for twelve 4H-thiazine-4-ones (**102**) (86PS327). The chemical shifts of four carbons from two 4,6-diphenyl-2H-thiazine-2-thiones (**103**) [82JCS(P1)2149] and the chemical shifts of the carbons from 2-substituted-4-methyl-6H-thiazine-6-thiones (**104**) are also shown (81BSB75). (See structures **24, 102–104.**)

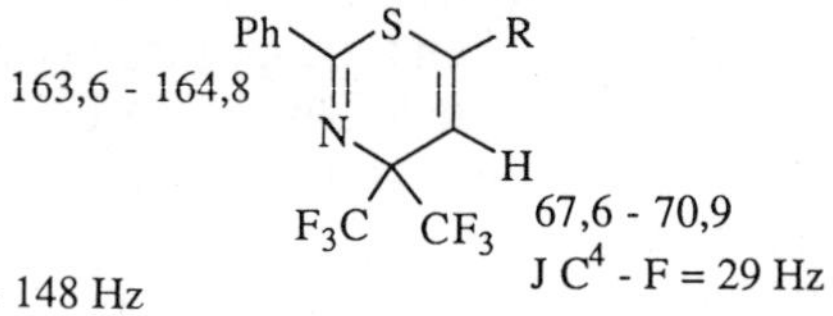

163,6 - 164,8

Ph — S — R

F$_3$C   CF$_3$ 67,6 - 70,9
J C^4 - F = 29 Hz

(**24**) R = Ph, H

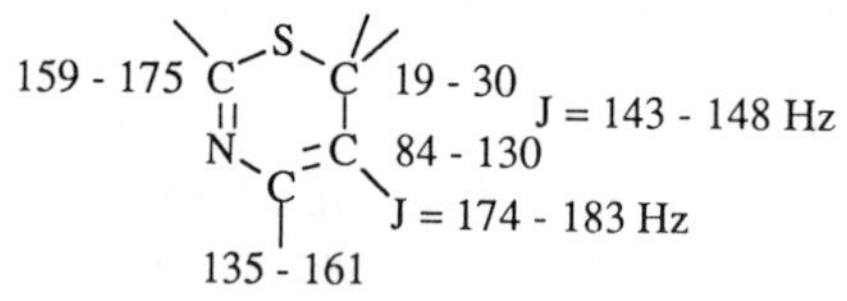

159 - 175   C — S — C   19 - 30
J = 143 - 148 Hz
N — C = C   84 - 130
J = 174 - 183 Hz
135 - 161

(**101**)

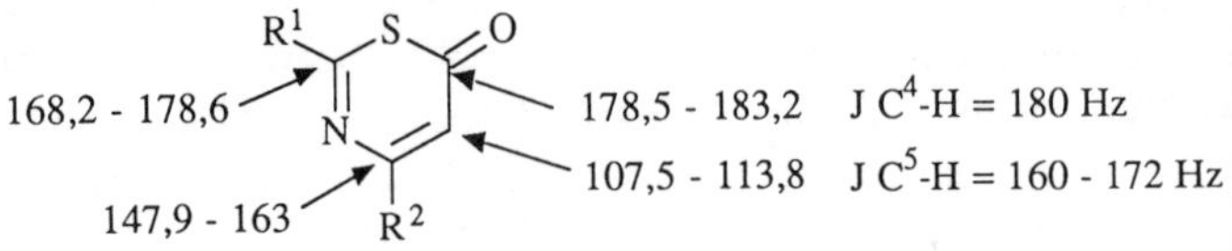

R^1 — S — O

168,2 - 178,6

178,5 - 183,2   J C^4-H = 180 Hz
107,5 - 113,8   J C^5-H = 160 - 172 Hz

147,9 - 163   R^2

(**102**)

S — S — Ph

N
R
Ph

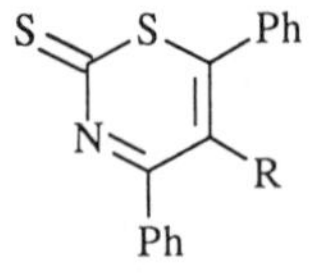

(**103**)

R = H, Me
111,4 - 122,2
136,7 - 138,2
163,8 - 164,3
168,3 - 169,1

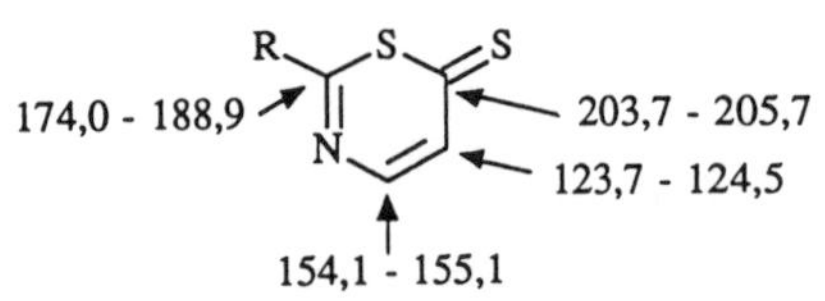

R — S — S

174,0 - 188,9
203,7 - 205,7
123,7 - 124,5

154,1 - 155,1

(**104**)

3. ^{15}N Spectra

^{15}N-NMR spectra display chemical shifts for the nitrogen of thiazines $\delta = (-)67.2–91.3$. The chemical shifts are expressed in ppm referenced to CH_3NO_2 used as an internal standard (80OMR479; 86CJC597).

D. Mass Spectra

The 6H-1,3-thiazines (**105**) studied are substituted thiazinyl-alanines, potential precursors of either cephems or cephalosporins. Scheme 43 shows the fragmentation founded on the study of metastable ions and "peak matching" (82TH1; 83CJC1169). The mass spectral fragmentation of 4H-1,3-thiazines, such as **106** (R^1 = Me, OMe, OEt, NO$_2$) occurred mainly by loss of one of the geminal methyl groups, whereas fragmentation of isomeric compound **107** yielded an intense [M-C$_6$H$_{10}$S]$^+$ ion resulting from cleavage of the thiazine ring (83MI1).

With NH compounds **108** and **109**, the amino/imino tautomer ratio could be determined by the ratio of the [M-Me]$^+$ and [M-C$_6$H$_{10}$S]$^+$ peaks. The amino form was more important for R^1 = Cl, H, and NO$_2$ than for R^1 = OMe and OEt. For 1,3-thiazine-6-ones such as **110,** the fragments noted where m/e 290 (M$^+$), 229 (M-COS-H), 275 (M-Me), and 151

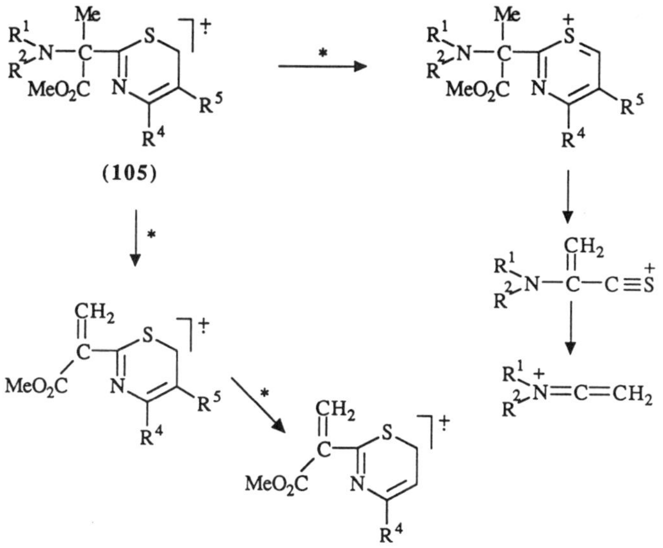

(**105**)

SCHEME 43

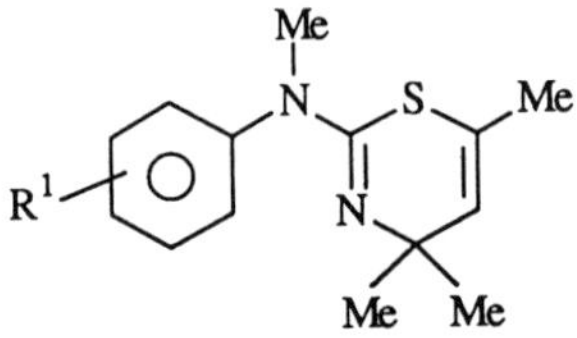

(106)

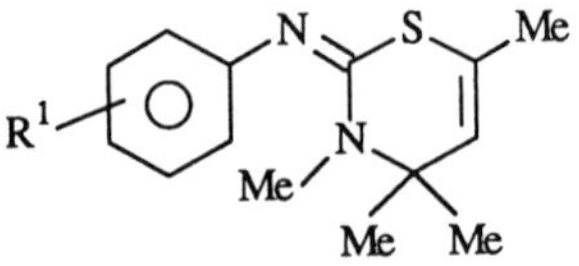

(107)

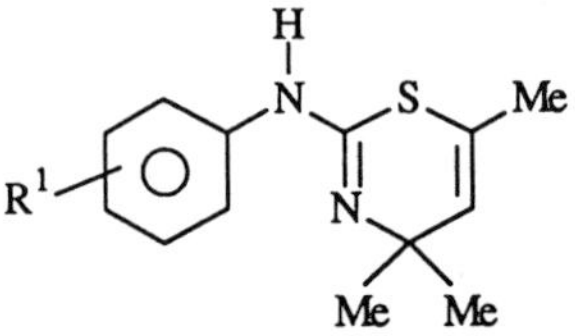

(108)

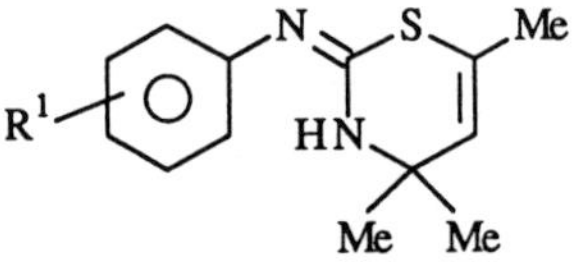

(109)

(MeOC$_6$H$_4$CS) (84JOC74). The fragmentation of one 1,3-thiazine-4-one (111) and a part of the scheme of fragmentation of one 2-imino-1,3-thiazine-4-one (112) are noted in Scheme 44 (82JOC1090). The same type of fragmentation has been found in the mass spectra of thiazinylergolines

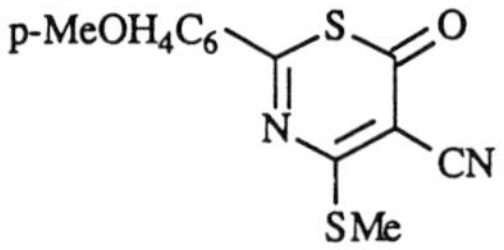

(110)

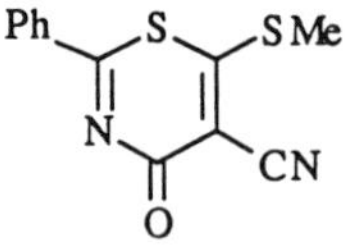

(111)

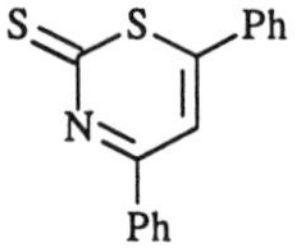

(114)

m/e   260 (M$^+$), 157 (M - PhCN)
      131 (PhCNCO), 110 (M - PhCN - SMe)
      103 (PhCN)

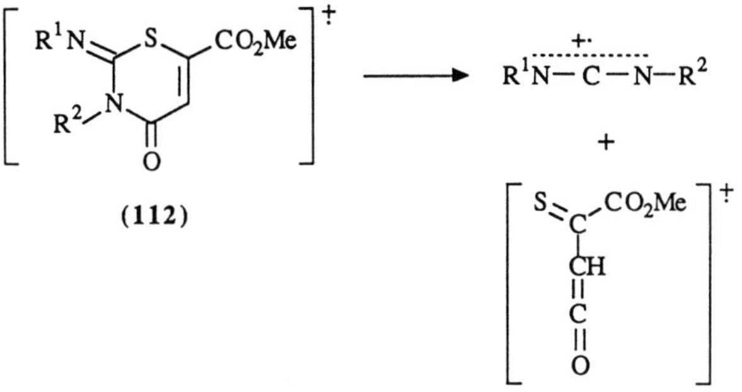

SCHEME 44

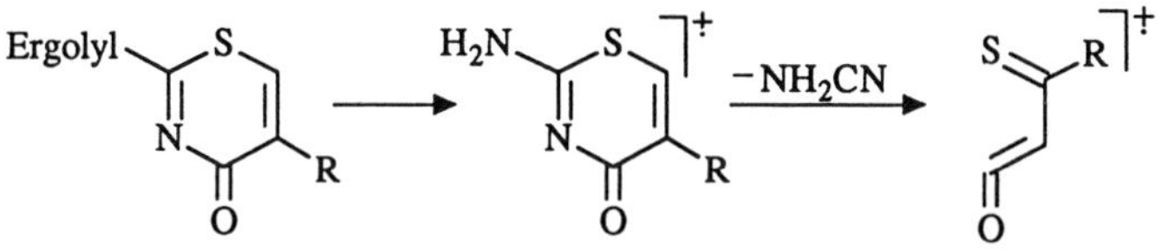

SCHEME 45

(113) (Scheme 45) (82MI1). The fragmentation of 4,6-diphenyl-1,3-thiazine-2-thione (114) under electron impact is noted here together with the relative abundance of the peaks: m/e 281 (M^+, 91%), 280 (M^+-H, 100%), 248 (M^+-H, 100%), 248 (M^+-SH, 21%), 237 (M^+-CS, 38%), 223 (M^+-NCS, 36%), 191 (M^+-NCS2, 26%, 178 (M^+-PhCN, 11%), 160 (M^+-PhCS, 17%) (83S827).

E. X-RAY

The X-ray analysis of compounds **115–118** is given in Table II. The

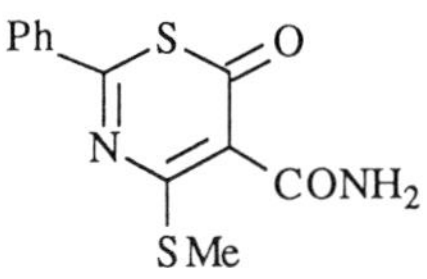

(115)

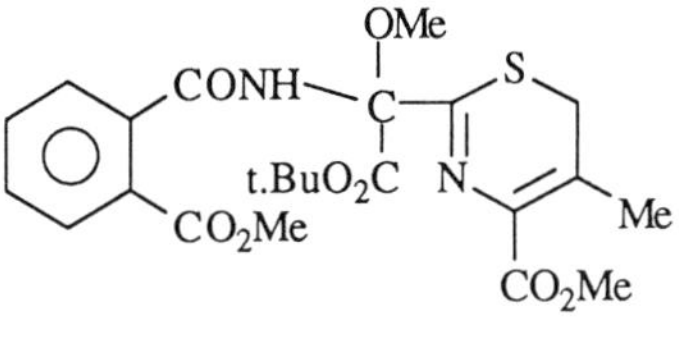

(116)

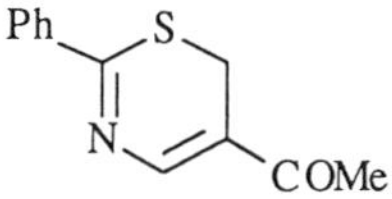

(117)

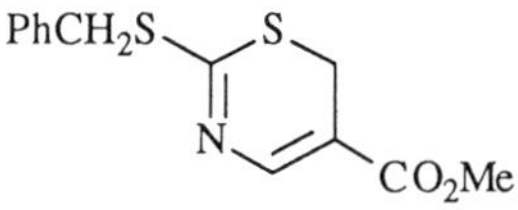

(118)

TABLE II
X-Ray Analysis of Compounds 115–118

Compound	Space group	a(Å)	b(Å)	c(Å)	B°	Z
115 (C$_{12}$H$_{10}$N$_2$O$_2$S$_2$)^a	P2₁/C$_{2h}^5$	7688(3)	5156(3)	31,473(8)	105.52(3)	4
116^b	P2₁/n	9783(4)	15,401(5)	16,938(9)	102.71(5)	4
117 (C$_{12}$H$_{11}$NOS)^c	P2₁/C	12,258(3)	9313(2)	9785(2)	98.90(2)	4
118 (C$_{15}$H$_{13}$NOS)^d	P2₁/C	14,638(2)	8844(1)	10,734(2)	—	4

^{a} Monoclinic crystal; data from Yokoyama *et al.* (82JOC1090).
^{b} Data from Bakasse *et al.* (88T139).
^{c} Monoclinic crystal.
^{d} Monoclinic system.

results of structural determination of 117a show that the benzene ring, the acetyl group, and the chain C-2—N—C-4—C-5 are almost coplanar. The bond lengths observed by X-ray are near the measured values for the same bonds from analogous systems (87BSF149).

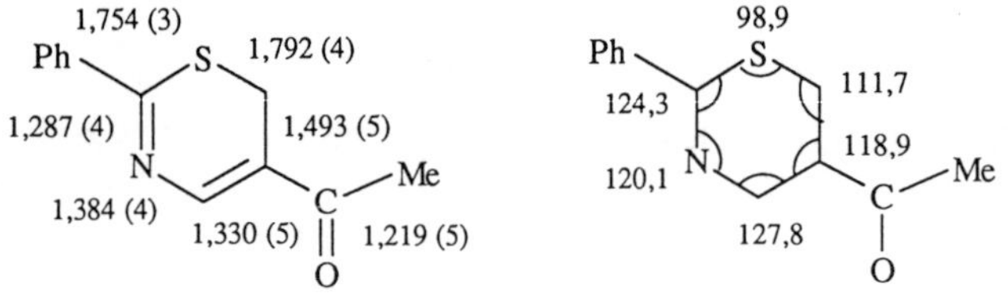

(117)

F. Miscellaneous Physical Properties

The basicities of the few 1,3-thiazines studied are located between those of sodium bicarbonate and ammonia. Ethyl 2-methylamino-1,3-thiazine-5 carboxylate (72) has pK$_a$ 5.41 (65JOC2290). Russian authors have determined the basicities of some compounds by potentiometric titration with perchloric acid in ethanol. The 2-methyliminothiazine (119), pK$_a$ 7.84, is more basic than the isomeric aminothiazine (120), pK$_a$ 7.12 (71KGS946).

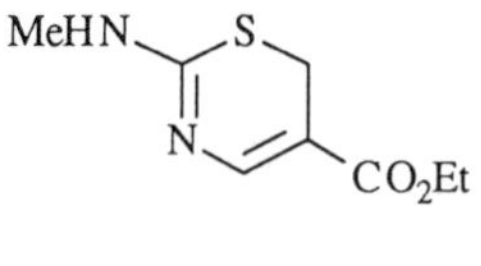

(72) 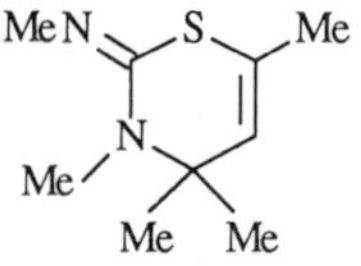 (119) 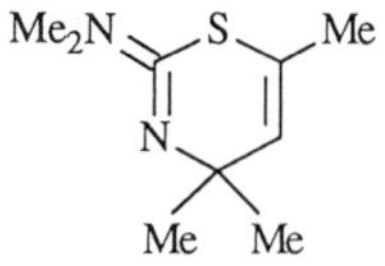 (120)

The pK_a of 2-amino-4-methyl-6H-1,3-thiazine-6-one has been spectro-photometrically determined and is high (13.30) similar to that of disodium sulfide (80CCC732).

The electrical resistance of complex salts of the 2,4,6-triphenylthia-zinium cation and tetracyanoquinodimethane (TCNQ) has been measured. For one thiazinium cation and 2 mol of TCNQ, the specific resistances were found to be in the range of 5–15 Ω/cm (82MI2).

IV. Chemical Properties

A. REDUCTION

1. *Chemical Reduction*

a. *Sodium.*   Reductive cleavage of the sp^3 carbon–sulfur bond in the alkyl- or aryl-substituted 2-mercapto-6H-1,3-thiazines (**121**) with sodium

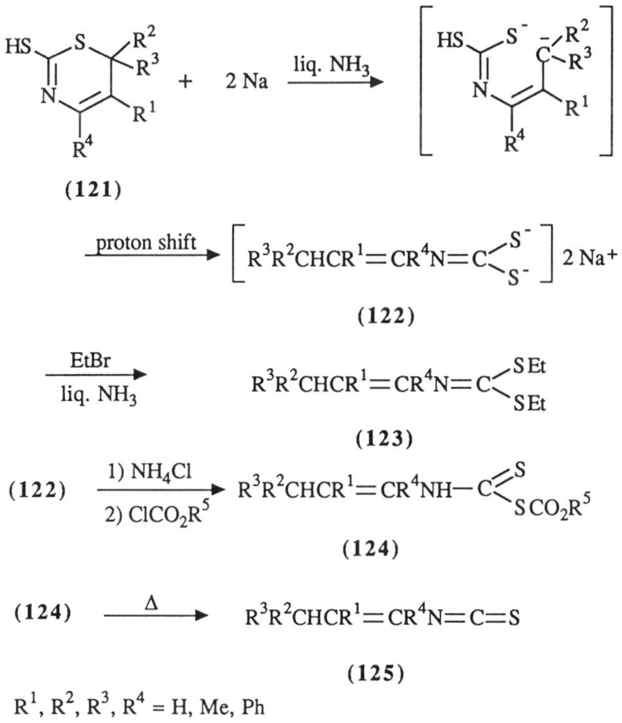

$$\text{(121)} + 2\,\text{Na} \xrightarrow{\text{liq. NH}_3} \left[\text{HS} \cdots \right]$$

(121)

$$\xrightarrow{\text{proton shift}} \left[R^3R^2CHCR^1{=}CR^4N{=}C{\stackrel{\displaystyle S^-}{\scriptstyle S^-}} \right] 2\,Na^+$$

(122)

$$\xrightarrow[\text{liq. NH}_3]{\text{EtBr}} \quad R^3R^2CHCR^1{=}CR^4N{=}C{\stackrel{\displaystyle SEt}{\scriptstyle SEt}}$$

(123)

$$\text{(122)} \xrightarrow[\text{2) ClCO}_2R^5]{\text{1) NH}_4Cl} R^3R^2CHCR^1{=}CR^4NH{-}C{\stackrel{\displaystyle S}{\scriptstyle SCO_2R^5}}$$

(124)

$$\text{(124)} \xrightarrow{\Delta} R^3R^2CHCR^1{=}CR^4N{=}C{=}S$$

(125)

$R^1, R^2, R^3, R^4 = H, Me, Ph$

SCHEME 46

in liquid ammonia affords almost quantitative yields of the corresponding disodium dithiolic compounds (122) (74RTC78). The dianion can be dialkylated with EtBr, giving the S,S'-diethyldithiocarbamates (123). The resulting compounds retain the cis configuration of the starting compounds. Protonation of 122 with one equivalent of ammonium chloride results in the formation of the corresponding monosodium salt which, when treated with an alkyl chloroformate at 0°C, affords the S-alkoxycarbonyl-N-alkenyldithiocarbamate (124). Heating induces decomposition of 124 to yield the N-alkenyl isothiocyanate (125). (Scheme 46). Similarly reductive cleavage of 2-amino-4,6,6-trimethyl-$6H$-1,3-thiazine (126) affords the N-substituted thiourea (127) (Scheme 47).

b. *Tin.* Tin and hydrochloric acid with thiazinylidenic derivatives (128) substituted by a methyl at the 4-position afford regioselective reduction of the exocyclic double-bond, resulting in compounds 129 (Scheme 48) (83PS143).

c. *Raney Nickel.* Desulfurization/dimerization of the 2,4-disubstituted-1,3-thiazines (130) by Raney nickel in benzene gave, in 6 hr, the

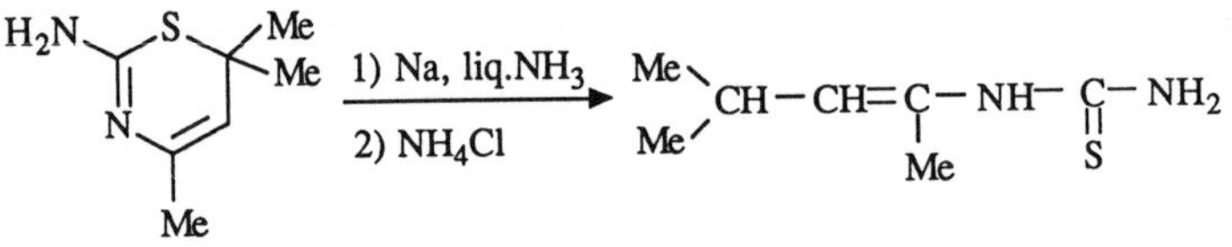

(126)   (127)

SCHEME 47

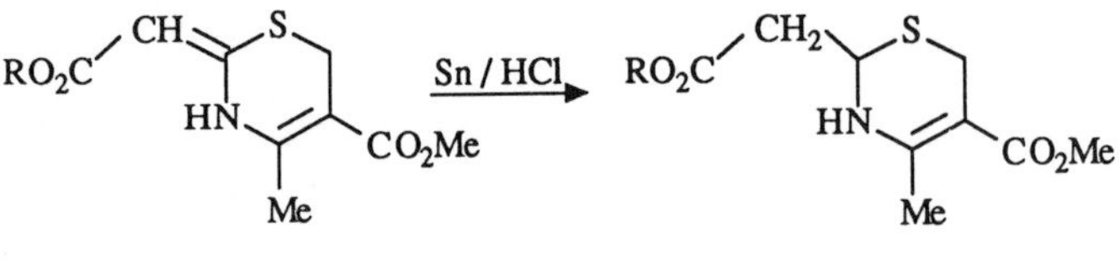

(128)   (129)

SCHEME 48

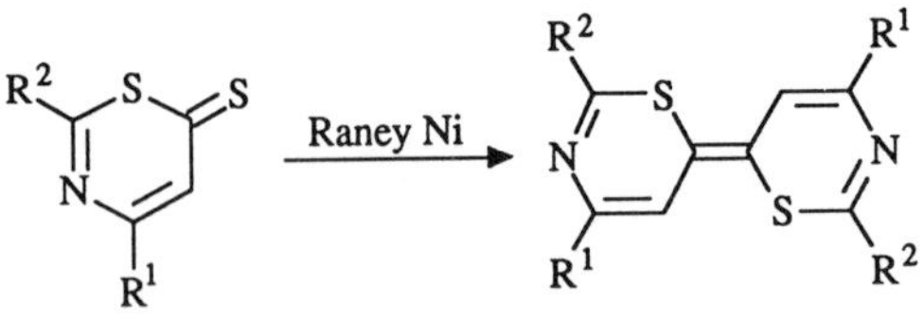

(131 E)

SCHEME 49

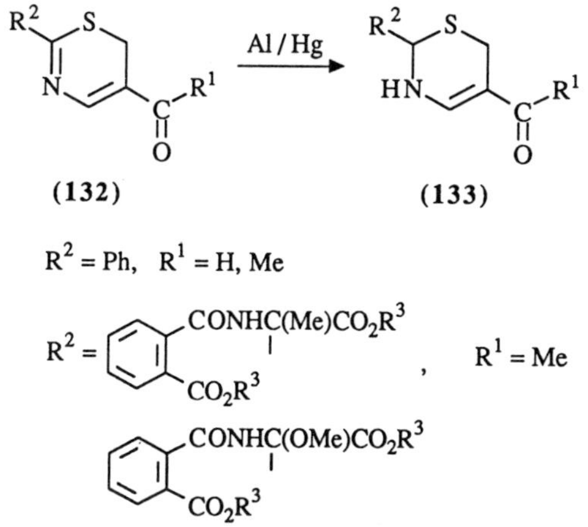

(132)   **(133)**

$R^2 = Ph, \quad R^1 = H, Me$

$R^2 =$ [structure] CONHC(Me)CO$_2$R^3 / CO$_2$R^3 , $R^1 = Me$

[structure] CONHC(OMe)CO$_2$R^3 / CO$_2$R^3

SCHEME 50

thermodynamically more stable 6,6′ E isomer of bis(6H-1,3-thiazinylidene) (**131**). A shorter time (2 hr) gave a mixture of both E and Z isomers (Scheme 49) (84MI1).

d. *Aluminum Amalgam.* Aluminum amalgam, in ethanolic solution, regioselectively reduces the endocyclic imine group of thiazines such as **132** (Scheme 50) [82JCR(S)72, 82TL2315; 83CJC1169]. With the thiazine-2-ylidenic compounds (**134**), aluminum reduces the exocyclic carbon–carbon double bond when the thiazine is substituted by an ester at the 5-position affording the 3,6-dihydro-2H-1,3-thiazine derivative (**135**). When the substituent in the same position is an acetyl (**136**), which is more

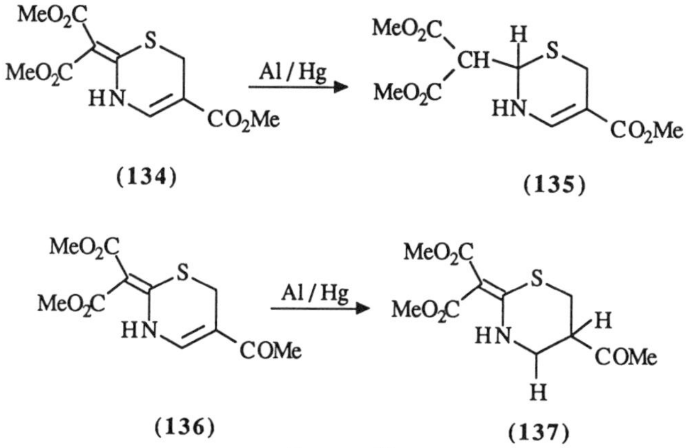

(134)   **(135)**

(136)   **(137)**

SCHEME 51

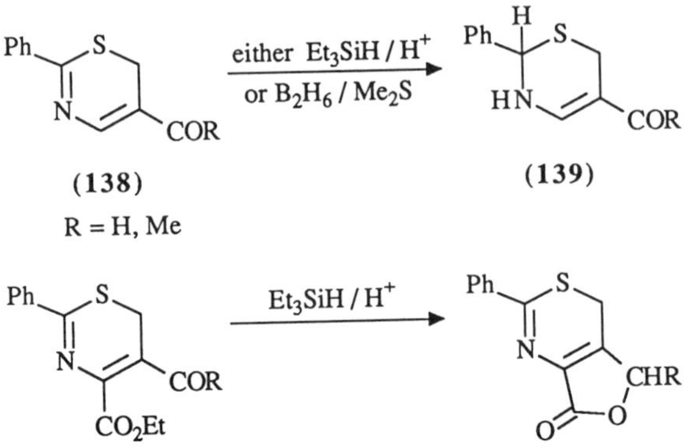

(138)

R = H, Me

(139)

(140)

(141)

Scheme 52

able to stabilize an anion, the reduction proceeds to compound **137**, in which the endocyclic carbon–carbon double bond is reduced. In both examples, the reduction is regioselective (Scheme 51) (84CC157).

e. *Borane and Triethylsilane.* In the examples investigated, the two reagents gave approximately the same result, i.e., saturation of the activated C=N double bond. When the bond was less activated (carbonyl and ester in adjacent positions), reaction was restricted to the formation of a lactone ring (Scheme 52) [82JCR(S)72; 85TL745]. The yield was up to 86% for saturation of the imino group in the example shown in Scheme 53.

f. *Sodium Borohydride.* With an electron-withdrawing substituent (ester) in the 4-position, the 6*H*-1,3-thiazine (**144**) was converted to the tetrahydro derivative (**145**) (Scheme 54) [73JAP(K)73-80575]. In contrast, the 6*H*-1,3-thiazines (**146**) and the thiazine-2-ylidene derivative (**147**), which had an electron-withdrawing group at the 5-position, were not reduced at the double bonds, but the aldehyde and ketone functions were

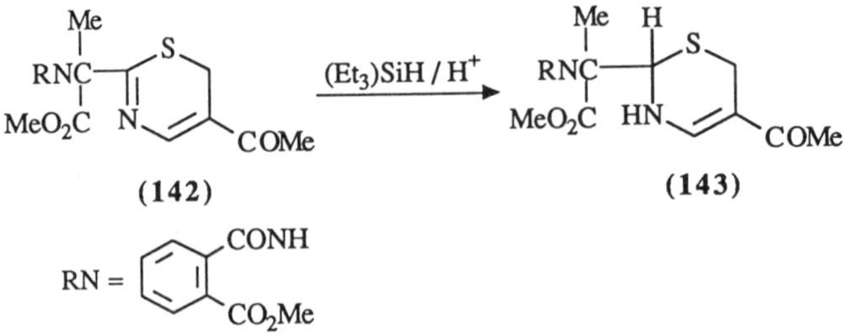

(142)

(143)

Scheme 53

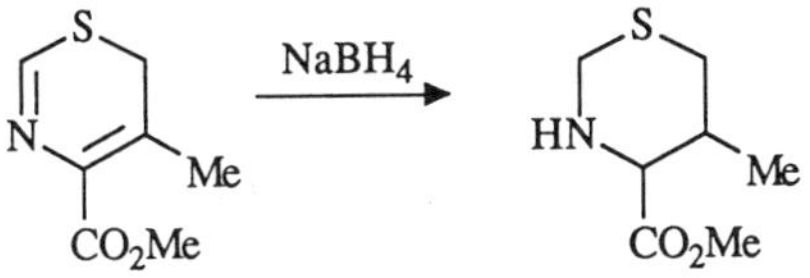

(144) (145)

SCHEME 54

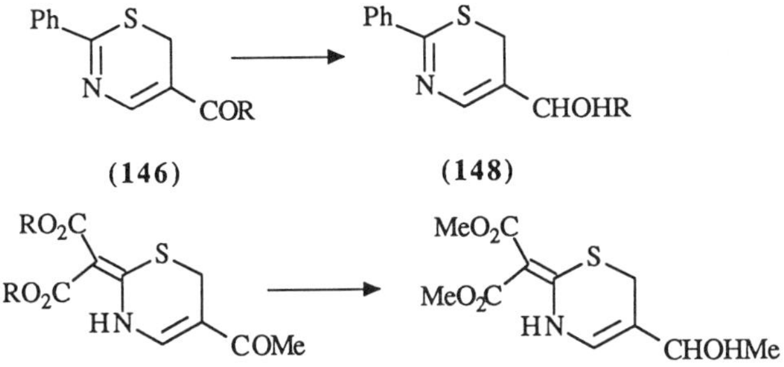

(146) (148)

(147) (149)

SCHEME 55

reduced to the corresponding alcohols **148** and **149** (Scheme 55) [82JCR(S)72; 83PS143].

Reduction of the 1,3-thiazine-6-ones (**150**) was easier, and 5-carbamoyl- or 5-cyano-4-methylthio-2-phenyl-2,3-dihydro-6*H*-1,3-thiazine-6-ones (**151**) were obtained (Scheme 56) (82JOC1090). The structure of compound (**151**) was determined by a single-crystal X-ray diffraction study. 2,6-Disubstituted-4*H*-1,3-thiazine-4-ones (**152**) with $NaBH_4$ afford 3,4-dihydro-2*H*-1,3-thiazine-4-one derivatives **153** (Scheme 57) (83CPB1929).

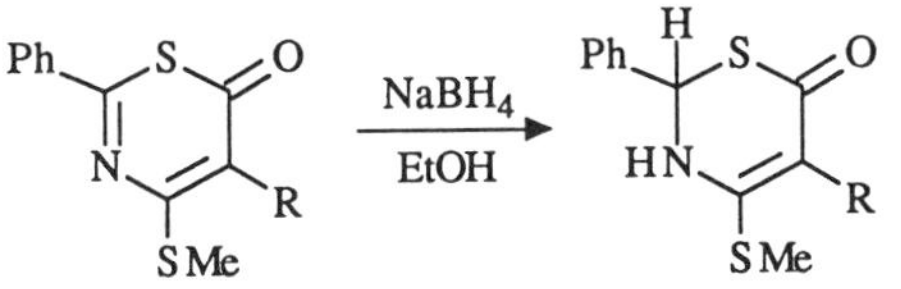

R = $CONH_2$, CN

(150) (151)

SCHEME 56

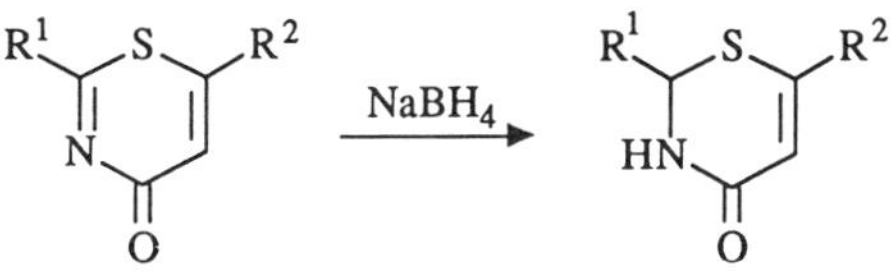

(152) (153)

SCHEME 57

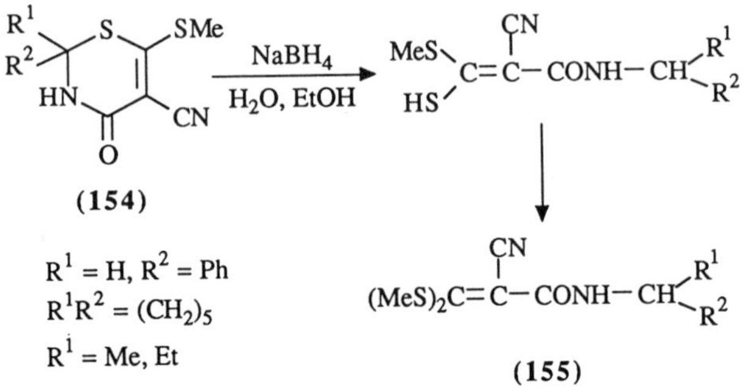

(154)

$R^1 = H, R^2 = Ph$
$R^1R^2 = (CH_2)_5$
$R^1 = Me, Et$

(155)

SCHEME 58

With an excess of sodium borohydride in water–ethanol, hydrogenolysis of the ring by scission of the S—C-2 bond was observed (Scheme 58) [75JCS(P1)1417].

g. *Sodium Cyanoborohydride.* With the 6*H*-1,3-thiazines (**156**), the nitrogen may be protonated. In this case, the hydride attacks at the 2-carbon. Thus, we have reduction of the endocyclic imine group in **156**, giving **157** (Scheme 59) (85TL745). When the thiazinylidenic derivative (**158**) is unsubstituted at position 4, there is a simple reduction of the exocyclic carbon-carbon double bond to yield **159**. Two enamino ester systems are present concurrently here. The better-stabilized carbocation (sulfur electron-releasing) is made preferentially and then reacts with H⁻. The double bond linked to sulfur is then reduced (Scheme 60) (83CJC1169, 83PS143). When the thiazinylidenic derivative (**160**) is substituted at the 4-position by a methyl, the carbocation linked to Me gains in stabilization. Both of the enamino ester systems are stabilized, and there is a double reduction to yield (**161**) (Scheme 61).

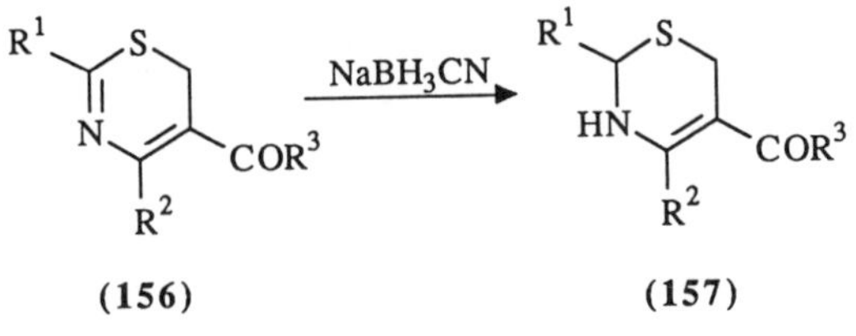

(156)   **(157)**

$R^1 = Ph, PhthC(Me)(CO_2Me)$
$R^2 = H, CO_2Et$
$R^3 = H, Me$

SCHEME 59

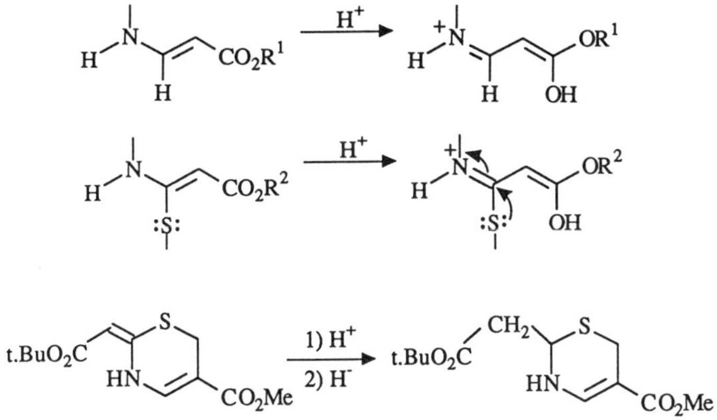

(158) (159)

SCHEME 60

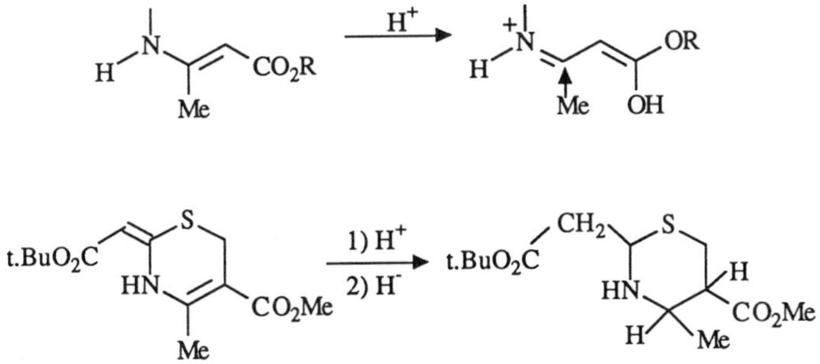

(160) (161)

SCHEME 61

h. *Lithium-Aluminum Hydride*. Results are often comparable with those obtained from $NaBH_4$. Compound **154** is, however, obtained in improved yield (74% instead of 52%). Hydrogenolysis may occur with breaking of the ring in the tetrahydrothiazine intermediate (**162**) (Scheme 62) (65JOC3092).

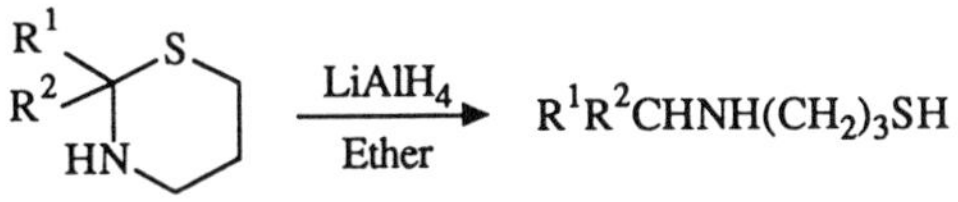

(162)

SCHEME 62

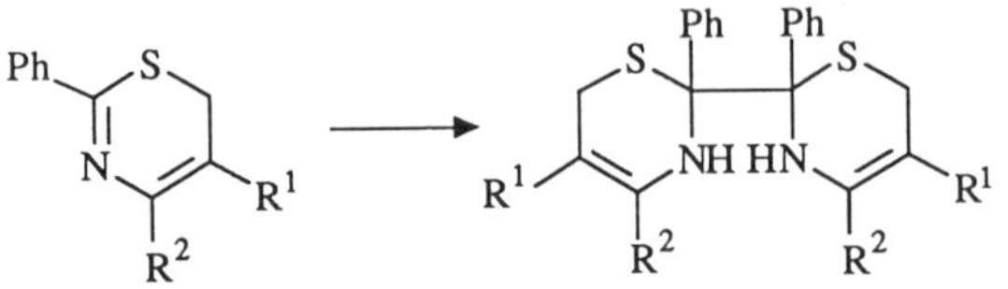

(163)   (164)

SCHEME 63

2. Electrochemical Reduction

Electrochemical reduction of 2-phenyl-6H-1,3-thiazines carried out at a mercury cathode in acetate buffer and ethanol (1 : 1) has been studied. In the case of compounds that are monoactivated at carbon 5, such as 163 (R^1 = CHO, COMe; R^2 = H), either hydrodimers (164) (resulting from coupling at C-2), 3,6-dihydro-2H-1,3-thiazines (reduction of the imine bond), or tetrahydrothiazines can be obtained (Scheme 63) (85TL745). Diactivated 6H-1,3-thiazines (163, R^1 = CHO, COMe, or R^2 = CO$_2$Et) successively lead to 5,6-dihydro-4H-1,3-thiazines (reduction of the ethylenic bond) and tetrahydrothiazines (87T2709). The reduction of a 6H-1,3-thiazine bearing only one withdrawing group at the 4-carbon (163) gives rise to ring opening and compound 165 (Scheme 64) (86MI1).

B. REACTIONS WITH NUCLEOPHILES

1. Hydroxides

With K$_2$CO$_3$ in aqueous ethanol, 1,3-thiazines (166) substituted at C-2 by an imino group rearrange to 1,2-dihydropyrimidine-2-thiones (167). The

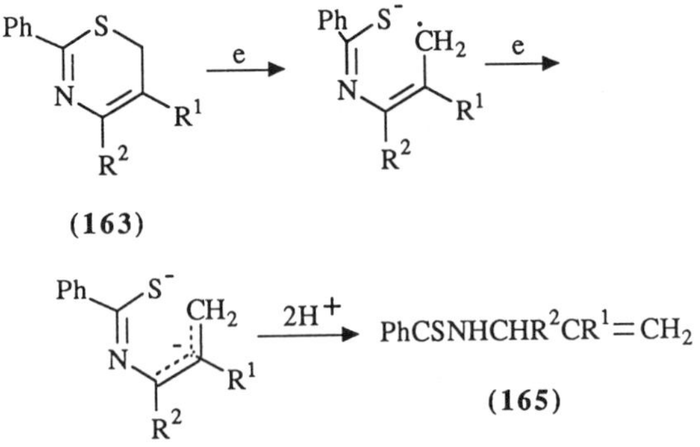

(163)

PhCSNHCHR^{2}CR^1=CH$_2$

(165)

SCHEME 64

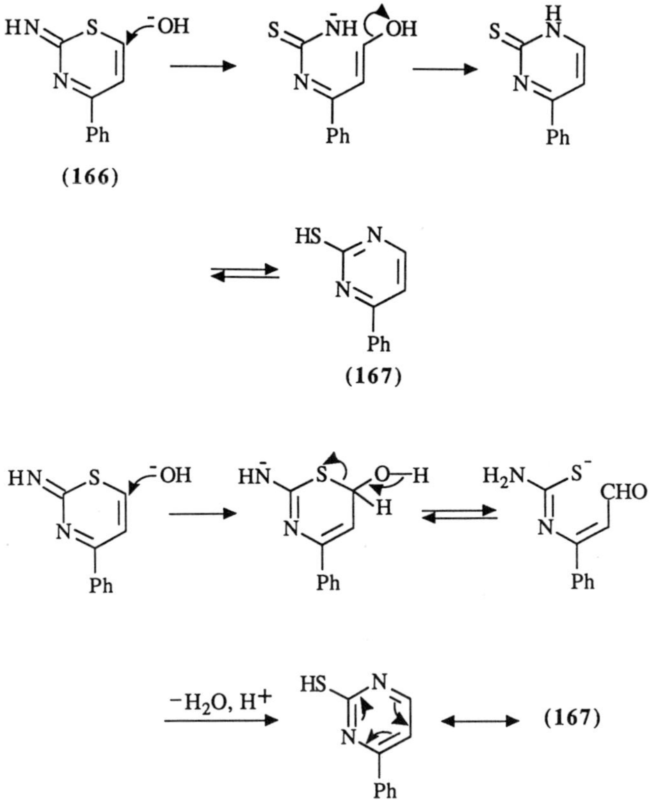

SCHEME 65

authors propose the following mechanism in Scheme 65. However, to avoid writing nucleofugal loss from an sp^2 carbon, we may take advantage of the extended conjugation (51JA2544).

2. Hydrosulfides

H_2S in the presence of $NaHCO_3$, leads to elimination of a nucleofugal group at the 2-position. However, attack at either the 2- or 6-positions may give the same results (Scheme 66) (83TL3713).

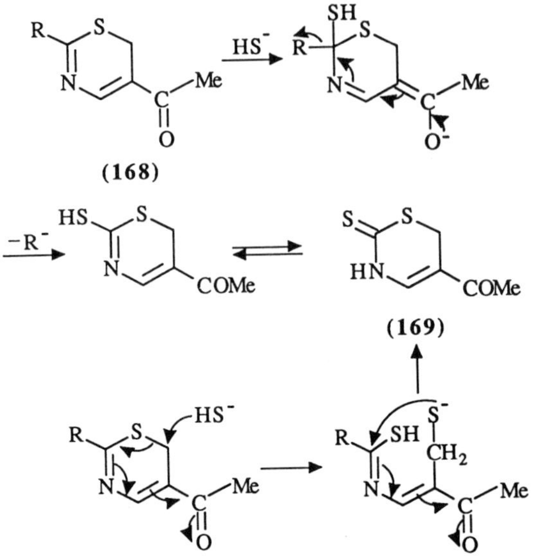

SCHEME 66

3. Hydrogensulfite

With HSO_3^-, the thiazine (170) gives an unstable adduct, which is hydrolyzed to 171 (Scheme 67) (87MI1).

4. Carbanions

Substitution by the soft anion derived from diethyl malonate occurs with ring opening at the 6-position of 172. Subsequent cyclization gives the corresponding thiocarbonyl compound 173 after elimination of the labile benzyl mercaptan (Scheme 68) (83TL3713). It has been shown that 173 is different from its isomer (174) which is obtained elsewhere (81UP1).

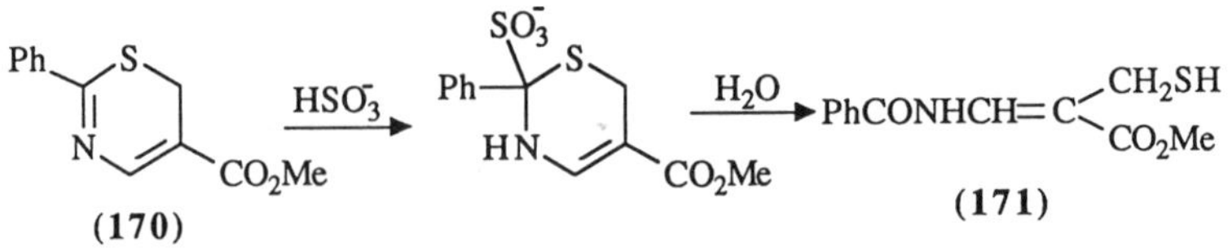

SCHEME 67

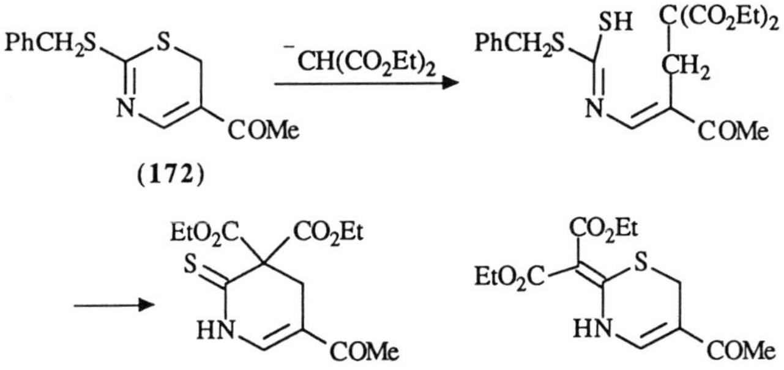

(172)

(173)   (174)

SCHEME 68

5. Hydrocyanic Acid

Hydrocyanic acid generated *in situ* (NaCN + CH$_3$CO$_3$H) adds to the imino group of thiazines **170** and **175**. The addition product **176** has actually been isolated (Scheme 69) (87MI1).

6. Water

2-Benzylmercapto-5-methoxycarbonyl-6H-1,3-thiazine (**177**) slowly hydrates with atmospheric moisture which leads, after elimination of benzyl mercaptan, to the 3,6-dihydro-2H-1,3-thiazine-2-one (**178**) (Scheme 70)

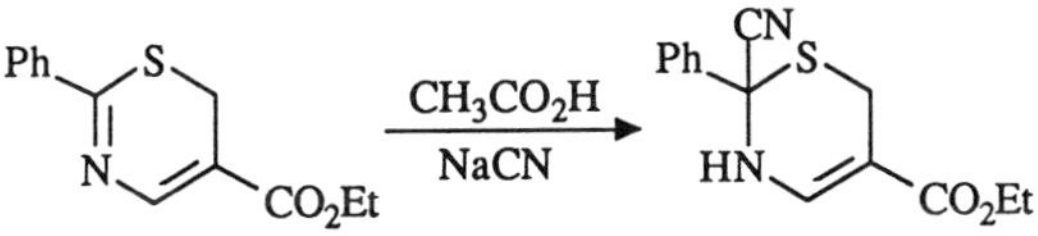

(175)   (176)

SCHEME 69

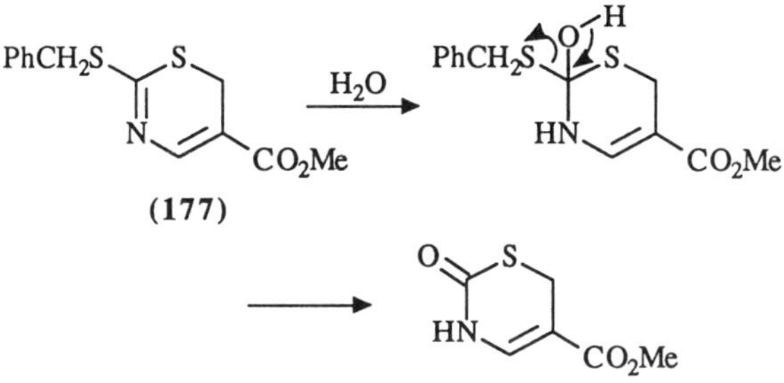

(177)

(178)

SCHEME 70

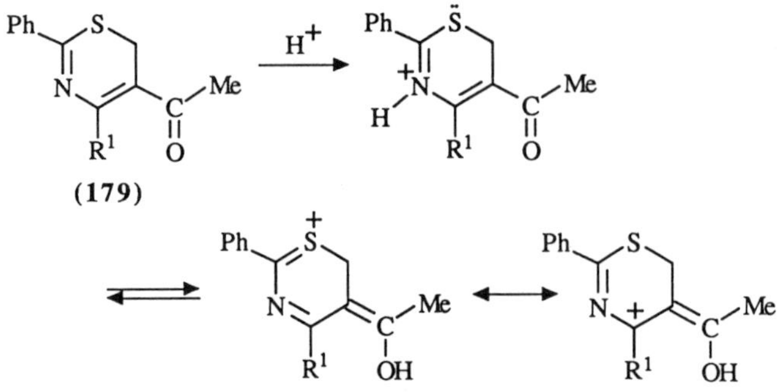

(179) (180) (181)

SCHEME 71

(87MI1). In acidic conditions (for example, in electrochemical reductions), water adds to either the 2-carbon or the 4-carbon. Protonation at either nitrogen or carbonyl can explain the results (Scheme 71) (87T2709).

When R^1 = Me (electron-releasing) the form **181** is more significant, and the stabilized carbocation at C-4 is attacked by H_2O (Scheme 72). With electron-withdrawing substituents R^1 = CO_2Et, form **180**, being more conjugated, is more significant, and H_2O attacks at the 2-position (Scheme

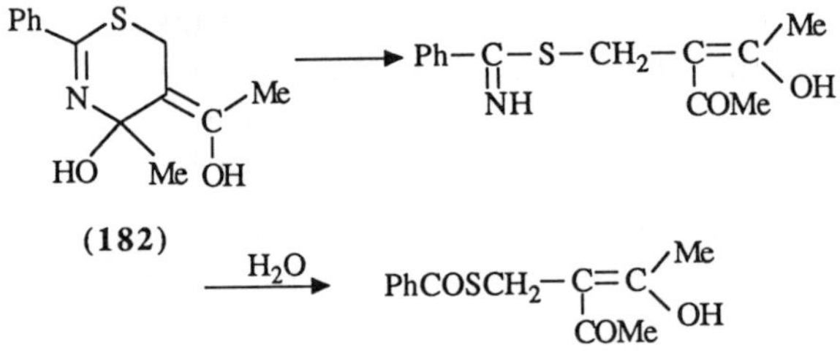

(182) (183)

SCHEME 72

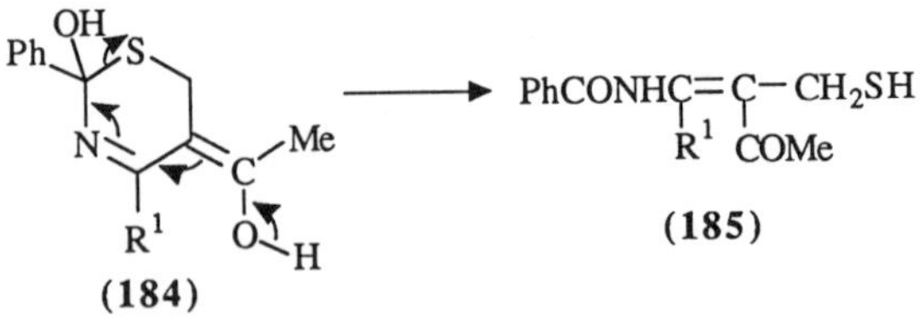

(184) (185)

R^1 = H, CO_2Et

SCHEME 73

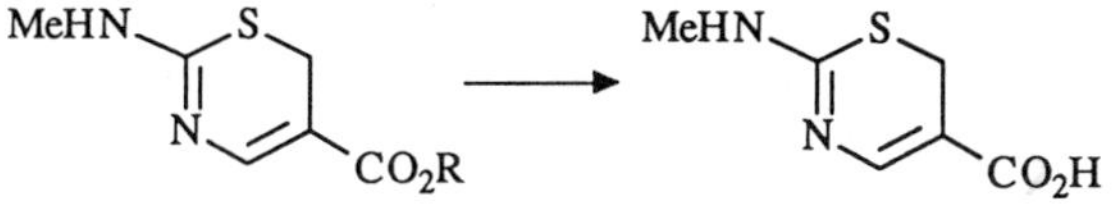

(186)   (187)

SCHEME 74

73). With 15% HCl–H$_2$O, compound **186** only undergoes hydrolysis of the ester function (Scheme 74) (65JOC2290).

In the case of 1,3-thiazine-2-ylidene compounds, hydrolysis of *N*-methylimino-5-ethoxycarbonyl-1,3-thiazine **188** (R = Me) is carried out in an acid medium. Formic acid does not react. Incorporation of water was found in the course of subsequent research (formic acid/aqueous triethylamine; either formic acid or acetic acid/water: 50–50; formic acid/aqueous formaldehyde) and resulted in the isolation of three compounds: 5-ethoxycarbonyl-2-oxo-2,3-dihydro-6*H*-1,3-thiazine **(189)**, 5-ethoxycarbonyl-1-methyl-2-thioxo-1,2,3,4-tetrahydropyrimidine **(190)** and 5-ethoxy-carbonyl-3-methyl-2-thioxo-1,2,3,4-tetrahydropyrimidine **(191)**. The authors have proposed a mechanism involving cleavage of the C$_{-6}$-S bond and reclosure of the six-membered ring by a Michael addition (Scheme 75).

Hydrolysis of **188** can also occur at the 4-carbon leading to **191** (Scheme 76) (Scheme 77). As in the case of **188** when the 2-position of the thiazine-4-one **(194)** is substituted by a nucleofugal group, the latter can also be substituted (Scheme 78) (67CB3671).

7. *Amines and Ammonia*

When 2*H*-thiazines are substituted at C-5 by an electron-withdrawing group, two electrophilic sites, the 2- and 6-carbons can be seen. When attack at the C-2 of 6*H*-thiazines is sterically disfavored, we observe

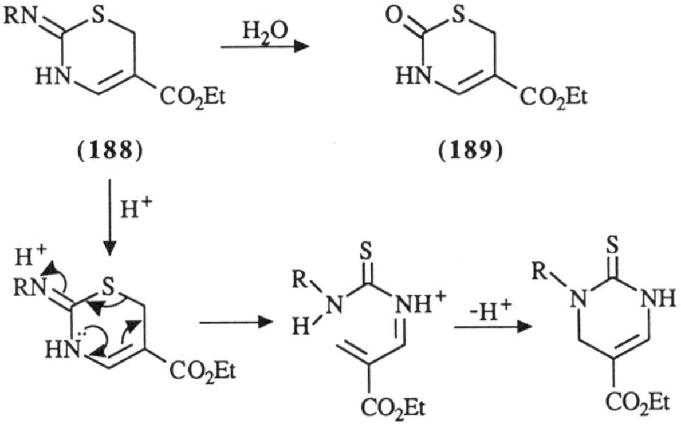

(188)   (189)

(190)

SCHEME 75

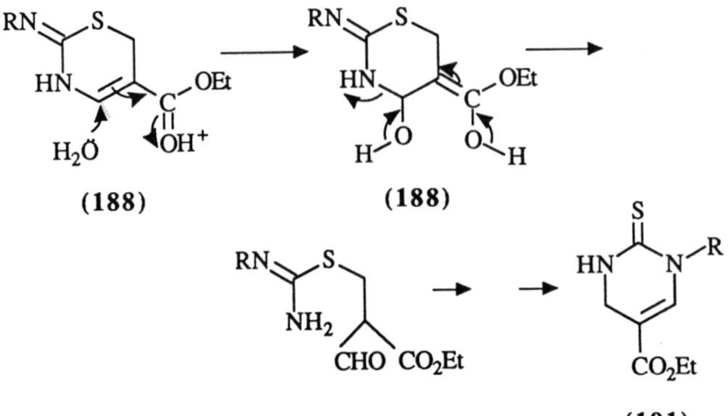

(188) (188) (191)

SCHEME 76

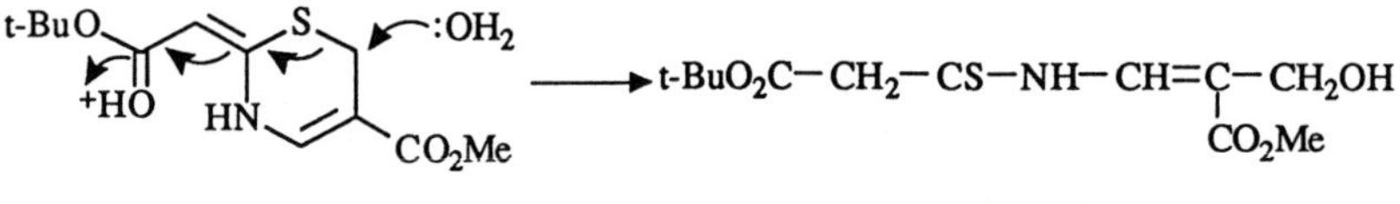

(192) (193)

SCHEME 77

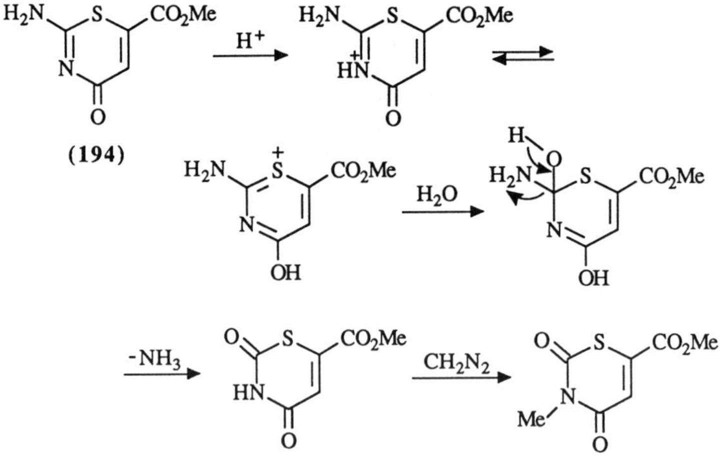

(194) (195) (196)

SCHEME 78

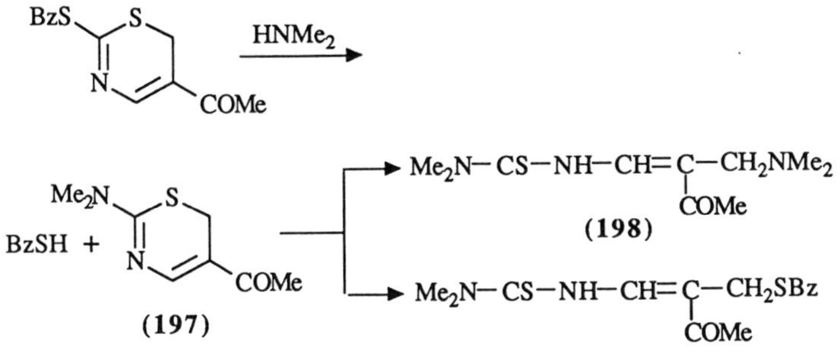

SCHEME 79

thiazine ring-opening by cleavage of the S–C-6 bond. Unsaturated dithio-carbamates, thioamides, and thioureas are obtained (Scheme 79) (83-TL3713).

In this sequence, substitution by 1 mol of dimethylamine first replaces the benzylmercaptan leading to 5-acetyl-2-dimethylamino-6H-1,3-thiazine (**197**). The thiazine then undergoes attack by dimethylamine excess at C-6, leading to the thiourea (**198**). The benzylmercaptan liberated in the first reaction may act as a nucleophile (BzS^-,$H_2NMe_2^+$), and a different thiourea substituted by dimethylamino and benzylthio groups is obtained. The action of pyrrolidine on 1,3-thiazine-4-ones (**194**) can be seen as a Michael addition followed by elimination of H_2S. In acidic media, the linear compound obtained is cyclized to the pyrimidone (**200**) (Scheme 80)

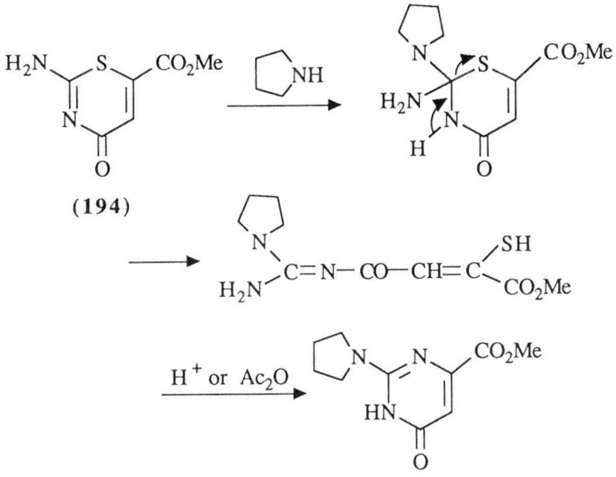

SCHEME 80

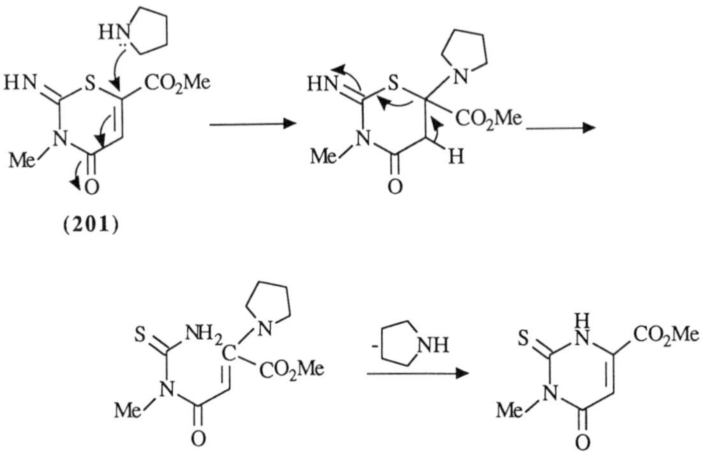

(201)   →

(202)

Scheme 81

(67CB3671). When the nitrogen of compound **194** is methylated, as in **201**, pyrolidine attack yields the pyrimidinethione (**202**) (Scheme 81).

Ammonia gave similar reactions (Scheme 82). 1,3-Thiazine-4-one derivatives **203** and **204** smoothly underwent ammonolysis with ethanolic ammonia, leading to the corresponding pyrimidine-4-one derivatives **205** in high yields (83CPB1929). 2-Chloro-thiazine-4-ones (**206**) react with amines undergoing substitution of chlorine, they also undergo subsequent addition of the amine to the carbonyl group followed by ring opening (Scheme 83) (77LA1249).

The electrophilic properties of substituted 6*H*-1,3-thiazine-6-ones (**208**) in solution also shows reactivity at C-2 in acidic conditions and at C-6 in basic conditions. The regioselective reaction of 4-ethoxycarbonyl-2-phenyl-6*H*-1,3-thiazine-6-one with dimethylamine leads, after ring opening and reclosure, to two diasteriomeric 5-dimethylcarboxamido-4-ethoxy-carbonyl-2-phenyl-Δ²-thiazolines **209** and **210**, whose structures were confirmed by X-ray diffraction studies on **210** (Scheme 84) (88BSF897). Com-

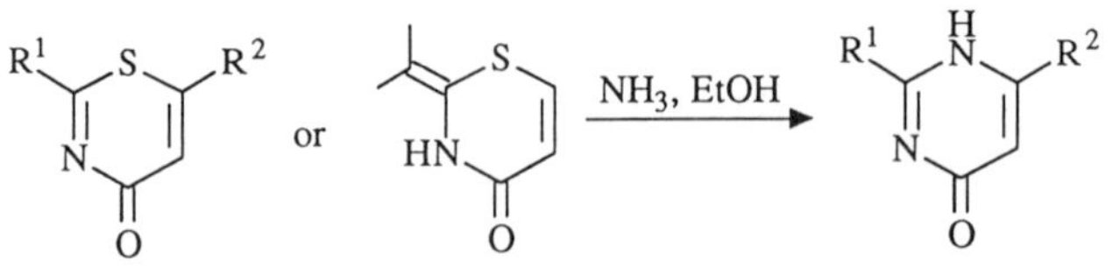

(203)   or   (205)

(204)

NH_3, EtOH

Scheme 82

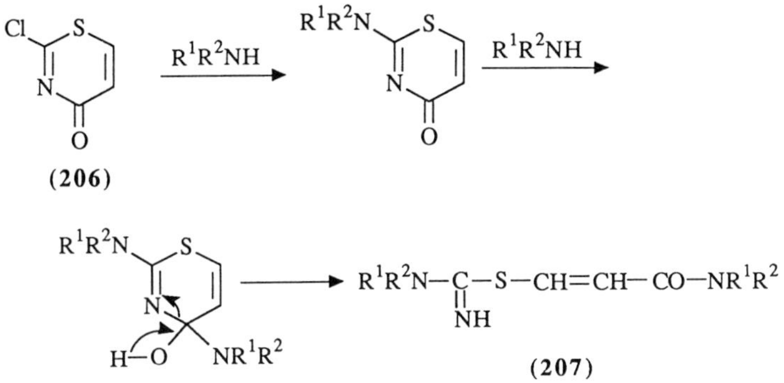

(206)

(207)

SCHEME 83

pound **210** can be selectively transformed into the trans substituted compound **209** by epimerization in the presence of dimethylamine in acetone.

Secondary amines (diethylamine, piperidine, or morpholine) attack the 1,3-thiazine-2,6-dithiones (**211**) at C-2 (Scheme 85) [80JCR(S)148].

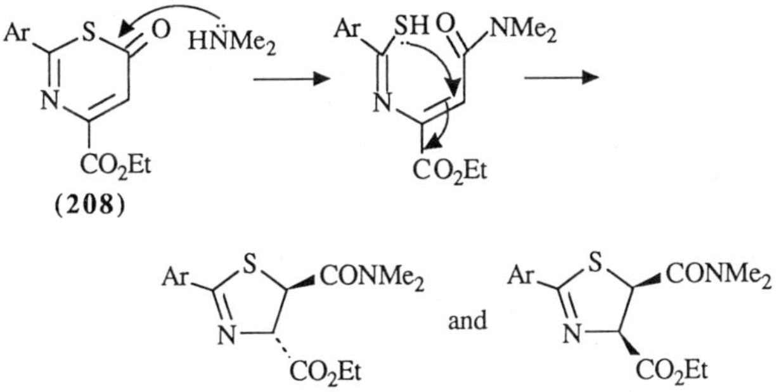

(208)

(209) (210)

SCHEME 84

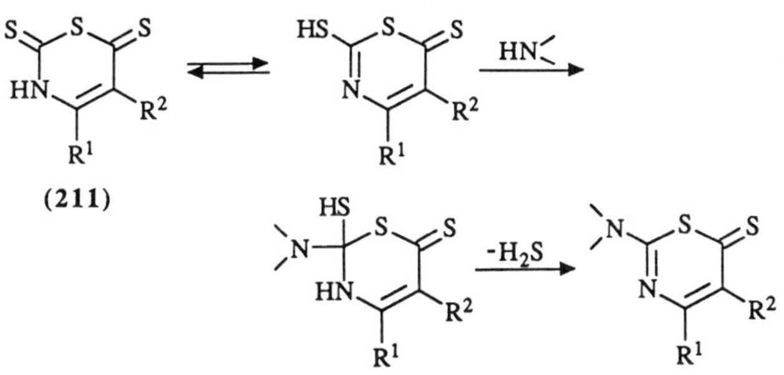

(211)

(212)

SCHEME 85

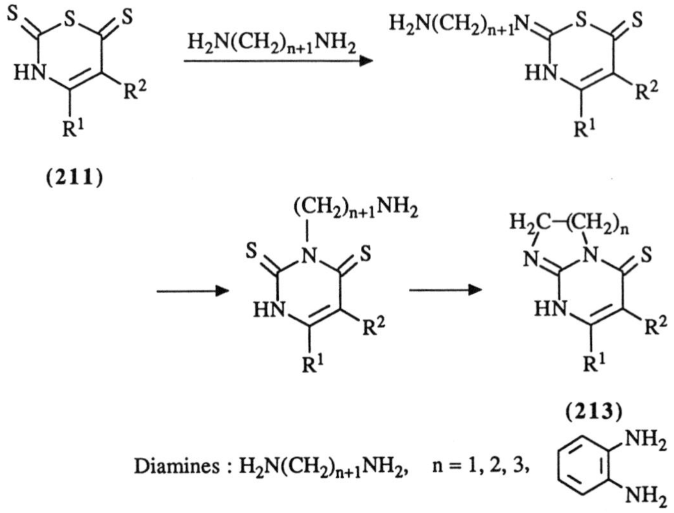

Diamines : $H_2N(CH_2)_{n+1}NH_2$,   n = 1, 2, 3,

SCHEME 86

This reaction appears to be a substitution of the mercapto group by the amine. It is claimed that primary amines do not afford 2-alkylamino-1,3-thiazine-6-thiones such as **212**, but instead give thioureas. Few details are given by the authors of this work. Diamines give bicyclic compounds **213** (Scheme 86).

The thiazine dithiones also react with hydrazine, hydroxylamine, semicarbazide, and thiosemicarbazide. Hydrazine gives 3-aminopyrimidine-2,4-dithiones (**214**) (Scheme 87) [80JCR(S)148]. Thiosemicarbazide reacts in the same way as diamines to give [1,2,4]triazolo[1,5-*a*]pyrimidines (**215**) (Scheme 88).

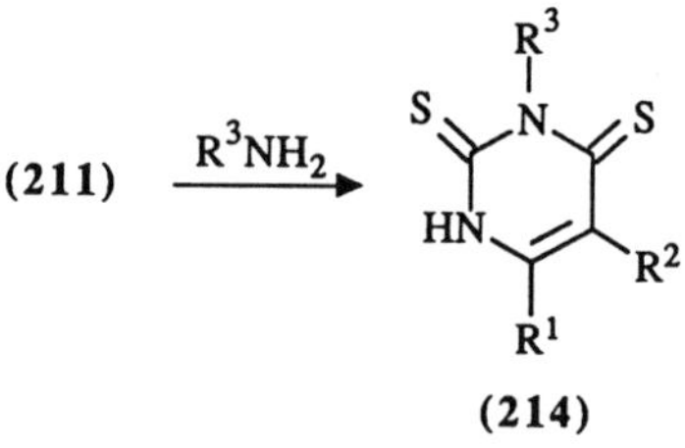

$R^3 = NH_2$, OH, $NHCONH_2$

SCHEME 87

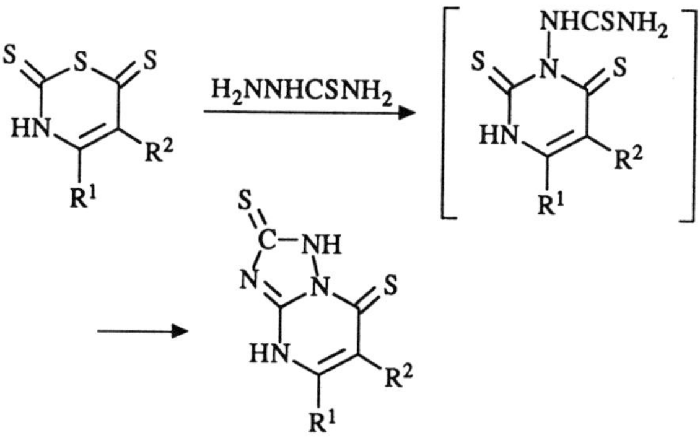

(215)

SCHEME 88

C. REACTIONS WITH ELECTROPHILES

1. *Alkylation*

With alkylating agents, 4-ethoxycarbonyl-2-N-methylaminothiazines appear to be ambident substrates. It seems the thiazines cannot be directly alkylated, and alkylations are made in the anionic form **(216)** (65JOC2290). In dioxan, alkylation occurs at the exocyclic nitrogen, in water, alkylation concerns the endocyclic nitrogen (Scheme 89). It has been noted that there is no methylation of NH by diazomethane.

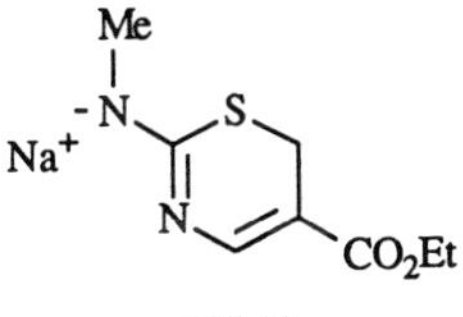

(216)

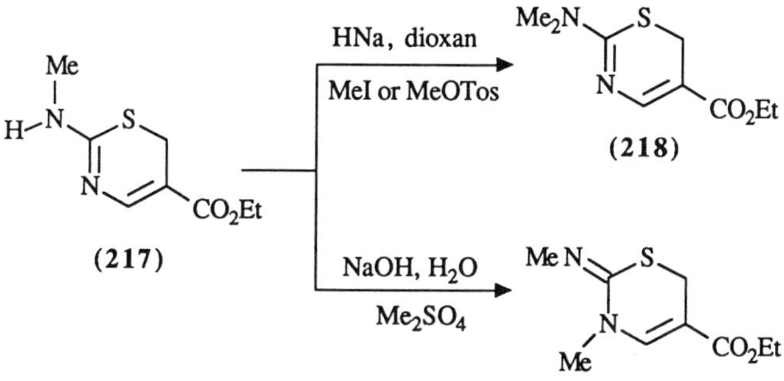

SCHEME 89

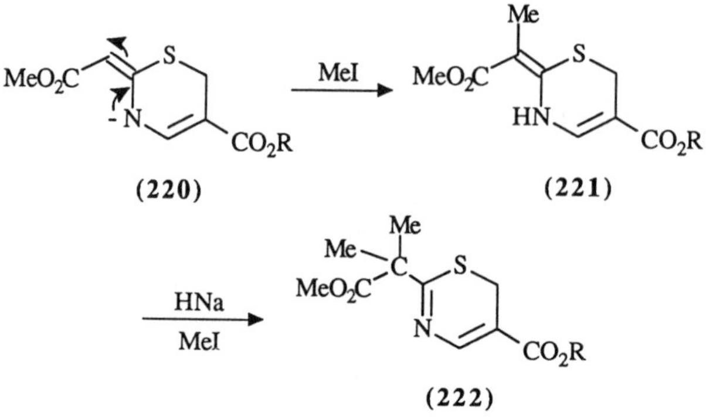

SCHEME 90

With alkyl iodides, alkylation of 1,3-thiazine-2-ylidene compounds (**220**) occurs to yield products such as **221** and **222** (Scheme 90) (87PS153). There are reports of some reductive alkylations. The thiazinedithiones (**223**), when treated with three molecular equivalents of thiolate anion, yield 2-alkylthio-2,3-dihydro-1,3-thiazine-6-thiones (**224**) (84JHC953). When the reaction is carried out with less than two molecular equivalents of the thiolate, the yield is extraodinarly low. Further, thiols without base never react with the thiazinedithiones (Scheme 91).

2. Acylation

Acetylation of methylamino-2-thiazines (**217**) (65JOC2290) and imino-2-thiazines (**226**) (70JAP37529) occurs on the exocyclic NH (in the latter case, the endocyclic NH is not acylated). Thiazine-2-ones (**228**)

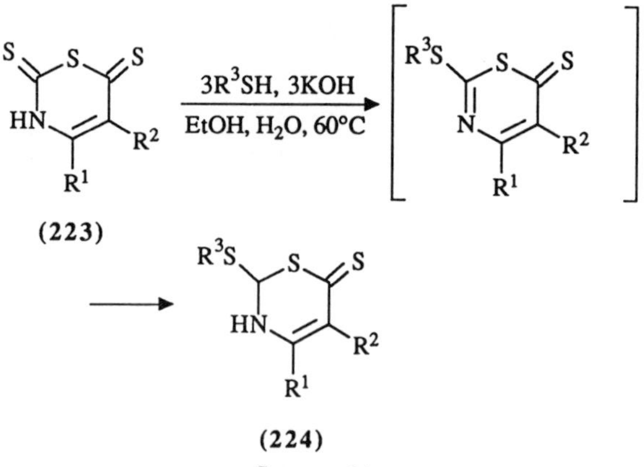

SCHEME 91

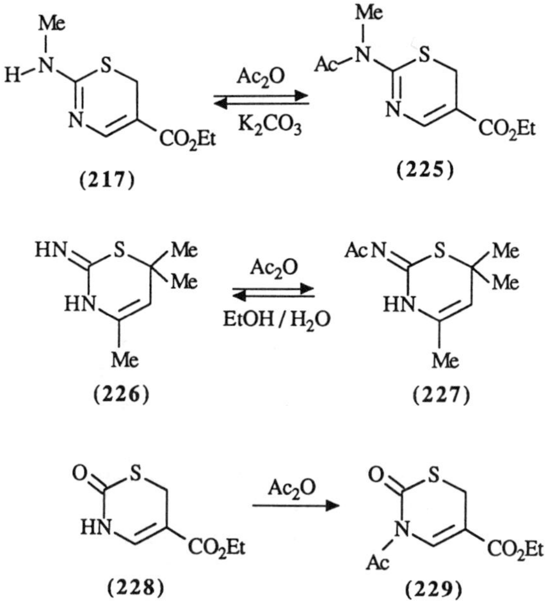

SCHEME 92

(67JAP7391) are normally acylated at the nitrogen to yield products such as **229** (Scheme 92). In aqueous ethanol with potassium carbonate, the *N*′-acetyl *N*-methyl-5-ethoxycarbonylthiazine (**227**) is quantitatively deacetylated to yield (**226**).

3. *Electrophilic Substitution*

The available spectral and calculated (Pariser–Parr–Pople method) data constitute evidence for a decrease in electron density in the benzene ring and activation of the 5-position to electrophilic attack in the 4-hydroxythiazinone (**203**) (79KGS44). Nitration was carried out in organic solvents at −60°C. The best results were obtained in the case of nitration with nitric acid (specific gravity = 1.42) in glacial acetic acid containing catalytic amounts of acetic anhydride at 40–50°C (Scheme 93). The mono-

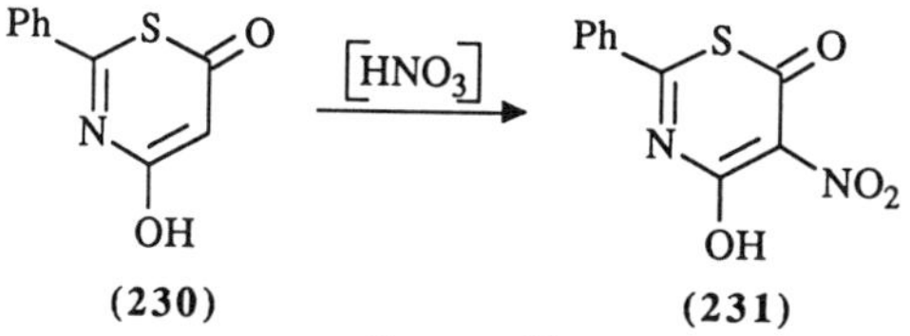

SCHEME 93

nitrated compound **231** was obtained in a yield of 75%. Bromination of **230** in organic solvents (dioxan, acetic acid, or carbon tetrachloride) also

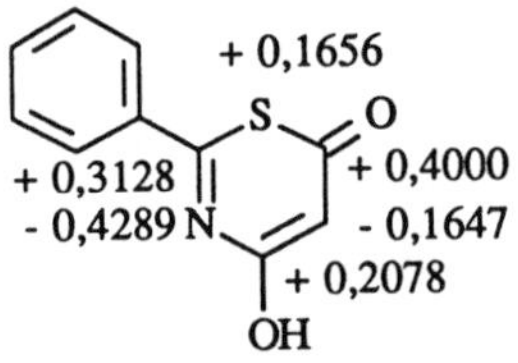

(230)

proceeded satisfactorily in all cases, but the highest yield of the bromo derivative (**232**) was obtained in glacial acetic acid. This is probably associated with the higher solubility of **230** in this solvent (Scheme 94).

The authors of this work were able to carry out iodination only when they used iodine monochloride. Compound **233** is unstable when stored in air and loses iodine quite rapidly. We were unable to isolate reaction products in the case of sulfonation with sulpuric acid. The most convenient sulfonating reagent was a solution of sulfur trioxide in dichlorethane. The sulfonic derivative **234** was very hygroscopic and was further converted into the disodium or bis(isobutylammonium) salt.

Diazo coupling has been carried out in a 10% solution of sodium carbonate, since NaOH solutions partially decompose the thiazinedione ring. Compounds **235** were only slightly soluble in organic sovents and water, but were quite soluble in alkali to give intensely red solutions (Scheme 95).

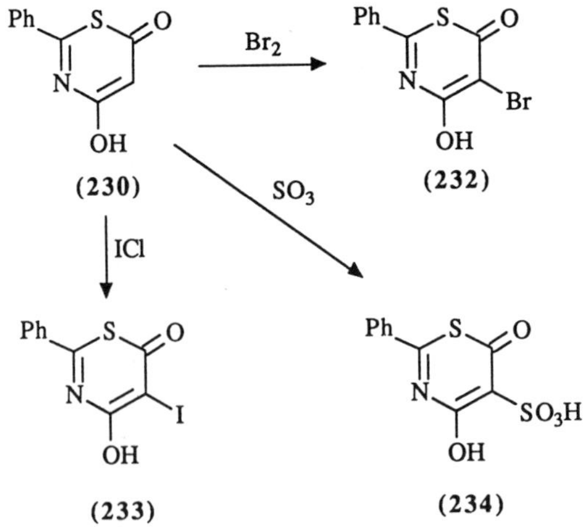

SCHEME 94

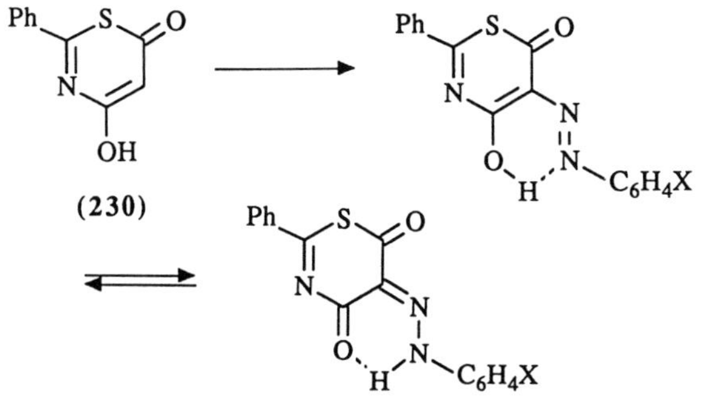

(230)

(235)   X = H, Br, NO$_2$

SCHEME 95

D. CYCLOADDITIONS

When the 2H-1,3-thiazine derivatives (236) were allowed to react with diphenylketene or with chloroacetyl chloride and a tertiary amine, β-lactam systems 237 and 238 were obtained (Scheme 96) (85BSB149). Instead of 3,4-cyclocondensation, as in the previous case, 2,3-cyclocondensation may be obtained by starting from 6H-1,3-thiazines (239) (Scheme 97)

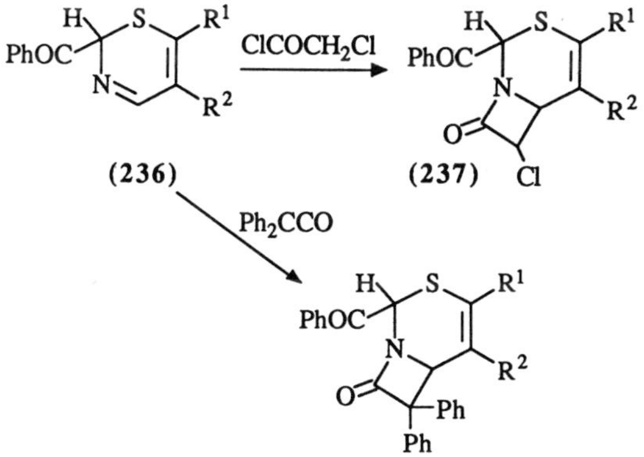

(236)   (237)   Cl

(238)

SCHEME 96

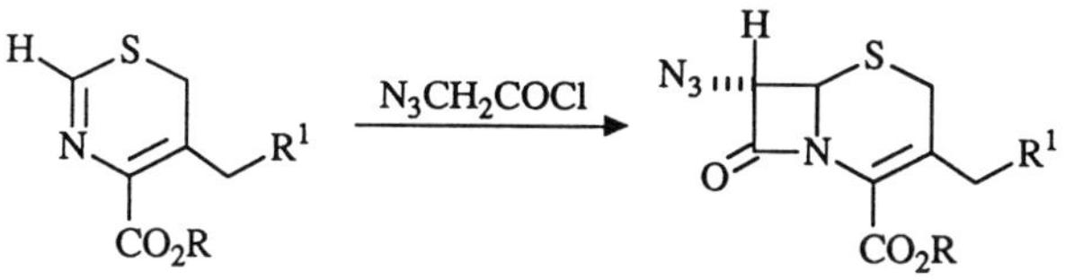

(239)   (240)

SCHEME 97

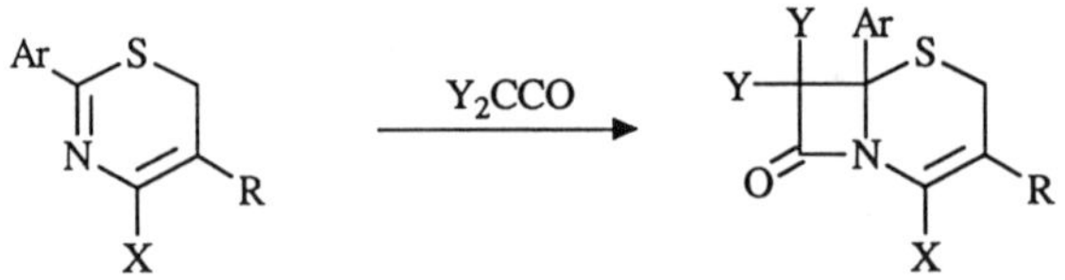

(241)   (242)

SCHEME 98

(73TL4649). The [2 + 2] cyclocondensation of thiazines (241) with ketenes has been studied. The reversibility of the reaction is a function of the nature of the substituents on positions 4 and/or 5 (Scheme 98) (86CJC597). As we have noted at the beginning of this review, this aspect of thiazine chemistry will not be discussed further here.

E. SIDE CHAINS

5-Cyano-3,6-dihydro-2H-1,3-thiazine-2-one (243) has been hydrolyzed in concentrated hydrochloric acid to give the amide 244, which was converted to the starting material 243 by the action of phosphorus oxychloride (64JOC1740). Another example of a carboxamido compound (245) giving a nitrile (246) using trimethylsilyl polyphosphate as a dehydrating agent has been described (Scheme 99) (82S591). 2-Amino-4,6,6-trimethyl-6H-1,3-thiazine (247), when mixed with dimethyl acetylenedicarboxylate at 0°C, gives 72% of 248 (Scheme 100) (74MI1).

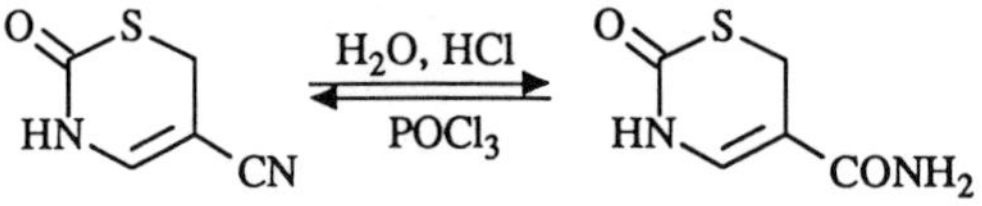

(243)   (244)

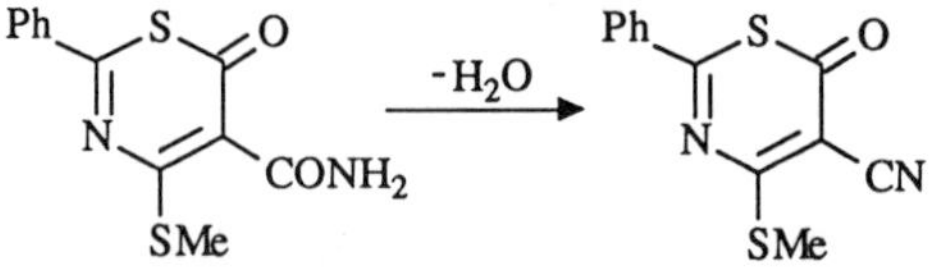

(245)   (246)

SCHEME 99

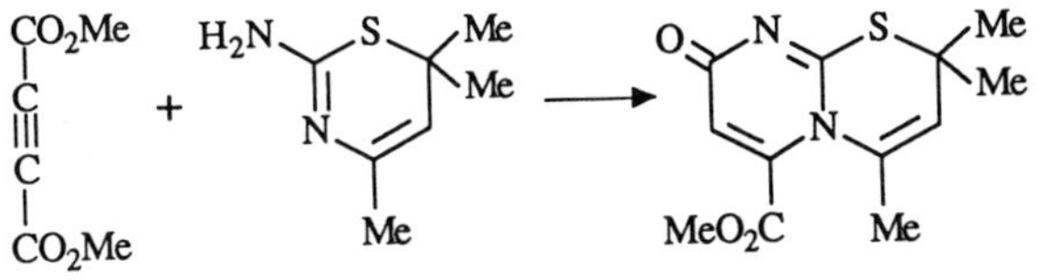

(247) (248)

SCHEME 100

Two modes of dimerization of the 2-alkyl-6-methyl-4*H*-1,3-thiazine-4-one (**249**) have been observed (83CC56). One mode gave the linearly combined dimer **250**, and the other led to a spiro compound (**251**). The structure of one of the spiro compounds was confirmed by X-ray crystallography (Scheme 101). The reactions occurred when the 2-alkyl group was methyl, ethyl, or *n*-propyl, but not when it was isopropyl. 4-Acyloxy- and 4-*p*-tosyloxy-1,3-thiazine-6-ones (**253**) have been prepared by acylation and sulfonylation starting from thiazine-4,6-diones (**252**), in a solvent, at temperatures from room temperature to reflux (Scheme 102) (79KGS44).

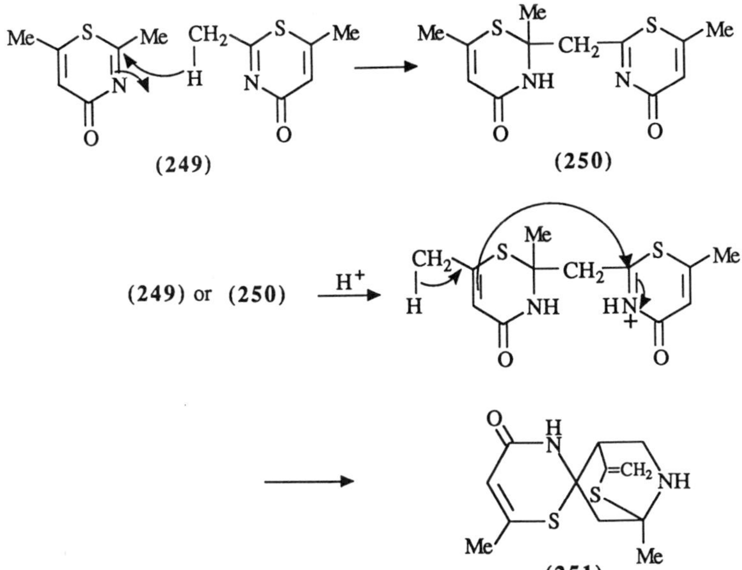

(249) (250)

(251)

SCHEME 101

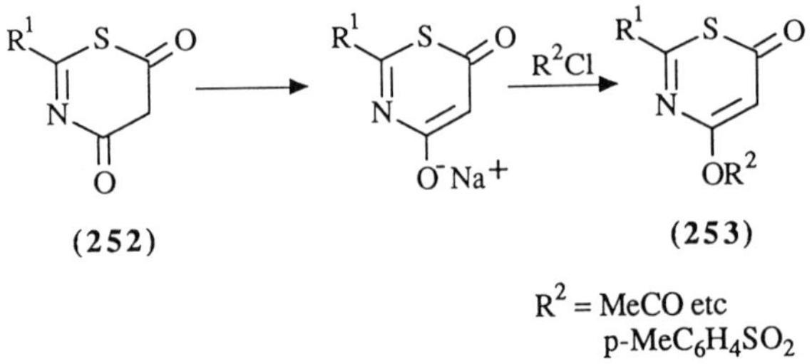

(252)   (253)

R^2 = MeCO etc
p-MeC$_6$H$_4$SO$_2$

SCHEME 102

Thiazine-5-carboxaldehyde (254) has been converted to esters 255 by the action of AcOH and NaCN, followed by MnO$_2$ (Scheme 103) (85MI1). Thiazine-5-carbonitrile (256) has also been obtained from compound 254. This is first converted to the oxime, which is then dehydrated by acetic anhydride. Alternatively, 254 can be treated with CF$_3$CONO$_2$CCF$_3$ in benzene and pyridine (Scheme 104), to also give 256.

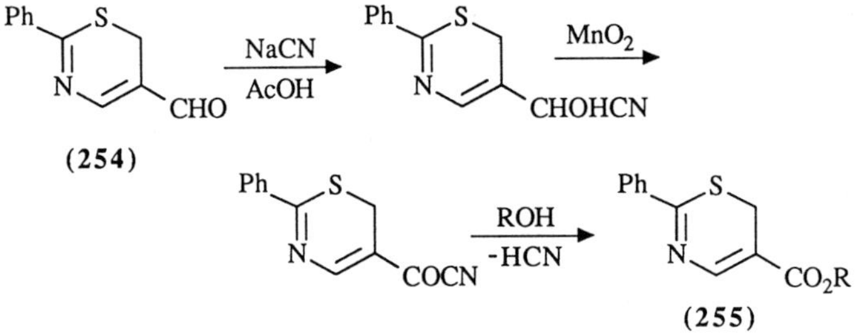

(254)

(255)

SCHEME 103

(254)

(256)

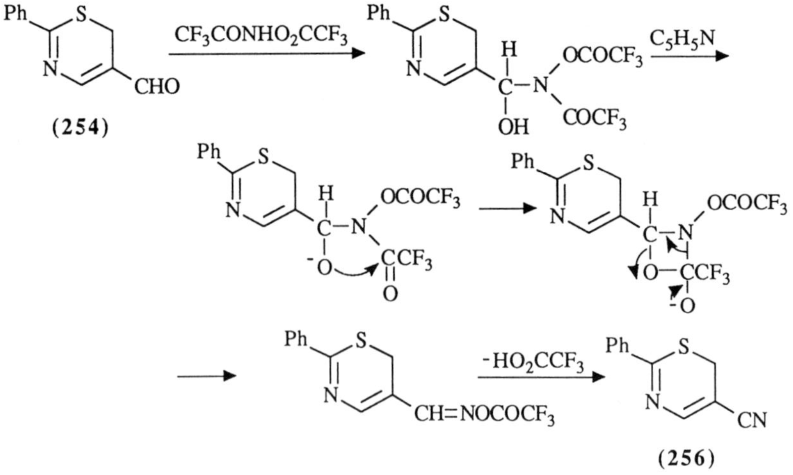

SCHEME 104

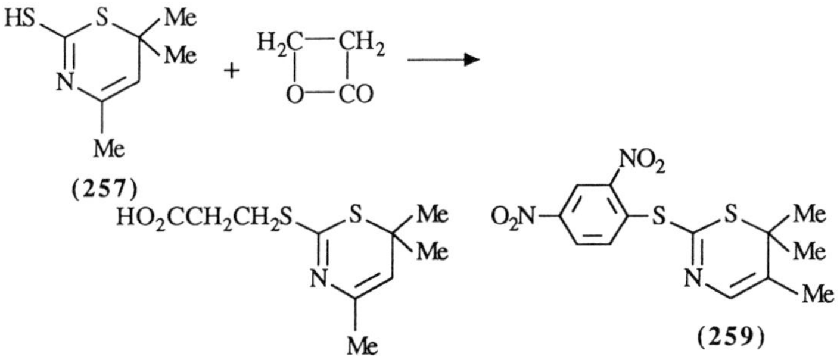

SCHEME 105

2-Mercapto-1,3-thiazine (**257**) can be alkylated or arylated (49USP2483416). Thus, with β-propiolactone, **257** gives (4,6,6-trimethyl-1,3-thiazin-2-yl) thiopropionic acid (**258**) (Scheme 105). The potassium salt of the thiazine (**257**), when suspended in benzene with $(EtO)_2POCl$ or EtI, gives the corresponding ethylated derivative (59IZV1037). *S*-Nitroaryl derivatives of 2-mercaptothiazines (**259**) are prepared by treating the sodium salt of the corresponding 2-mercapto-thiazine with 2,4-dinitrochlorobenzene (51U2P2547682).

F. REARRANGEMENTS

Dimroth rearrangement of 5-ethoxycarbonyl-2-methylamino-6*H*-1,3-thiazine (**260**) gives 2-thioxotetrahydropyrimidine (**261**) on heating with aqueous formic or acetic acid. A mixture of **261** and its 3-methyl isomer (**262**) is obtained (Scheme 106) (65JOC2290). Other related Dimroth rearrangements have been observed. Starting from trimethylthiazines (**263**) (Scheme 107), rearrangement takes place with thermal heterolysis of the 1,6-bond (75M1469).

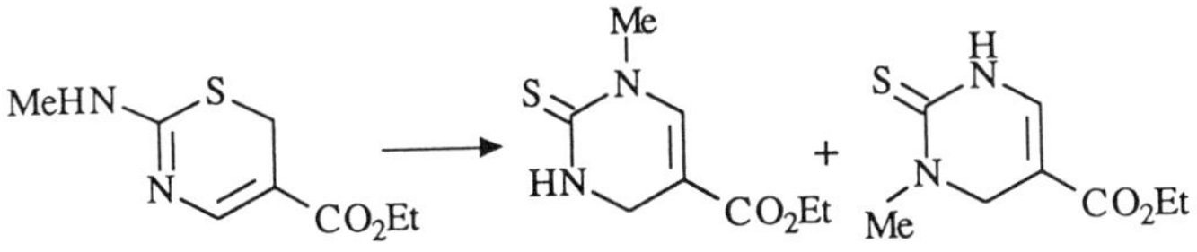

SCHEME 106

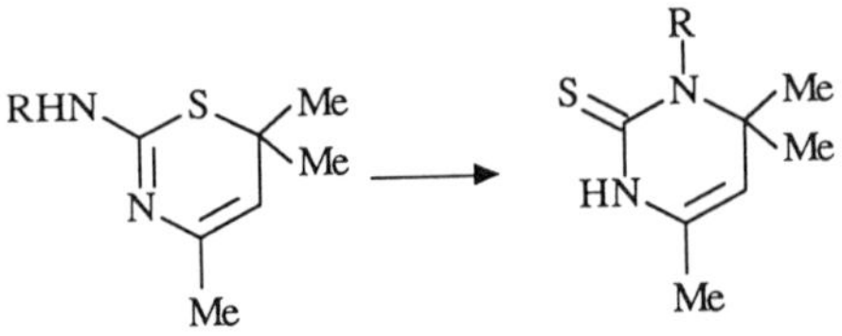

(263)   (264)

R = H, Me, Ph

SCHEME 107

Reaction of 6-morpholino- and 6-phenylthio-6H-1,3-thiazines (265) with BuLi at low temperature is followed by ring contraction to a pyrrolic derivative (266) (Scheme 108) (85TL3971). 5-Acyl-4-ethoxycarbonyl-2-phenyl-6H-1,3-thiazines rearrange with contraction to give 3-mercaptopyrroles [85JCS(P1)1875)]. These latter compounds were isolated only as their oxidation products; the corresponding disulfide 268 and products of subsequent condensation, 269 or 270, were obtained when an excess of acrylic reagent was present (Scheme 109). When N-thioacylfor-

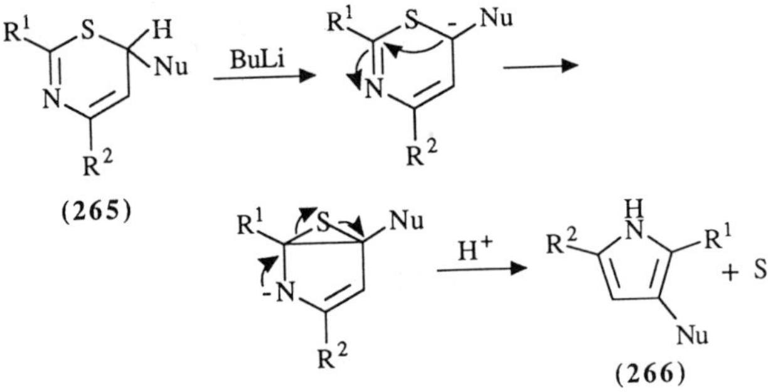

(265)

(266)

SCHEME 108

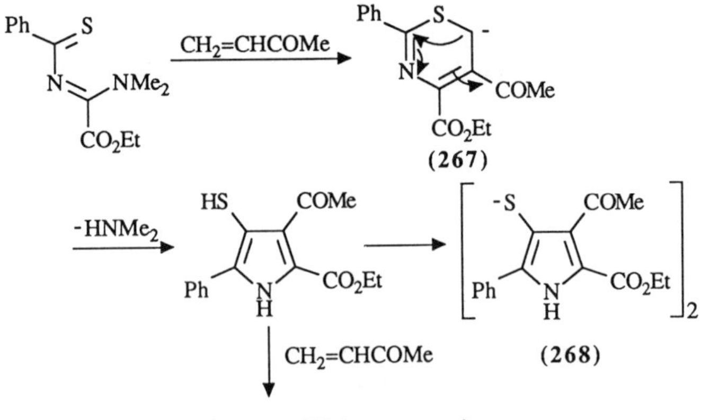

(267)

(268)

SCHEME 109 (CONTINUED)

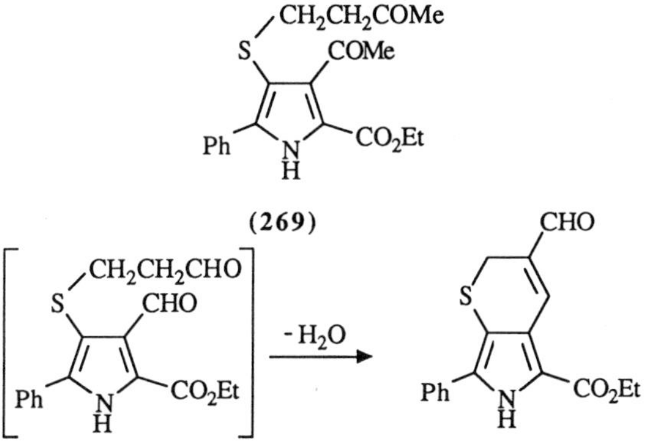

(269)

(269)   −H₂O   →   (270)

SCHEME 109

mamidines (267) are starting materials, it is possible to obtain 6H-1,3-thiazines in acidic medium and functionalized sulfides in basic medium.

Thermolysis of 4H-1,3-thiazines (271) leads to thioamide vinylogues (272) by ring opening (Scheme 110) (86SC79). At more elevated temperature (toluene reflux), compound 273 is not isolated. A heterocyclic isomer (274) with a 3,4-dihydro-2H-1,3-thiazine structure is obtained (Scheme 111).

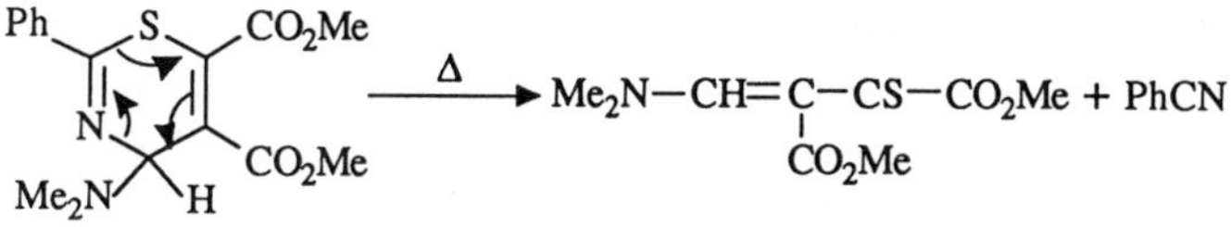

Δ → Me₂N—CH=C—CS—CO₂Me + PhCN
                        |
                      CO₂Me

(271)   (272)

SCHEME 110

$[1,5]$ H
Δ

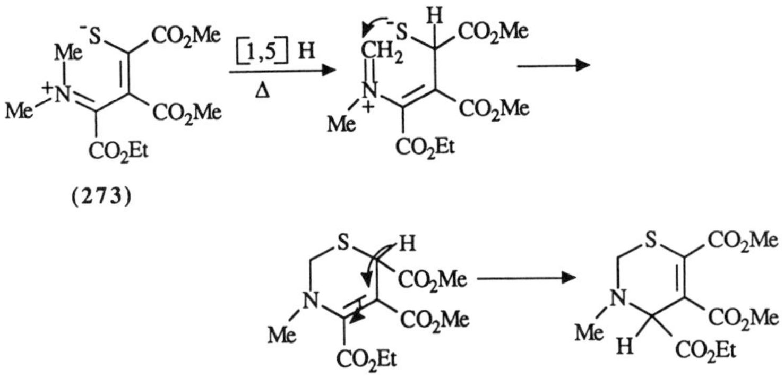

(273)

(274)

SCHEME 111

V. Uses

There are three uses of 1,3-thiazines: (a) medicinal applications, (b) 1, 3-thiazines as pesticides and growth factors, and (c) miscellaneous.

A. Medicinal Applications

4,6,6-Trimethyl-2-mercaptothiazine (**257**) was slightly active in producing hyperplasia of the thyroid and in impairment of the ability to fix administered ^{131}I in rats (54MI1). 2-Amino-1,3-thiazines increase the survival time in mice irradiated with X-rays (74MI3). Thiazines are also sedative, anticancer, antimicrobial, and antiviral drugs (69JAP06820; 74JAP(K)93534, 74JMC609) and central nervous system depressants (63USP3098071).

B. 1,3-Thiazines As Pesticides And Growth Factors

2-(2,6-Dichlorophenyl)-4H-1,3-thiazine-4-ones behave as herbicidal products (70GEP2025339; 70NEP6903457;73MI1), as do dinitro-arylesters of 2-mercaptothiazines (51USP2547682). 4-Aryl-6-thiazines have bactericidal, fungicidal, and algicidal properties that are useful in agriculture, spinning mixtures, and manufacturing papers and paints (75GEP2426653). 2-Aminothiazine-6-thiones have strong antibacterial activity and low toxicity. Antifungal activity is also observed for 1,3-thiazine-2,4-diones (63MI1). 2-Mercaptothiazines, sometimes linked by the sulfur atom to propionic acid in the 3-position, are stimulants for plant growth, especially in rooting (50USP2535876).

C. Miscellaneous

2-Mercapto-4,6,6-trimethyl thiazines (**257**), used as additives in extreme pressure lubricants, inhibit the staining of copper (58USP2836561), copper alloys (57GEP1013820), silver, and other sulfur-sensitive metals (53USP2620303). Addition of these compounds to a lubricating oil containing selenium inhibits the formation of silver selenide, and this maintains the effectiveness of the lubricant (52USP2597838). The sodium and potassium salts of the sulfuric acid reaction products (perhaps a 2-sulfo-derivative) of 2-mercapto-4,6,6-trimethyl-6H-1,3-thiazines are useful as wetting agents (48USP2439810). This product and its dihydro derivative are good stabilizers of silver halide emulsions (58BEP571916). It has also been observed to be an accelerator of vulcanization (47USP2426855).

ACKNOWLEDGMENT

The authors thank Professor D. Young, the University of Sussex, for reviewing the manuscript. The authors are grateful to Christine Michaut and Anne Laigo for typing and Mina Bakasse for drawing formulas and schemes.

References

1886LA1	R. List, *Justus Liebigs Ann. Chem.* **236**, 1 (1886).
12CB1557	P. Brigl, *Chem. Ber.* **45**, 1557 (1912).
28JPJ802	E. Miyamichi, *J. Pharm. Soc. Jpn.*, 802 (1928).
37JA1486	D. E Worrall, *J. Am. Chem. Soc.* **59**, 1486(1937).
47USP2426855	A. J. Beber (B. F. Goodrich Co.), U.S. Pat. 2,426,855 (1947) [*CA* **42**, 6572(1948)].
48MI1	H. M. Randall, R. G. Fowler, N. Fuson, and J. R. Dangl, "Infrared Determinations of Organic Structures." Van Nostrand, New York, 1948.
48USP2439810	J. E. Jansen (B. F. Goodrich Co.), U.S. Pat. 2,439,810 (1948) [*CA* **42**, 4791(1948)].
48USP2440095	J. E. Jansen, U.S. Pat. 2,440,095 (1948) [*CA* **42**, 5473 (1948)].
49USP2483416	J. E. Jansen and R. A. Mathes, U.S. Pat. 2,483,416 (1949)[*CA* **44**, 1544 (1950)].
50JA1879	R. A. Mathes and E. D. Stewart, *J. Am. Chem. Soc.* **72**, 1879 (1950).
50USP2535876	W. D. Stewart (B. F. Goodrich Co.), U.S. Pat. 2,535,876 (1950) [*CA* **45**, 3986 (1951)].
51LA140	H. Behringer and P. Zillikens, *Justus Liebigs Ann. Chem.* **574**, 140 (1951).
51JA2544	C. J. Cavallito, C. M. Martini, and F. C. Nachod, *J. Am. Chem. Soc.* **73**, 2544 (1951).
51USP2547682	L. L. Baumgartner (B. F. Goodrich Co.), U.S. Pat. 2,547,682 (1951) [*CA* **45**, 5872 (1951)].
51USP2568633	J. E. Jansen, U.S. Pat. 2,568,633 (1951) (*CA* **46**, 3574, (1952)].
52USP2597838	W. Lowe, W. T. Stewart, and J. O. Clayton (California Research Corp.), U.S. Pat. 2,597,838 (1952) [*CA* **46**, 7760 (1952)].
53USP2620303	W. Lowe and J. O. Clayton, U.S. Pat. 2,620,303 (1952) [*CA* **47**, 2480 (1953)]
54MI1	L. F. Hartmann, A. Portela, and A. F. Cardeza, *Rev. Soc. Argent. Biol.* **30**, 87 (1954).
55JA2866	J. E. Jansen and R. A. Mathes, *J. Am. Chem. Soc.* **77**, 2866 (1955).
55JA5431	J. E. Jansen and R. A. Mathes, *J. Am. Chem. Soc.* **77**, 5431 (1955).
55UKZ732	L. K. Mushkalo and G. Ya. Yangol, *Ukr. Khim. Zh.* **21**, 732 (1955).
56GEP954417	H. Behringer and P. Zillikens, Ger. Pat. 954,417 (1956) [*CA* **53**, 11418 (1959)].
56JCS3847	M. R. Atkinson, G. Shaw, K. Schaffner, and R. N. Warrener, *J. Chem. Soc.*, 3847 (1956).

57GEP1013820   J. S. Elliott and E. D. Edwards (C. C. Wakefield and Co. Ltd.), Ger. Pat. 1,013,820 (1957) [*CA* **54**, 11465 (1960)].

57MI1   R. C. Elderfield and E. E. Harris, *in* "Heterocyclic Compounds" (R. C. Elderfield, ed.), Vol. 6, Chapter 13. Wiley, New York, 1957.

57MI2   G. W. Kenner and A. Todd, *in* "Heterocyclic Compounds" (R. C. Elderfield, ed.), Vol. 6, p. 236. Wiley, New York, 1957.

58BEP571916   L. P. Roosens (Gevaert Photoproducter N. V.), Belg. Pat. 571,916 (1958) (*CA* **56**, 8216(1962)].

58USP2836561   J. S. Elliott and E. D. Edwards, U. S. Pat. 2,836,561(1958) [*CA* **52**, 19112 (1958)].

59IZV1037   B. A. Arbuzov and V. M. Zoroastrova, *Izv. Akad. Nauk SSSR, Otdl. Khim. Nauk,* 1037 (1959) [*CA* **54**, 1498 (1960)].

60MI1   G. R. Ramage, E. H. Rodd, and J. K. Landquist, *in* "Chemistry of Carbon Compounds" (E. H. Rodd, ed.), Vol 4C, p. 1502. Elsevier, Amsterdam, 1960.

62CPB13   H. Tanaka and A. Yokoyama, *Chem. Pharm. Bull.* **10**, 13 (1962).

62M26   E. Ziegler, U. Rossmann, F. Litvan, and H. Meier, *Monatsh. Chem.* **93**, 26 (1962).

62USP3062816   E. Ziegler, U.S. Pat. 3,062,816 (1962) [*CA* **59**, 6841 (1963)].

63MI1   G. A. Carter, J. L. Garraway, D. M. Spencer, and R. L. Wain, *Ann. Appl. Biol.* **51**, 135 (1963).

63USP3098071   H. M. Blatter and G. De Stevens (Ciba Corp.), U.S. Pat. 3,098,071 (1963) [*CA* **60**, 1766 (1964)].

64CPB683   A. Yokoyama and H. Tanaka, *Chem. Pharm. Bull.* **12**, 683 (1964).

64JCS4008   J L. Garraway, *J. Chem. Soc.,* 4008 (1964).

64JOC1740   A. Takamizawa, K. Hirai, Y. Sato, and K. Tori, *J. Org. Chem.* **29**, 1740 (1964).

64M147   E. Ziegler and E. Steiner, *Monatsh. Chem.* **95**, 147 (1964).

64M1061   E. Ziegler and R. Wolf, *Monatsh. Chem.* **95**, 1061 (1964).

64M1550   E. Steiner and E. Ziegler, *Monatsh. Chem.* **95**, 1550 1964).

65JOC2290   A. Takamizawa and K. Hirai, *J. Org. Chem.* **30**, 2290 (1965).

65JOC3092   E. L. Eliel and J. Roy, *J. Org. Chem.* **30**, 3092 (1965).

65M411   E. Ziegler and H. D. Hanus, *Monatsh. Chem.* **96**, 411 (1965).

66TL3225   R. N. Warrener and E. N. Cain, *Tetrahedron Lett.,* 3225 (1966).

67CB3671   E. Winterfeld and J. M. Nelke, *Chem. Ber.* **100**, 3671 (1967).

67JAP7391   A. Takamizawa, K. Hirai, and T. Ishinami (Shionogi and Co, Ltd.), Jpn. Pat. 7391 (1967) [*CA* **68**, 29706 (1968)].

68CB1428   H. Behringer, D. Bender, J. Falkenberg, and R. Wiedenmann, *Chem. Ber.* **101**, 1428 (1968).

68USP3408348   J. C. Martin and R. H. Meen, U.S. Pat. 3,408,348 (1968) [*CA* **70**, 11709 (1969)].

69JAP06820   A. Takamizawa and K. Hirai, Jpn. Pat. 06,820 (1969) [*Ca* **71**, 3395 (1969)].

69KGS896   B. V. Unkovskii and L. A. Ignatova, *Khim. Geterosikl. Soedin.,* 896(1969).

69ZOR621   E. G. Kataev, L. K. Konovalova, and E. G. Yarkova, *Zh. Org. Khim.* **5**, 621 (1969).

70AJC51   E. N. Cain and R. N. Warrener, *Aust. J. Chem.* **23**, 51 (1970).

70GEP2025339   N. V. Philips' Gloeilampenfabrieken, Ger. Pat. 2,025,339 (1970) [*CA* **74**, 53819 (1971)].

70JAP37529   K. Ueno, Y. Hirose, and F. Tada (Mitsui Toatsu Chemicals Co.,
                Ltd.), Jpn. Pat. 37529 (1970) [*CA* **74,** 88009 (1970).
70KGS1690   B. V. Unkovskii, L. A. Ignatova, P. L. Ovechkin, and A. I. Vino-
                gradova, *Khim. Geterosikl. Soedin.,* 1690 (1970).
70NEP6903457   N. V. Philips' Gloeilampenfabrieken, Neth. Pat 6,903,457 (1970)
                [*CA* **74,** 125711(1971)].
71GEP2010558   G. Simchen and G. Entenmann, Ger. Pat. 2,010,558 (1971) [*CA* **76,**
                3881 (1972).
71KGS946   P. L Ovechkin, L. A. Ignatova, and B. V. Unkovskii, *Khim.
                Geterosikl. Soedin.,* 946 (1971).
71MI1   H. S. Kimmel and W. H. Snyder, *Spectrosc. Lett.* **4,** 15 (1971).
72KGS937   P. L. Ovechkin, L. A. Ignatova, A. E. Gekhman, and B. V. Un-
                kovskii, *Khim. Geterosikl. Soedin.,* 937 (1972).
72MI1   S. M. Deshpande and A. K. Mukerjee, *Curr. Sci.* **41,** 139 (1972).
73JAP(K)73-80575 S. Terao, T. Matsuo, S. Tsushima, N. Matsumoto, and T. Miya-
                waki, Jpn. Kokai 73-80,575 [*CA* **80,** 59951 (1974)].
73JHC223   R. Ketcham, T. Kappe, and E. Ziegler, *J. Heterocycl. Chem.* **10,**
                223 (1973).
73JOC802   M. Yokoyama, Y. Sawachi, and T. Isso, *J. Org. Chem.* **38,** 802
                (1973).
73MI1   T. J. Monaco, *HortScience,* 308 (1973).
73TL4649   R. W. Ratcliffe and B. G. Christensen, *Tetrahedron Lett.,* 4649
                (1973).
74EGP106649   W. Schroth and G. Dill, Ger. (East) Pat. 106,649 (1974) [*CA* **82,**
                112091 (1975)].
74EGP106650   W. Schroth and G. Dill, Ger. (East) Pat. 106,650 (1974) [*CA* **82,**
                73011 (1975)].
74G849   C. Giordano, *Gazz. Chim. Ital.* **104,** 849 (1974).
74GEP2356388   B. G. Christensen and R. W. Ratcliffe (Merck and Co., Inc.), Ger.
                Pat. 2,356,388 (1974) [*CA* **83,** 28 249 (1975)].
74JAP(K)93534   T. Misato, K. Ko, Y. Honma, T. Shida, I. Kusumoto, J. Sano, and
                Y. Iwashita, Jpn. Kokai 93534 (1974) [*CA* **82,** 165873 (1975)].
74JMC609   E. Akerblom, *J. Med. Chem.* **17,** 609 (1974).
74LA1753   J. Lepschy, G. Hoefle, L. Wilschowitz, and W. Stelich, *Liebigs
                Ann. Chem.,* 1753 (1974).
74MI1   M. V. Sharma, *Curr. Sci.* **43,** 147 (1974).
74MI2   L. A. Ignatova, A. E. Gekhmann, P. L. Ovechkin, and B. V.
                Unkovskii, *Tezisy. Dokl. Nauchn. Sess. Khim. Tekhnol. Org.
                Soedin. Sery Sernistykh Neftei, 13th, 1974,* 249 (1974) [*CA* **85,**
                160007 (1976)].
74MI3   Y. Takaji, M. Shikita, and S. Akaboshi, *J. Radiat. Res.* 116 (1974).
74RTC78   S. Hoff and A. P. Blok, *Recl. Trav. Chim., Pays-Bas* **93,** 78
                (1974).
74TL3567   N. G Steinberg, R. W Ratcliffe, and B. G Christensen, *Tetrahedron
                Lett.,* 3567 (1974).
75CR(C)119   J. P. Pradère, *C. R. Hebd. Seances Acad. Sci., Ser. C* **281,** 119
                (1975).
75GEP2426653   W. Paulus, H. Scheinpflug, and H. Genth (Bayer A. G.), Ger. Pat.
                2,426,653 (1975) [*CA* **84,** 121876 (1976)].
75JCS(P1)1417   M. Yokoyama, *J. C. S., Perkin Trans. 1,* 1417 (1975).
75KGS1614   E. M. Peresleni, L. A Ignatova, P. L. Ovechkin, A. P. Engoyan,

	Y. N. Sheinker, and B. V. Unkovskii, *Khim. Geterosikl. Soedin.,* 1614 (1975).
75M1469	G. Zigeuner, T. Strallhofer, F. Wede, and W. B Lintschinger, *Monatsh. Chem.* **106,** 1469 (1975).
75MI1	G. Prota, "Organic Compounds of Sulphur, Selenium and Tellurium," Vol. 3. Chemical Society, London, 1975.
75MI2	G. Prota, "Organic Compounds of Sulphur, Selenium and Tellurium, Vol. 4. Chemical Society, London, 1977.
75T3055	J. C. Meslin and H. Quiniou, *Tetrahedron* **31,** 3055 (1975).
76BCJ2828	O. Tsuge and A. Inaba, *Bull. Chem. Soc. Jpn.* **49,** 2828 (1976).
76ZOR904	A. N. Mirskova, G. G. Levkoskaya, and A. S. Atavin, *Zh. Org. Khim.* **12,** 904 (1976).
77ABC65	M. Shiozaki and T. Hiraoka, *Agric. Biol. Chem.* **41,** 65 (1977).
77EGP128124	K. Schulze, M. Mühlstädt, and W. Mai, Ger. (East) Pat. 128,124 (1977) [*CA* **88,** 105391 (1978)].
77LA1249	G. Simchen and G. Entenmann, *Liebigs Ann. Chem.,* 1249 (1977).
77PHA461	W. Schroth, J. Herrmann, C. Feustel, S. Schmidt, and K. M. Jamil, *Pharmazie* **32,** 461 (1977).
78AJC2307	H. Singh, P. Singh, and R. K. Mehta, *Aust. J. Chem.* **31,** 2307 (1978).
78JAP(K)78-111080	T. Hirayama, S. Fukatsu, A. Seki, H. Goi, and Y. Yamada, Jpn. Kokai 78-111,080 (1978).
78JCS(P1)1428	L. I. Giannola, S. Palazzo, P. Agozzino, L. Lamartina, and L. Ceraulo, *J. C. S., Perkin Trans. 1,* 1428 (1978).
78MI1	M. Sainsbury, *in* "Rodd's Chemistry of Carbon Compounds" (S. Coffey, ed.), 2nd ed., Vol. 4H, p. 488. Elsevier, Amsterdam, 1978.
79BSF347	J. C. Meslin and H. Quiniou, *Bull. Soc. Chim. Fr.,* 347 (1979).
79CB3939	M. Kloft and D. Hoppe, *Chem. Ber.* **112,** 3939 (1979).
79JAP(K)79-20504	K. Tomita and T. Murakami, Jpn. Kokai 79-20,504 (1979).
79JCR(S)384	T. Yamamoto, M. Muraoka, and T. Takeshima, *J. Chem. Res., Synop.,* 384 (1979).
79KGS44	V. G. Beilin, V. A. Gindin, V. N. Kuklin, and L. B. Dashkevich, *Khim. Geterosikl. Soedin.,* 44 (1979).
80CB2699	K. Burger, R. Ottlinger, H. Goth, and J. Firl, *Chem. Ber.* **113,** 2699 (1980).
80CCC732	J. Kavalek, S. El Bahaie, V. Machacek, and V. Sterba, *Collect. Czech. Chem. Commun.* **45,** 732 (1980).
80JCR(S)148	T. Yamamoto, M. Muraoka, M. Takahashi, and T. Takeshima, *J. Chem. Res., Synop.,* 148 (1980).
80JCS(P1)1013	H. Singh and P. Singh, *J. C. S., Perkin Trans. 1,* 1013 (1980).
80MI1	I. N. Azerbaev, L. A. Tsoi, A. D. Salimbaeva, S. T. Cholpankulova, G. A Ryskieva, L. T. Kalkabaeva, and M. Zh. Aitkhozhaeva, *Tr. Inst. Khim. Nauk, Akad. Nauk Kaz. SSR* **52,** 128(1980)[*CA* **94,** 208766 (1981)].
80OMR479	C. Rabiller, G. J. Martin, J. P. Pradère, J. C. Meslin, and H. Quiniou, *Org. Magn. Reson.* **14,** 479 (1980).
80S453	J. C. Meslin, A. Reliquet, F. Reliquet, and H. Quiniou, *Synthesis,* 453 (1980).
81BSB75	R. Shabana, J. B. Rasmussen, and S. O. Lawesson, *Bull. Soc. Chim. Belg.* **90,** 75 (1981).

81EGP147359   K. Seifert, S. Johne, and S. Haertling, Ger. (East) Pat. DD147,359
                [*CA* **96**, 6926 (1982)].
81EGP149807   K. Seifert, S. Johne, and A. Schaks, Ger. (East) Pat. DD149,807
                (1981) [*CA* **96**, 52326 (1982)].
81H851   Y. Yamamoto, Y. Azuma, and S. Ohnishi, *Heterocycles* **15**, 851
                (1981).
81UP1   A. Reliquet, unpublished data.
81ZC326   M. Mühlstädt, B. Schulze, and I. Schubert, *Z. Chem.* **21**, 326 (1981).
82CCC3268   J. Imrich and P. Kristian, *Collect. Czech. Chem. Commun.* **47**, 3268
                (1982).
82JCR(S)72   J. P. Pradère, J. C. Rozé, and G. Duguay, *J. Chem. Res., Synop.*, 72
                (1982).
82JCS(P1)1905   R. M. Acheson and J. D. Wallis, *J. C. S., Perkin Trans. 1*, 1905
                (1982).
82JCS(P1)2149   T. Nishio and Y. Omote, *J. C. S., Perkin Trans. 1*, 2149 (1982).
82JOC1090   M. Yokoyama, M. Nakumura, H. Ohteki, T. Imamoto, and K.
                Yamaguchi, *J. Org. Chem.* **47**, 1090 (1982).
82MI1   J. Shmidt, K. Seifert, S. Haertling, S. Johne, and H. J. Veith,
                Biomed. Mass. Spectrom. **9**, 174 (1982).
82MI2   I. Sibuya and M. Kurabayashi, *Tokyo Kogyo Shikensho Hokoku* **77**,
                597 (1982) [*CA* **99**, 5583 (1983)].
82MI3   X. Wang, Z. Wang, and R. Lian, *Huaxue Xuebao*, 459 (1982) [*CA*
                97, 127334 (1982)].
82S591   M. Yokoyama, S. Yoshida, and T. Imamoto, *Synthesis*, 591 (1982).
82TH1   A. Guevel, Thesis 3ᵉ cycle, University of Nantes (1982).
82TL2315   J. C. Roze, J. P. Pradère, G. Duguay, and H. Quiniou, *Tetrahedron
                Lett.*, 2315 (1982).
83CC56   Y. Yamamoto, S. Ohnishi, R. Moroi, and A. Yoshida, *J. C. S.,
                Chem. Commun.*, 56 (1983).
83CC945   K. Burger, E. Huber, W. Schoentag, and R. Ottlinger, *J. C. S.,
                Chem. Commun.* **17**, 945 (1983).
83CJC1169   J. C. Rozé, J. P. Pradère, G. Duguay, A. Guevel, H. Quiniou, and
                S. Poignant, *Can. J. Chem.* **61**, 1169 (1983).
83CPB1929   Y. Yamamoto, S. Ohnishi, and Y. Azuma, *Chem. Pharm. Bull.* **31**,
                1929 (1983).
83EGP156908   M. Mühlstädt, B. Schulze, I. Schubert, E. Kleinpeter, and G. Kir-
                sten, Ger. (East) Pat. DD156,908 [*CA* **98**, 143440 (1983)].
83MI1   A. E. Lyuts, V. V. Zamkova, O. V. Agashkin, and L. A. Ignatova,
                Izv. Akad. Nauk Kaz. SSR, Ser. Khim., 46 (1983) [*CA* **98**, 160052
                (1983)].
83MI2   J. P. Pradère, J. C. Roze, G. Duguay, A. Guevel, C. G. Tea, and
                H. Quiniou, *Sulfur Lett.* **1**, 115 (1983).
83PS143   A. Reliquet, F. Reliquet, J. C. Meslin, and H. Quiniou, *Phosphorus
                Sulfur* **15**, 143 (1983).
83S827   W. Schroth, R. Spitzner, and J. Freitag, *Synthesis*, 827 (1983).
83TL3713   C. G. Tea, J. P. Pradère, J. Villieras, and H. Quiniou, *Tetrahedron
                Lett.*, 3713 (1983).
84CC157   M. Lees, M. Chehna, M. A. Riahi, G. Duguay, and H. Quiniou,
                J. C. S., Chem. Commun., 157 (1984).
84JHC953   M. Muraoka, A. Yamada, and T. Yamamoto, *J. Heterocycl. Chem.*
                21, 953 (1984).

84JOC74	M. Yokoyama, M. Kodera, and T. Imamoto, *J. Org. Chem.* **49,** 74 (1984).
84JPR101	K. Schulze, C. Richter, F. Richter, E. Mzorek, R. Weisheit, and F. Achenbach, *J. Prakt. Chem.* **326,** 101 (1984).
84MI1	M. Ebel. and P. Mothes, *Sulfur Lett.* **2,** 127 (1984).
84ZC435	W. Schroth, U. Burkhardt, P. Thiess, and R. Spitzner, *Z. Chem.* **24,** 435 (1984).
85BSB149	M. E. Hassan, *Bull. Soc. Chim. Belg.* **94,** 149 (1985).
85EGP222310	D. Briel, G. Wagner, and U. Schubert, Ger. (East) Pat. DD222,310 (1985) [*CA* **104,** 88573 (1986)].
85H1225	S. Coen, B. Ragonnet, C. Vieillescazes, and J. P. Roggero, *Heterocycles* **23,** 1225 (1985).
85JCS(P1)1875	C. G. Tea, J. P. Pradère, H. Quiniou, and L. Toupet, *J. C. S., Perkin Trans. 1,* 1875 (1985).
85JOC1545	C. G. Tea, J. P. Pradère, and H. Quiniou, *J. Org. Chem.* **50** 1545 (1985).
85MI1	A. Reliquet, F. Reliquet, J. C. Meslin, and F. Sharrard, *Sulfur Lett.* **3,** 143 (1985).
85TL745	M. Jubault, A. Tallec, B. Bujoli, J. C. Roze, and J. P. Pradère, *Tetrahedron Lett.* **26,** 745 (1985).
85TL3971	R. Spitzner and W. Schroth, *Tetrahedron Lett.* **26,** 3971 (1985).
85ZC324	J. Bernat, L. Kniezv, P. Kristian, and A. Dondoni, *Z. Chem.* **25,** 324 (1985).
85ZC327	D. Briel and G. Wagner, *Z. Chem.* **25,** 327 (1985).
86CJC597	J. P. Pradère, J. C. Roze, H. Quiniou, R. Danion-Bougot, D. Danion, and L. Toupet, *Can. J. Chem.* **64,** 597 (1986).
86KGS3	G. A. Mironova, V. N. Kuklin, E. N. Kirillova, and B. A. Ivin, *Khim. Geterosikl. Soedin.,* 3 (1986).
86MI1	B. Bujoli, M. Chehna, M. Jubault, and A. Tallec, *J. Electroanal. Chem.,* 461 (1986).
86PS327	C. G. Tea, M. Chehna, J. P. Pradère, G. Duguay, and L. Toupet, *Phosphorus Sulfur* **27,** 327 (1986).
86SC79	C. G. Tea, J. P. Pradère, and H. Quiniou, *Synth. Commun.,* 79 (1986).
87BSF149	C. G. Tea, J. P Pradère, B. Bujoli, H. Quiniou, and L. Toupet, *Bull. Soc. Chim. Fr.,* 149 (1987).
87MI1	J. C. Meslin, A. Reliquet, and F. Reliquet, *Sulfur Lett.* **5,** 117 (1987).
87PS153	A. Reliquet, F. Reliquet, J. C. Meslin, F. Sharrard, and H. Quiniou, *Phosphorus Sulfur* **32,** 153 (1987).
87SC1971	M. Chehna, J. P. Pradère, and A. Guingant, *Synth. Commun.,* 1971 (1987).
87T2709	B. Bujoli, M. Jubault, J. C. Roze, and A. Tallec, *Tetrahedron* **43,** 2709 (1987).
88BSF897	M. Chehna, J. P. Pradère, J. Vicens, L. Toupet, and H. Quiniou, *Bull. Soc. Chim. Fr.,* 897 (1988).
88MI1	M. Chehna, J. P. Pradère, M. Perrin, S. Lecocq, F. Baert, and J. Vicens, *Mol. Cryst. Liq. Cryst., Inc. Nonlin. Opt.* **161,** 55 (1988).
88T139	M. Bakasse, G. Duguay, H. Quiniou, and L. Toupet, *Tetrahedron* **44,** 139 (1988).

Benzo[c]Pyrylium Salts: Syntheses, Reactions, and Physical Properties

EVGENII V. KUZNETSOV AND
IRINA V. SHCHERBAKOVA

Research Institute of Physical and Organic Chemistry, Rostov on Don University, 344006 Rostov on Don, USSR

ALEXANDRU T. BALABAN

Polytechnic Institute, Organic Chemistry Department, Bucharest, Romania

I. Introduction

Among the numerous classes of heterocyclic cations that are now actively studied, pyrylium salts are distinguished by the extraordinary variety of their transformations, making them good synthons and widely applicable. An exhaustive review on this subject has been published by an international group of chemists dealing with pyrylium chemistry [82AHC(Suppl.)].

Two benzoannelated analogues of the pyrylium cation are known: benzo[b]pyrylium (chromylium or 1-benzopyrylium) and benzo[c]pyrylium (isochromylium or 2-benzopyrylium) salts. So far, the 1-benzopyrylium system is considered to be the more interesting (79MI2; 84MI1) because it is the basic heterocyclic system of important plant pigments (anthocyanins), and because there are major differences between properties of its heterocyclic ring and those of monocyclic pyrylium systems as a result of benzoannelation. Thus, for instance, 1-benzopyrylium salts do not possess the ability to have recyclization reactions with heteroatom exchange.

2-Benzopyrylium salts have not been found in nature, and in spite of their ability to take part in recyclization reactions, until recently it was considered that there were no specific features in their transformations relative to monocyclic pyrylium salts (79MI2), and the scope of these transformations has been considered as rather restricted (71CB2984). The reason for such a conclusion was based on the incorrect interpretation of the effects of benzoannelation on reactions of 2-benzopyrylium cations; so far no review on this subject has been published.

The major advantage of 2-benzopyrylium salts is their easy $O \rightarrow N$ exchange, which results in isoquinolines, unlike 1-benzopyrylium salts which do not afford quinolines under similar reaction conditions. Many isoquinolines are known (78MI1) to possess high biological activity and are the structural units of various natural alkaloids and their analogues. Thus, the use of benzo[c]pyrylium salts in isoquinoline synthesis could be very promising in competition with known methods. However, such an approach to isoquinolines was not developed, evidently because of the lack of reliable methods for synthesizing the initial 2-benzopyrylium salts.

The search for benzo[c]pyrylium systems and the elaboration of synthetic approaches to its development could lead to the discovery of novel regularities that might be applied in syntheses of different types of heteroaromatic and aromatic compounds. On the other hand, this route could give a possible systematic study of the influence of benzo[c]annelation on reactions of the pyrylium cation.

The active development of 2-benzopyrylium chemistry was begun in the 1960s by two research groups: one by Vajda and Ruff and Müller and co-workers in Hungary and the other by and Dorofeenko and co-workers in the USSR [64ACH(40)217; 66ACH(50)387] [66DOK(166)359]. The Hungarian chemists were interested chiefly in the investigation of 1-aryl-substituted benzo[c]pyrylium salts, whereas the Russian group successfully developed the synthesis of different substituted cations, as well as studying their transformations. At the moment, only the USSR group, in cooperation with Romanian chemists, works in this field, and the overwhelming majority of their results has been published in Russian.

The goal of the present review is to demonstrate the wide potential for using newly developed methods to synthesize 2-benzopyrylium salts and to discuss in detail the numerous and unusual transformations of this class of oxygen cations in comparison with monocyclic pyrylium salts, i.e., the study of influence of benzo[c]annelation.

On the basis data known so far, one can conclude confidently that the chemistry of benzo[c]pyrylium salts is very promising, not only because of its practical outcome leading to isoquinolines, but mainly because of a multiplicity of unexpected transformations.

We shall use both names for this class of compounds, 2-benzopyrylium and benzo[*c*]pyrylium, indiscriminately, preferring them to isochromylium or isobenzopyrylium. The term benzo[*c*]pyrylium is especially suitable when the positions of substituents are indicated numerically. It is common to designate by the 1-position α and the 3-position by α'. In subsequent formulas, unstable or nonisolated intermediates are enclosed in square brackets, whereas resonance formulas denoting the same compound are enclosed within braces { }. In formulas, "Ver" will denote veratryl(3,4-dimethoxyphenyl) and "An" will denote *para*-anisyl substituents.

II. Syntheses

A. FROM INDENE OR INDANE DERIVATIVES

The unsubstituted 2-benzopyrylium cation **1** was first synthesized by Blount and Robinson (33JCS555) starting with indene. Its oxidation by lead tetraacetate affords homophthalic dialdehyde, which is cyclized to 2-benzopyrylium tetrachloroferrate on treatment with hydrochloric acid and ferric chloride.

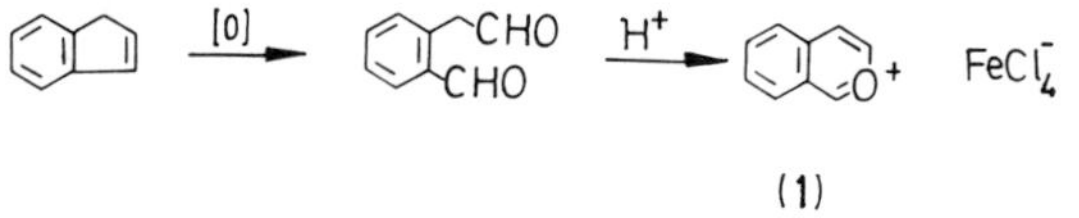

(1)

This approach was applied later to synthesize substituted 2-benzopyrylium salts **4** (85LA2116) via 1,5-diketones **3,** and a few dioxetanes **2** were isolated as intermediates on using singlet oxygen photochemically (74JA4339).

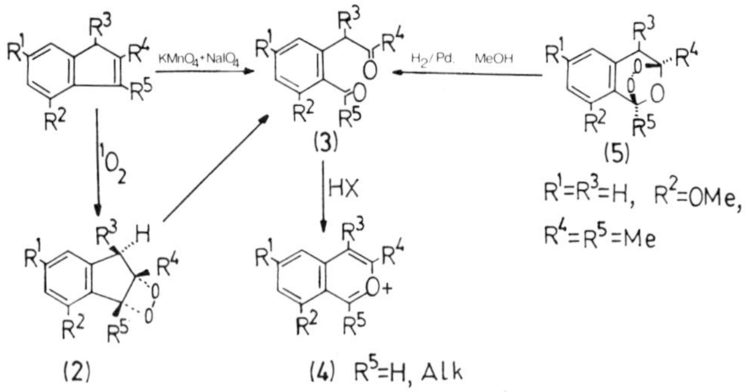

The only example of an ozonide (**5**) was reported (85LA2116) when ozone was used instead of photooxidation; **5** was also converted into 2-benzopyrylium via the corresponding *o*-acetyl benzyl ketone **3,** but not

(5)   (6)   (7)

through the independently obtained isochromene **7**. At the same time, isochromanes of type **6** were oxidized to the corresponding 2-benzopyrylium salts **4** by the Jones reagent (86MIP1).

The structural elucidation by Doering and Berson for the oxidized form of diisoeugenol **8,** which was found to be a 1,5-diketone **9** (50JA1118), triggered investigations by Hungarian groups of the synthesis of various 1-aryl-6,7-dimethoxybenzo[c]pyrylium salts **(10)** through oxidation of 1-aryl indanes and indenes [64ACH(40)217; 64ACH(40)225; 64ACH(40)295; 64ACH(41)451; 65M369; 66ACH(50)381; 66ACH(50)387; 67ACH(51)107; 67ACH(52)261; 68ACH(57)181; 75ACH(85)79; 78ACH(96)373].

(8)   (9)

(10) R^1, R^2=Alk,H;
R^3, R^4, R^5, R^6, R^7=H, OH, OAlk

Cyclization of diketones **9** to 2-benzopyrylium salts **10** occurs in the presence of strong acids (HClO$_4$, H$_2$SO$_4$, HCl, HBF$_4$). On heating 1,5-diketones in formic acid, such a cyclization also takes place, but so far only a single example of the isolation of a stable 2-benzopyrylium formate has been reported [66ACH(50)381]. 2-Benzopyrylium acetates were not isolated at all in spite of electronic-absorption spectral data (cf. Section III,E,2) on their formation in low yield from 1,5-diketones of type **9** in acetic acid at elevated temperature [66ACH(50)381].

Perchlorates are the most often used anions in syntheses and investigations of all 2-benzopyrylium salts. In the following sections, only anions different from perchlorate will be depicted in structural schemes. However, taking into account the danger of explosion inherent in working with solid perchlorates, it would be cautious to use BF$_4^-$, CF$_3$SO$_3^-$, or other anions instead.

Photolysis of 2-hydroxy-1-indanones **11** leads to a mixture of pseudo-base ether **12** and 1,5-dicarbonyl compound **13** (76JA5581), which are precursors of 2-benzopyrylium salts and should be convertible to the latter compounds under strongly acidic conditions.

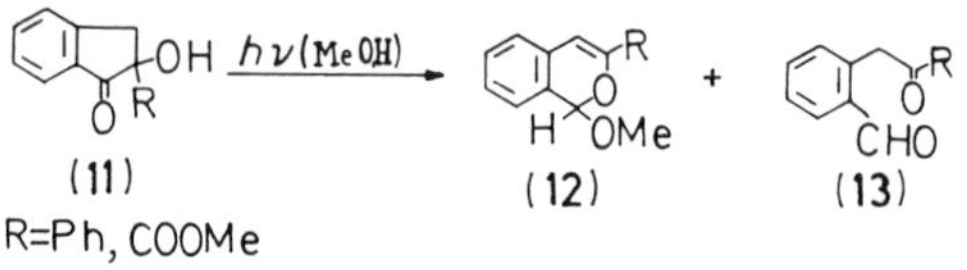

(11)
R=Ph, COOMe

(12)

(13)

At the same time, photolysis of 2,3-diaryl-2,3-epoxyindanones **14,** as well as thermolysis (62JA1315), includes the reversible formation of red inner salts benzo[*c*]pyrylium-4-oxides **15** (63JA3529; 64JA2084; 84KGS167; 86NKK1622).

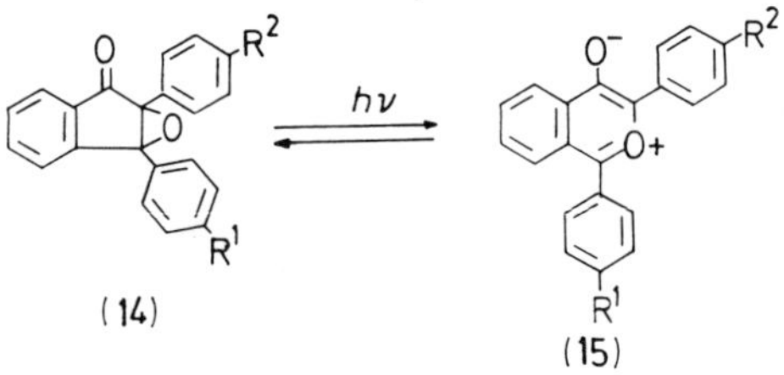

(14)

(15)

The photostationary state concentration of **15** depends on the wave-length used for excitation, but a sufficient concentration could not be obtained to allow isolation of **15** (62JA1315). Besides, the formation of beta-ines **15** is also complicated by their interaction with the initial epoxyinda-nones **14** (Section III,E,2). The high reactivity of benzo[*c*]pyrylium-4-oxides as dipolarophiles (Section III,E,2) initiated the search for new approaches to their synthesis, one of which is the intramolecular carbenecarbonyl reaction which yields the unstable 1-methoxy-4-oxide **16** (81BCJ240).

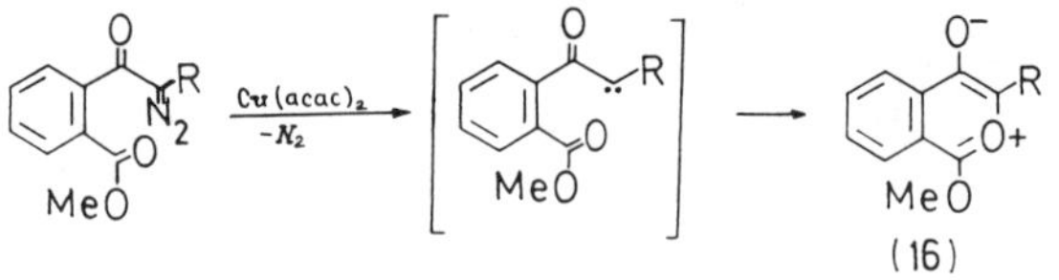

(16)

The unsubstituted benzo[*c*]pyrylium-4-oxide **19** was formed from iso-chroman-4-one derivatives **17** [76ACS(B)619] or from **18** (84CC702). All attempts to isolate oxide **19,** for instance as a fluoroboric acid salt, failed because it reacted further by polymerization and dimerization [76ACS(B)619]. Even in substituted systems **15,** in spite of added stabilization from aryl substituents in comparison with **19**, equilibrium exists

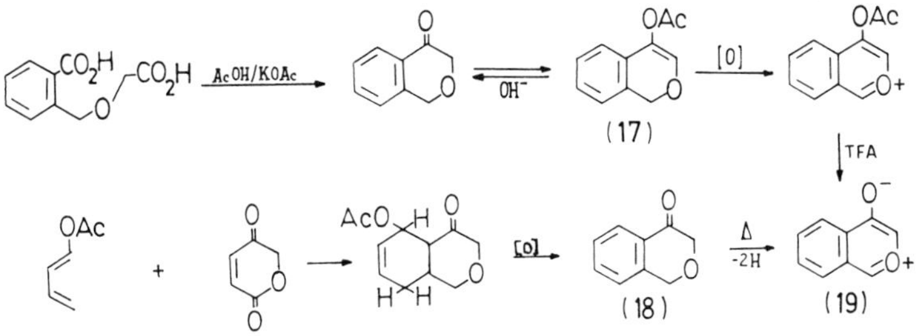

between the pyrylium and the epoxide valence isomers; the equilibrium position depends on conditions, but favors the epoxide structure **14** (64JA2084; 66JA4942).

B. FROM ISOCOUMARIN DERIVATIVES

The unstable 1-hydroxybenzo[c]pyrylium salts **21**, which are structural isomers of the protonated form of **19**, result when isocoumarins **20** are protonated with strong acids (49JOC204; 51JOC1064; 60JIC379).

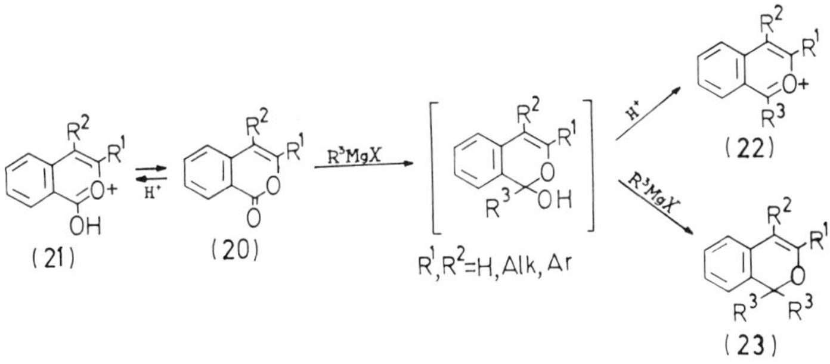

The application of the Grignard reaction to isocoumarins **20** (49JOC204; 51JOC1064; 71CB2984) may be called the traditional method for synthesizing 2-benzopyrylium salts **22**. However, since the synthesis of isocoumarins with different substituents in the benzo and lactone rings is a complicated problem (64CRV29), this method is not wide-spread (69TL711; 69ZOR191; 71CB2984; 78KGS1320; 80KGS193). Interestingly, excess of the Grignard reagent in this reaction in some cases causes formation of stable geminally substituted isochromenes **23** (71JIC707) (cf. Section III,C,2).

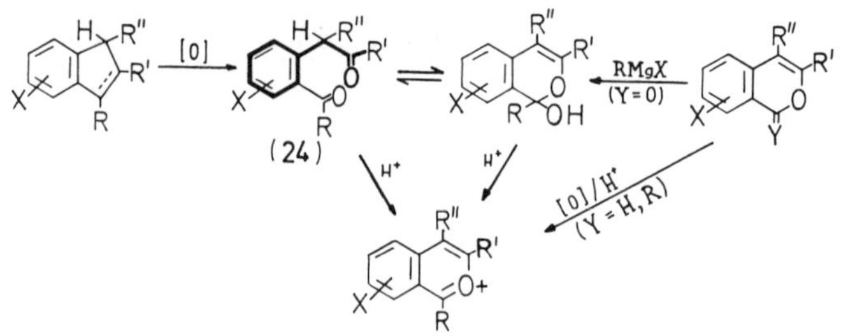

SCHEME 1

C. Acid-Catalyzed Acylation Of Benzyl Carbonyl Compounds And Their Derivatives

1. Survey

One can see that the main point of most methods described earlier is a construction of 1,5-dicarbonyl compounds or cyclic pseudobases by a transformation of the readily available indene (indane) or isocoumarin molecular carbon skeleton (Scheme 1).

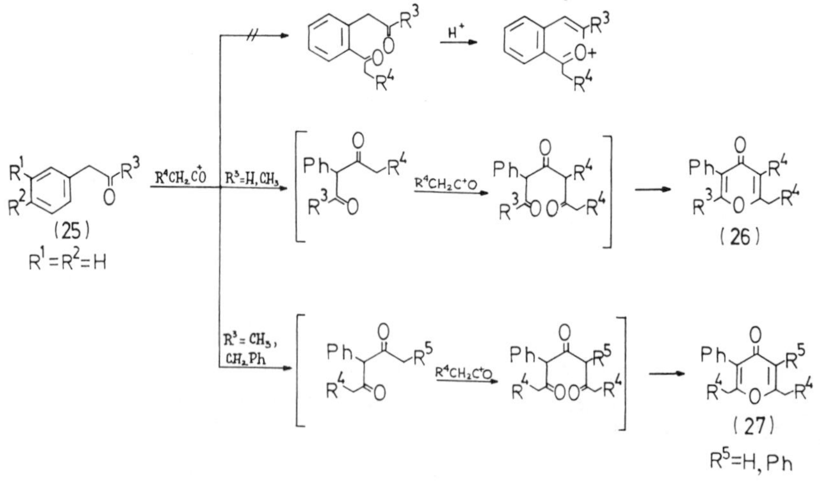

Such a construction of 1,5-dicarbonyl compounds **24** may also be reached by using the benzyl ketone framework marked out in **24**. *ortho*-Acylation under acidic conditions of the aromatic ring could lead to the formation of **24** and to cyclization *in situ* into a 2-benzopyrylium cation as proposed in 1961 (61RRC295). However, the acylation of benzyl ketones unsubstituted in aromatic ring **25,** or of phenylacetic aldehyde under conditions of the Friedel–Crafts reaction in the presence of different catalysts, led to attack on the methylene or methyl group and was completed by the formation of γ-pyrone derivatives **26** or **27** (61JA193; 61JOC92; 61JOC2993; 61RRC295; 72KGS152). The structure of the final γ-pyrone (**26** vs. **27**) is determined by group R^3 in the initial carbonyl compound **25**.

The acylation of benzyl ketones **25** to yield 2-benzopyrylium salts was successfuly applied for the first time by Dorofeenko *et al.* (65URP176592; 66ZOR1499) to structures with excess electron density in the ortho position of the aromatic ring, e. g., as in **25**, if $R^1 = R^2 = OMe$ [66DOK(166)359].

(25)

The same authors used this approach to synthesize pyrylium salts condensed at the *c*-site with different π-excess heterocycles (72MI1). However, these systems with heterocyclic rings replacing the benzo ring are outside the scope of this review and will not be discussed.

2. Mechanism of Acid-Catalyzed Acylation

The formation of 2-benzopyrylium cations **30** by acid-catalyzed acylation of benzyl carbonyl compounds **28** may be described by several equilibrium processes as shown in Scheme 2.

The extent of kinetically controlled formation of the carboxonium ions **31** depends on the nature of R^1 and Y^1. The possible existence of **31** allows formation of acylated enols **32** ($Y = R^3CO$), which are analogous with ω-acylaminostyrene derivatives. As is known, the latter compounds easily undergo an intramolecular acid-catalyzed cyclization to isoquinolines (the Pictet–Gams reaction) (80T1279).

Special experiments show that enol acetates **32** ($R^1 = Ar$, $R^2 = H$, $R^3 = Me$, $R^4 = OMe$) form 2-benzopyrylium salts under acidic conditions

SCHEME 2. $Y = R^3CO,Z$; $Z = H^+$, Lewis acid.

---

[1] The experimental support for such a conclusion is the synthesis of isoquinolines **33** from benzyl ketones **25,** which obviously takes place by a new version of the Ritter reaction [71JCS(C)3590; 78CPB245; 82MI1; 89TH1]. Detailed discussion of this conversion is outside the scope of this review.

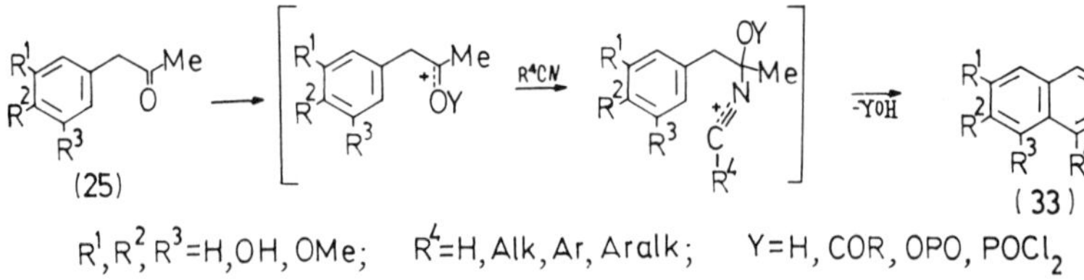

(25)

$R^1, R^2, R^3 = H, OH, OMe;$ $R^4 = H, Alk, Ar, Aralk;$ $Y = H, COR, OPO, POCl_2$

(76KGS50), and that the formation of the latter compounds does not occur by the course analogous to the Pictet–Gams reaction, but via the preliminary cleavage of the oxygen–vinyl bond, according to Scheme 2 (79TH1). Interestingly, the rearrangement of enol acetates of type **32** into 1,3-diketones (52JA3228), which may be converted to γ-pyrone derivatives (**26, 27**) does not take place if the aromatic ring of the enol acetate is activated by donor substituents.

Note that 1,5-diketones **29** (Scheme 2) are cyclodehydrated as soon as they are formed, and this process evidently occurs via protonation of the carbonyl group. The proton affinity in this case is probably determined by the substituent at each carbonyl group in **29**. So far, there are no synthetic investigations of the problem, whereas quantum chemical calculations reveal the equal possibility for proton addition to both oxygen atoms in homophthalic dialdehyde (86KGS460).

3. Application of Acid-Catalyzed Acylation to the Synthesis of 2-Benzopyrylium Salts

Various alkoxy-substituted phenylacetones **34** are used to synthesize a series of 2-benzopyrylium salts **35** by acid-catalyzed acylation [66DOK(166)359; 66ZOR1492; 70KGS(2)196, 70KGS(2)213; 70ZOR1118]. Interestingly, in all cases the alternative pathway of acylation of **34** (R^1 = H and R^2, R^3 are not equal to H) was not observed, and isomeric 2-benzopyrylium salts **36** were not obtained.

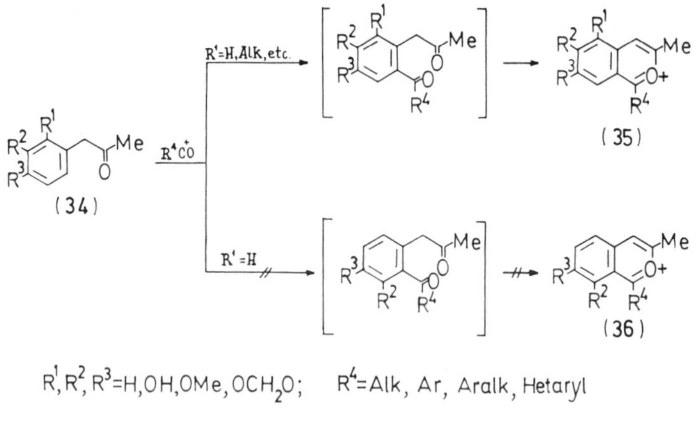

(34)

(35)

(36)

$R^1, R^2, R^3 = H, OH, OMe, OCH_2O;$ $R^4 = Alk, Ar, Aralk, Hetaryl$

Only the coordinated action of at least one donor group leads to acylation of **37** in the ortho position relative to the acetonyl group ($R^1 = R^3 = H$, $R^2 = R^4 = OMe$) (85LA2116). The reaction does not occur when $R^1 = R^3 = OMe$, $R^2 = R^4 = H$ (68UP1). Acylation of **37** when $R^1 = R^4 = OMe$, $R^2 = R^3 = H$ also does not take place (68UP1); this can be explained by the same reasons found in the case of **36.**

The various benzyl ketones **38** with two methoxy groups in positions 3 and 4 have become easily accessible, since polyphosphoric acid (PPA) and polyphosphoric acid ester (or ethyl polyphosphate) (PPE) were applied on a large scale in their synthesis by the Friedel–Crafts reaction (**38**, $R^1 = Ar, R^2 = H$; $R^1 = R^2 = Ar$; $R^1 + R^2 = $ indeno) [70KGS(2)207; 70ZOR578; 74KGS181; 75KGS25; 77TH1]. The acylation of the methylene group in homoveratric acid with homoveratric(3,4-dimethoxyphenyl-acetic) or benzoic anhydrides in the presence of sodium acetate also leads to benzyl ketones **38** ($R^1 = CH_2C_6H_3(OMe)_2$, $R^2 = H$ and $R^1 = Ph$, $R^2 = H$, respectively) (77ZOR631; 79TH1).

The need for different catalysts in the synthesis of 2-benzopyrylium salts (**39**) from compound **38** is determined by the introduced substituent R^3, which is introduced at the first position of the cation. Thus, formylation of **38** ($R^1 = Me, Ar$; $R^2 = H$) with dichloromethyl butyl ether in the presence of $AlCl_3$ [according to Rieche (60CB88)], followed by treatment with perchloric acid leads to 1-unsubstituted salts **39** ($R^1 = Me$, Ar; $R^2 = R^3 = H$) (70KGS278; 70KGS1013).

In the case of benzyl ketones (**40**) containing two meta situated hydroxy groups, orthoesters can be successfully used for formylation [71-JCS(C)3559, 71JCS(C)3571].

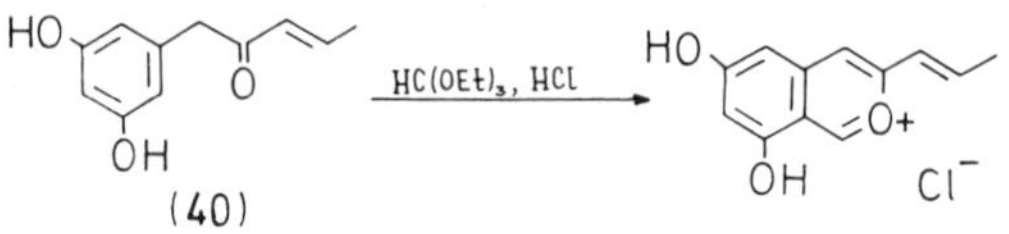

(40)

Similiarly, the deoxybenzoin (**41**), substituted in the β-aromatic ring, forms a 2-benzopyrylium salt (**42**) under Gattermann reaction conditions [71JCS(C)3590].

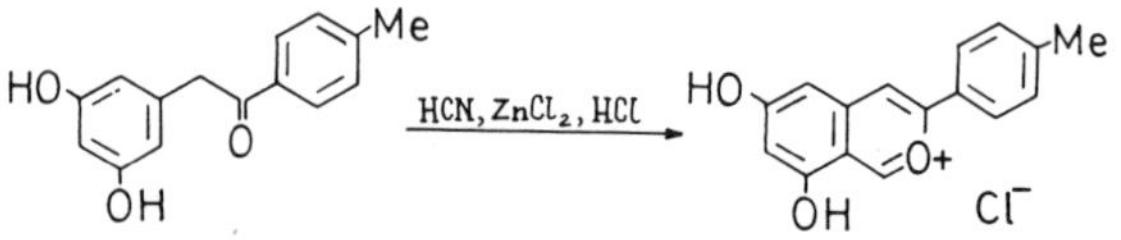

(41)   (42)

When **41** contains a methyl group instead of the α-aromatic (*p*-tolyl) ring, isoquinolines result [71JCS(C)3590] (cf. Section II,C,2).

1-Alkyl-substituted perchlorates **39** (R^3 = Alk) are the result of acylation of **38** with $(AlkCO)_2O$ + $HClO_4$, whereas 1-aryl or 1-aralkyl-substituted salts **39** (R^3 = Ar, Aralk) were synthesized by acylation of **38** with acid chlorides in the presence of $AlCl_3$ or with the corresponding acids in PPA or PPE followed by treatment with perchloric acid.

Depending on the initial carbonyl derivatives **38,** several types of 2-benzopyrylium perchlorates were synthesized; they are shown in Scheme 3.

Acylation of benzyl ketones **25** with dicarboxylic acids give bis-2-benzopyrylium salts **43** (72KGS454; 77KPS714).

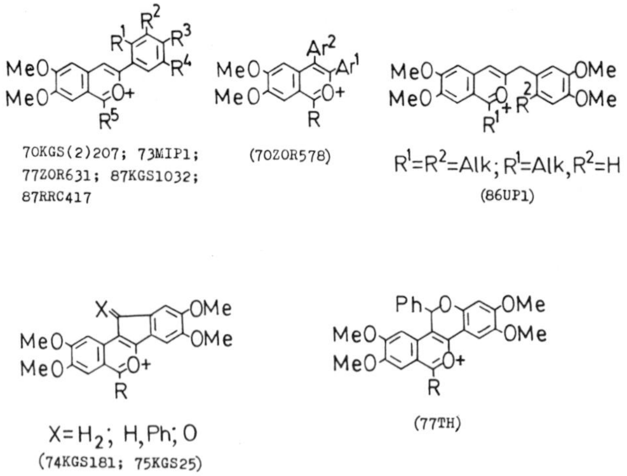

7OKGS(2)207; 73MIP1;
77ZOR631; 87KGS1032;
87RRC417

(7OZOR578)

$R^1 = R^2 =$ Alk; $R^1 =$ Alk, $R^2 =$ H

(86UP1)

X = H_2; H, Ph; O
(74KGS181; 75KGS25)

(77TH)

SCHEME 3

Another approach to bis-2-benzopyrylium salts **45** involves acylation of bis-benzyl ketone **44** (87KGS600).

The acylation of homoveratric aldehyde gives rise to the stable perchlorates **46** (84KGS702; 85KPS815) in spite of the prevailing opinion about the instability of 2-benzopyrylium cations with an unsubstituted 3-position [64ACH(40)225].

Salts **49** with the 3-carboxylic ester group in position 3 were synthesized from both glycidic **47** (70ZOR1429) and pyruvic **48** (73KGS1458) esters.

The latter method gives higher yields of **49** than the synthesis from **47**, but all attempts to use the free acid **48** (R = H) in this reaction failed.

The first example of acylation of homoveratryl nitrile was reported to result in 1-alkyl-3-acylaminobenzo[c]pyrylium salts **50** (71KGS730).

The structure of the final 2-benzopyrylium cation depends on the nature of the substituents at the methylene group in the initial nitrile (85MI2; 86KGS999).

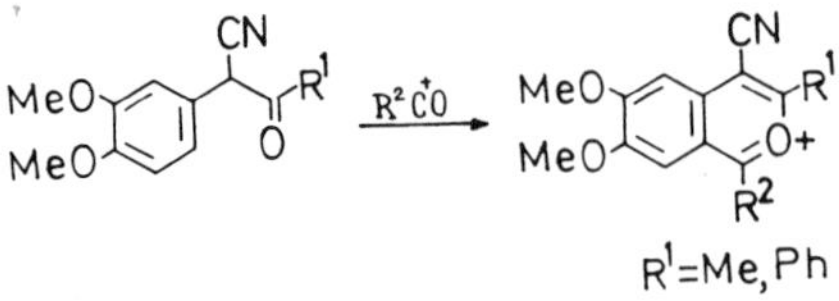

$R^1 = Me, Ph$

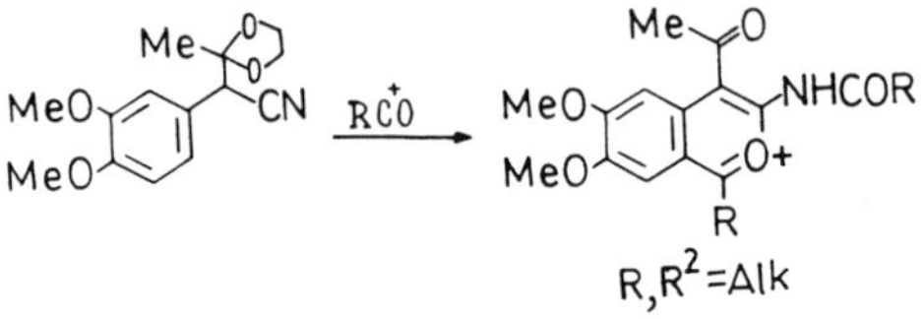

$R, R^2 = Alk$

Acylation of homoveratric esters **51** leads to the stable 3-alkoxy-substituted cations **52**, whereas the use of homoveratric acid **51** (R = H) gives the unstable 3-acyloxy-substituted salts **53**, which are easily hydrolyzed to ring-opened ketoacids **54** [70KGS(2)200].

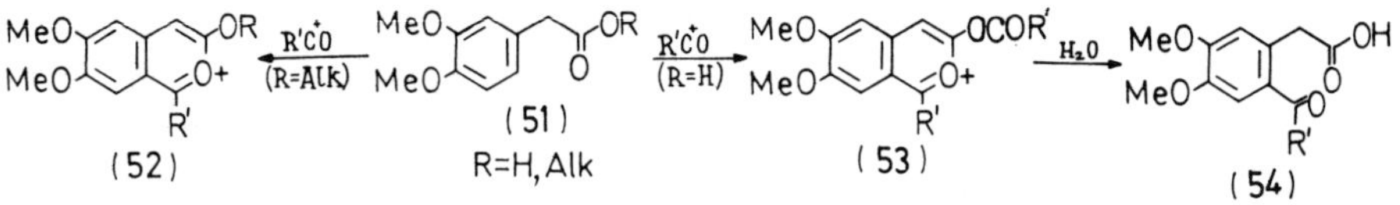

Interestingly, acylation in PPA of **51** (R = H) with fatty-aromatic acids, with the goal of obtaining **54**, also gave, in some cases, salts **56** in a yield of 3–5% [70KGS(2)200]. Obviously, self-acylation of **51**, resulting in ketoacid **55**, is the first step of the reaction (71KGS1601), then acylation of **55** with R'COOH occurs.

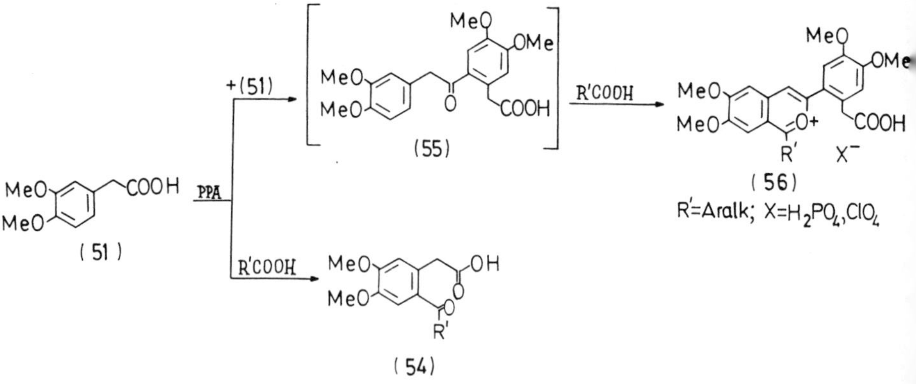

Indeed, isolated ketoacid **55,** as well as its lactone **57,** forms the corresponding salts **56** (R' = H, Alk) under the conditions of acid-catalyzed acylation (77KGS1481; 81KGS313).

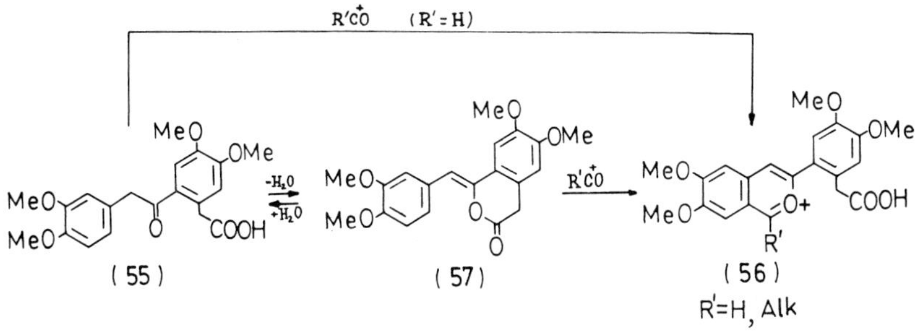

The five-membered lactone **58** reacts similarly under the same conditions (77KGS1479).

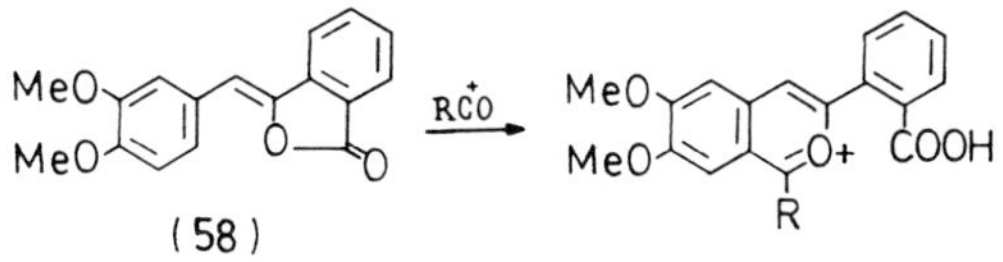

Interestingly, a side product in the formylation of **55** or **57** with dichloromethyl butyl ether was 5-oxoniachrysene **59** (81KGS313) (cf. Section II,C,6).

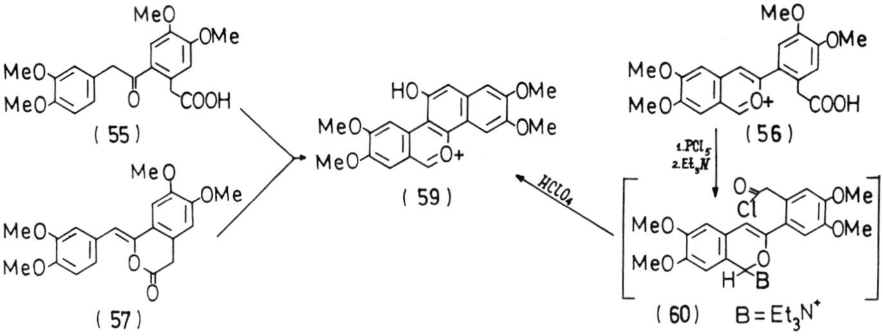

This polycyclic 2-benzopyrylium salt was also obtained in good yield from the 2-benzopyrylium salt **56,** through intramolecular acylation of the intermediate isochromene derivative **60.** The same approach was used later by Elliott and co-workers (84CJC2435) in the synthesis of some benzo[c]-phenanthridine analogs.

4. Use of Aromatic Aldehydes

The direct interaction of benzyl carbonyl compounds **38** and **48** with aromatic aldehydes in the presence of PPA or Ac_2O + $HClO_4$ (86KGS125; 87TH1), or with acetals or acylals of aliphatic or aromatic aldehydes in the presence of triphenylmethyl perchlorate (73KGS881; 75ZOR1962), gives rise to benzo[c]pyrylium salts **62** in yields of 15–70 %.

(38) R^1=Me,Ar
(48) R^1=COOH

(61)

(62)

R^2=H, 3(4)-NO_2, 3-Br, 3(3,4)-OMe
R^1=Me,Ar,COOH

The mechanism of the reaction was not investigated more closely, but the possible existence of 1-H-isochromenes **61** as intermediates was suggested (87TH1).

Contrary to the synthesis described earlier (Section II,A; C,3) of 1-aryl-2-benzopyrylium salts with acylation agents, the use of aromatic aldehydes has the obvious advantage of allowing the synthesis of cations **62** with electron-attracting groups in the 1-aryl substituent (87TH1) or with a carboxylic group in position 3 (86KGS125).

5. Competing Reactions in the Formation of 2-Benzopyrylium Salts upon Acid-Catalyzed Acylation

Upon acylation of some benzyl carbonyl compounds (**25**, R = H, Me; **51**, R = OH) dibenzo[a,d]tropylium salts **65** have been isolated in low yields (5–15 %) along with the major products, 2-benzopyrylium salts. Veratryl acetone **25** (R = Me) as well as homoveratric aldehyde **25** (R = H) (or carboxonium ions **31** which are formed from them) may undergo an oxidative α-cleavage, resulting in the benzyl cation **64**. The formation of the same cation from homoveratric acid **51** is the result of decarbonylation of the acylium ion **63**. Further interaction of the benzyl cation **64** with the substrate, followed by cyclization and oxidation, results in the polycyclic tropylium salts **65** (82ZOR589).

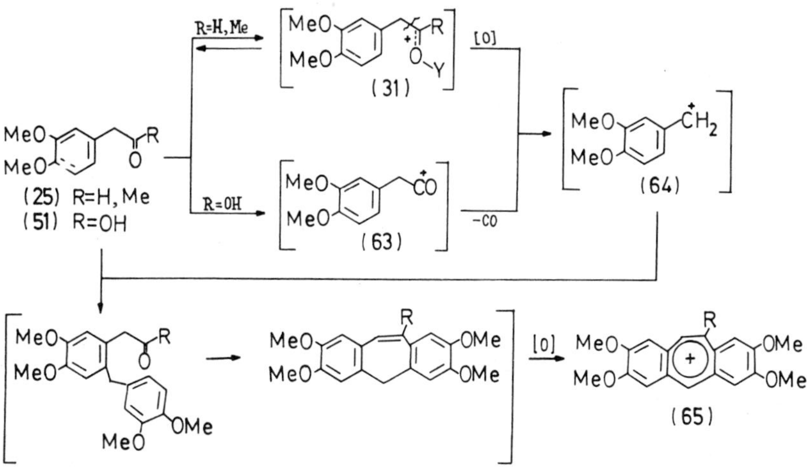

Independent syntheses of these interesting compounds are based on this scheme and are described in several references (79ZOR439; 79ZOR588; 81ZOR440; 84ZOR224; 87TH1).

6. Polycyclic Derivatives of 2-Benzopyrylium Salts

The acid-catalyzed acylation was successfully applied to the synthesis of two types of polycyclic 2-benzopyrylium salts **67** and **69**. The initial compounds in these cases were ortho-(3,4-dimethoxyphenyl)-substituted α-naphthols (**66**) or β-naphthols (**68**), and the final salts were 5-oxoniachrysenes **67** (76KGS745; 77KGS1176) and 6-oxoniabenz[a]anthracenes **69** (89ZOR164), respectively.

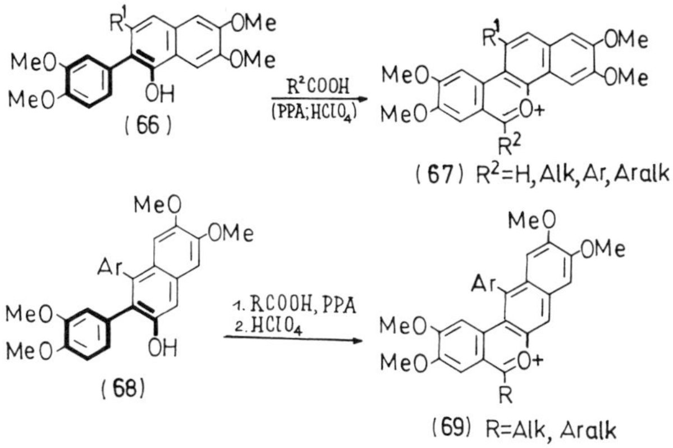

The aim of these syntheses of polycyclic oxonia cations was twofold. On one hand, naphthols **66** and **68** may be considered as benzyl carbonyl derivatives with a fixed enolic function (the marked-out framework), and the acid-catalyzed acylation in this case could have some differences in comparison with the same process for benzyl ketones **28** (Section II,C,2). On the other hand, the chemistry of the novel salts **67** and **69** could be very promising, since the O → N exchange process could lead simply to benzophenanthridine derivatives (71CB2984; 78MI1). Acylation of **66** and **68** with $(AlkCO)_2O + HClO_4$ leads not only to salts **67** and **69**, respectively, in a poor yield, but also to unidentified deeply blue-colored compounds, which may be the result of easy oxidation of naphthols with perchloric acid. This behavior is similar to that of some polycondensed aromatic compounds (67MI1).

However, the use of PPA as acylation catalyst, followed by treatment with $HClO_4$, gave in all experiments perchlorates **67** and **69** in good yield. Moreover, α-unsubstituted 5-oxoniachrysenes **67** ($R^2 = H$) were obtained by the formylation of **66** with 98% HCOOH in PPA in a 70–75% yield (76KGS745; 77KGS1176). Phenol ethers do not possess the ability to be C-formylated under the conditions used here (79TH1).

On the basis of data obtained in the previous paragraphs, one can propose that formation of 5-oxoniachrysenes **67** ($R^2 = H$) occurs via the primary O-formylation of **66** followed by cyclization of esters **70** ($R^2 = R^3 = H$), which is similar to the Pictet–Gams reaction (80T1279). Obviously, this mechanism can also be applied to the acylation of **66** with R^2COOH (R^2 does not equal H) in PPA in contrast to the scheme of formation of 2-benzopyrylium cations **30** from benzyl ketones **28** (Section II,C,2).

This suggestion is supported by the high stability of the ester bond in the easily obtained acetates **70** (R^2 = Me), under conditions of both acidic and basic hydrolysis. Also, an *ipso*-substitution of the bromine atom in the acetate **70** ($R^1 = R^2$ = Me, R^3 = Br) takes place in PPA, resulting in 5-oxoniachrysene **67** ($R^1 = R^2$ = Me). This behavior is contrary to that of the enol acetate **71**, which does not form any 2-benzopyrylium salt in PPA, but breaks down to deoxybenzoin **72** (77ZOR631; 79TH1).

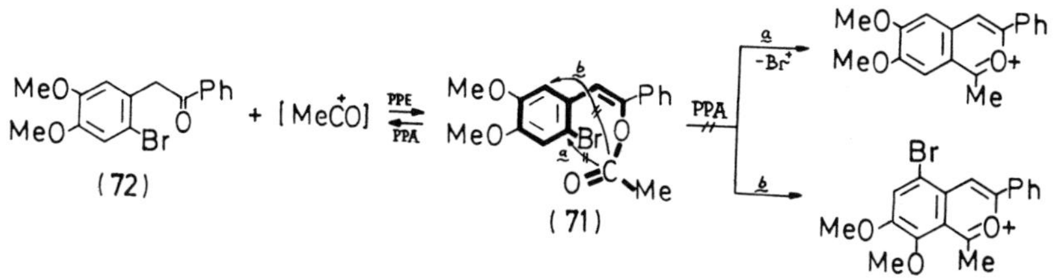

D. Rearrangement of *o*-Benzhydryl Homoveratric Acids

1,3-Diarylbenzo[c]pyrylium salts **76** are formed by rearrangement in PPA of *o*-benzhydryl homoveratric acids **73** (81ZOR440).

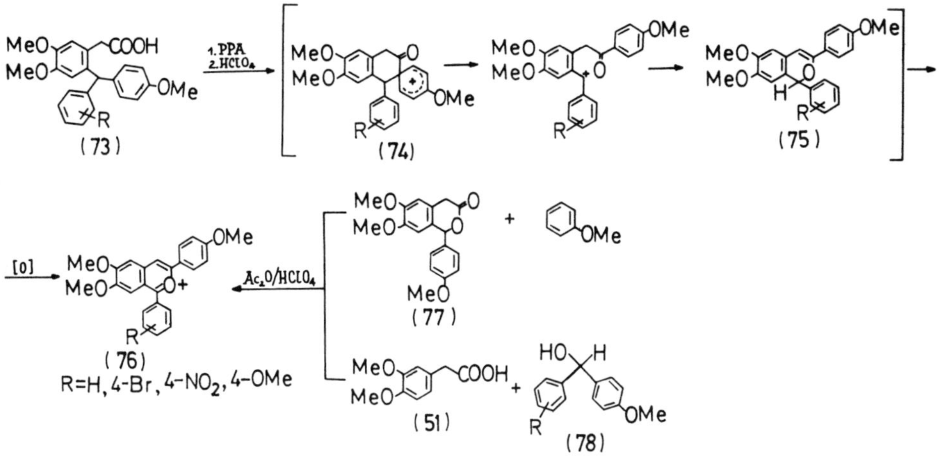

The prerequisite to this process is the presence of an anisyl substituent in **73**, necessary because this substituent is involved in *ipso*-substitution via intermediate **74**.

Interestingly, such migrations through spiro compounds (62JA788) are known for *peri*-benzhydryl naphthoic acids (59JA935), whereas *ortho-*

benzhydryl phenylacetic acids of type **73** without donor substituents do not undergo the rearrangement (61JOC92). The intermediate formation of 1-*H*-isochromenes **75** is postulated as the next to the last stage of rearrangement. The clarification of the key role of *o*-benzhydryl homoveratric acids **73** in this process has led to the synthesis of salts **76** starting from isochromanone **77** and anisole, or from homoveratric acid **51** and benzhydrols **78** (87TH1).

E. FROM ACYLVERATROLES

The intermediate 1,5-dicarbonyl compounds of type **24** (Scheme 1) can be constructed not only on the basis of *meta*-alkoxy-substituted benzyl ketones ($C_4 + C_1$ synthesis, Section II,C), but also under definite conditions starting from aryl ketones ($C_2 + C_3$ synthesis). Thus, in a molecule of acylveratrole derivatives of type **79,** the excess of π-electron density due to the presence of two *ortho*-methoxy groups allows such compounds to be involved in electrophilic substitutions with benzoin (73URP2; 74KGS1575).

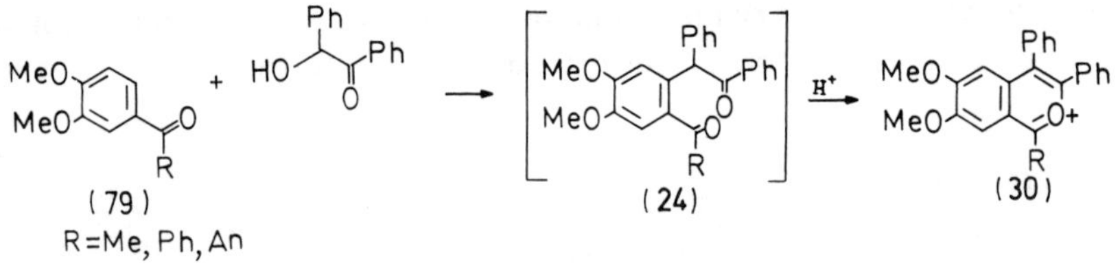

A formal and rather remote analogy to the method previously described for salts **30** can be seen in another $C_2 + C_3$ synthesis of the C_5-chain, which cyclizes to yield the pyrylium ring of 1-aminobenzo[*c*]pyrylium bromide **80** (78JOC3817).

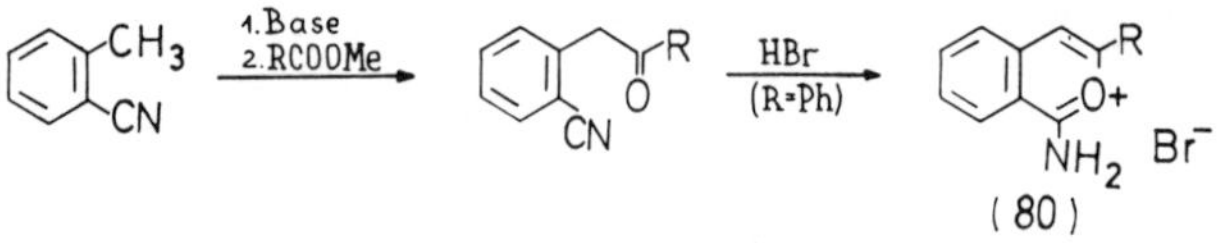

III. Reactions of Benzo[c]pyrylium Salts

A. INTRODUCTION

The reactions of 2-benzopyrylium salts can be divided into five groups:

(1) Reactions of substituents conserving the benzo[c]pyrylium system.

(2) Nucleophilic addition reactions which are the most common. Depending on the nature of nucleophile, the structure of 2-benzopyrylium cation, and the reaction conditions, such additions may result in isolable adducts, in products of ring opening, or in formation of a new ring.

(3) Deprotonation reactions if the 2-benzopyrylium cation has an α-alkyl substituent or a group with a mobile hydrogen atom in any other fragment of the molecule. This conversion takes place over a wide range of conditions, and it often accompanies nucleophilic additions. Moreover, the problem of the acid–base and nucleophilic–electrophilic interactions for 2-benzopyrylium salts concerns not only the primary step of their reactions with nucleophiles, but also the ring-open forms, as shown in Section III,C,4.

(4) [4 + 2]-Cycloaddition reactions with inverse electron demand, which become real because of the high electronegativity of the oxonium atom and the fixation, to a considerable extent, of the double bonds in the heterocyclic part of the cation.

(5) Dimerization and further transformation of dimers. This reaction type has to be considered separately since it includes practically all reactions of 2-benzopyrylium salts mentioned here.

B. REACTIONS THAT CONSERVE THE BENZO[c]PYRYLIUM CATION

Since the deactivating effect of the positive charge is obvious for the 2-benzopyrylium ring system, no electrophilic substitution is known for it. However, substituents on the heterocyclic fragment of the cation, as well as substituents on the annelated benzenoid ring, may undergo electrophilic attack. The expected easy acylation of hydroxy groups on the annelated ring of 2-benzopyrylium salts **30** leads to acyloxy-substituted derivatives **81** [70KGS(2)213].

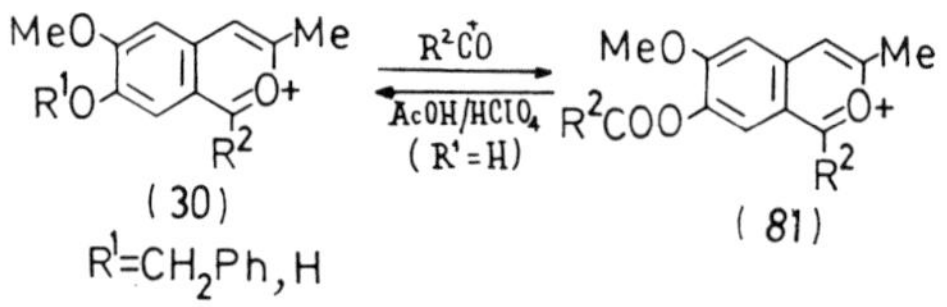

If R^1 is benzyl in **30**, the formation of **81** takes place by replacing the benzyloxy group with acyloxy. The acylation of hydroxy groups belonging to aryl substituents in positions 1 or 3 in **82** results in acyloxy derivatives **83** (87KGS1032) and **84** (87RRC417), respectively.

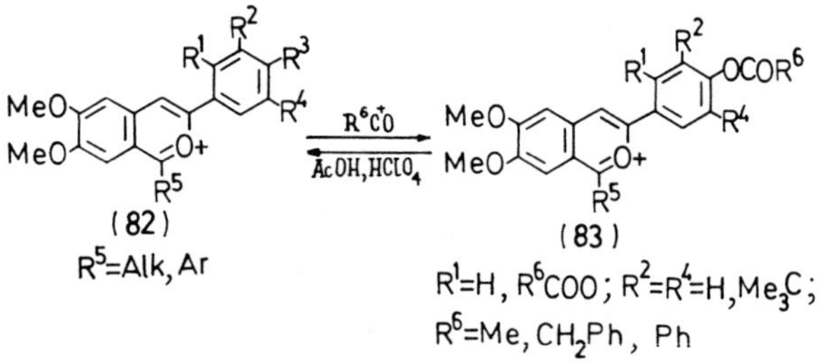

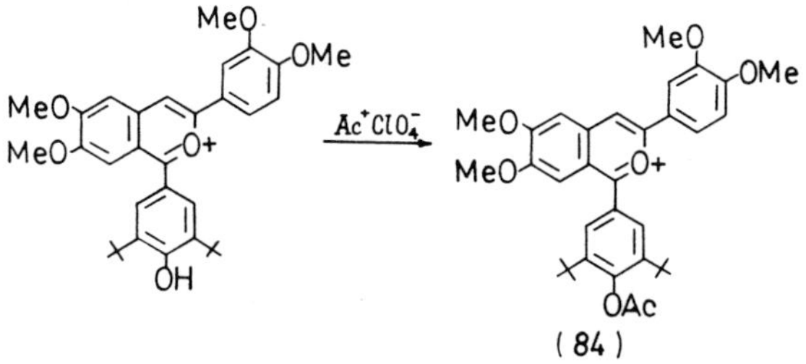

The acylated phenolic hydroxyl surrounded by *t*-butyl groups, as in **84** or **83** ($R^2 = R^4 = Me_3C$) is stable under conditions of both acidic and basic hydrolysis, whereas other acyloxy derivatives **81** and **83** ($R^2 = R^4 = H$) easily undergo hydrolysis to the initial hydroxy-substituted salts in solutions of hot acetic acid with a few drops of perchloric acid [70KGS(2)213]. Examples of aromatic C-acylation are known, for instance, 2-benzopyrylium salts having 1- or 3-benzyl substituents activated for electrophilic attack by two methoxy groups (**86** and **87**, respectively) (86UP1, 86UP2).

The acylated 3-(3,4-dimethoxybenzyl) substituted salts (**86**) or salts **81** can be directly obtained by the acid-catalyzed acylation of diveratryl-acetone (86UP1) or the corresponding hydroxyl-containing benzyl ketone

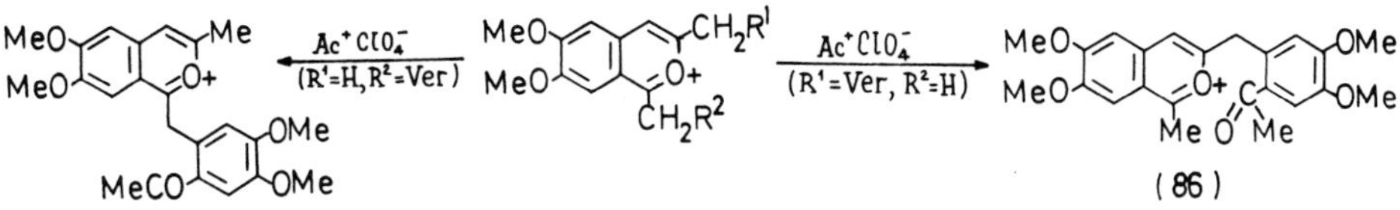

(85)   (86)

[70KGS(2)213)], respectively, with an excess of aliphatic anhydride as described in Section II,C,3. The unsubstituted benzyl ring is not activated for electrophilic attack in **87**, and the methylene group takes part in acylation, but only in alkaline medium [90KGS(ip1)]. The reaction then continues with easier deprotonation and formation of acylated anhydrobases **88** because of the considerable acidifying effect from the acyl group.

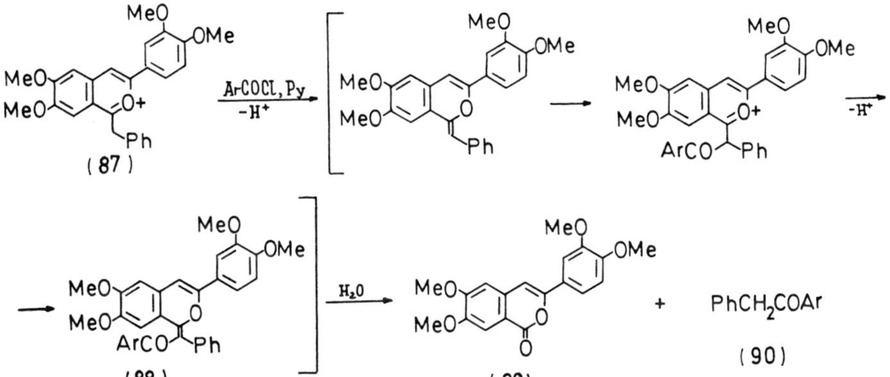

(87)

(88)   (89)   (90)

The latter compounds **88** are unstable and readily hydrolyzed to isocoumarin **89** and deoxybenzoin **90** by analogy with β-diketonic enol ethers (cf. Section III,F,2,d). The only example of electrophilic substitution in an aromatic ring conjugated with the cationic ring was described for the nitration regioselectively yielding the *meta* isomer **91** (84MI1).

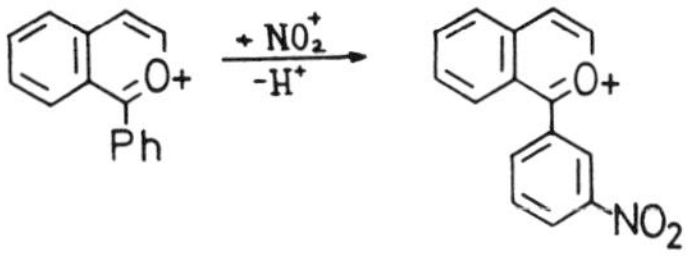

(91)

The condensation of the 1-methyl group in 2-benzopyrylium salts **92** with carbonyl derivatives such as aromatic aldehydes (70KGS1308) in acetic acid gives rise to 1-styryl-substituted salts **94**. The reaction proceeds similarly with orthoesters (74KGS37), isocoumarin derivatives, and dimethylformamide (DMF) (80KGS193) or azomethines (82KGS465). The formation of intermediate anhydrobases **93** is postulated for this reaction by analogy with a similar type of conversion for monocyclic pyrylium salts [82AHC(Suppl)].

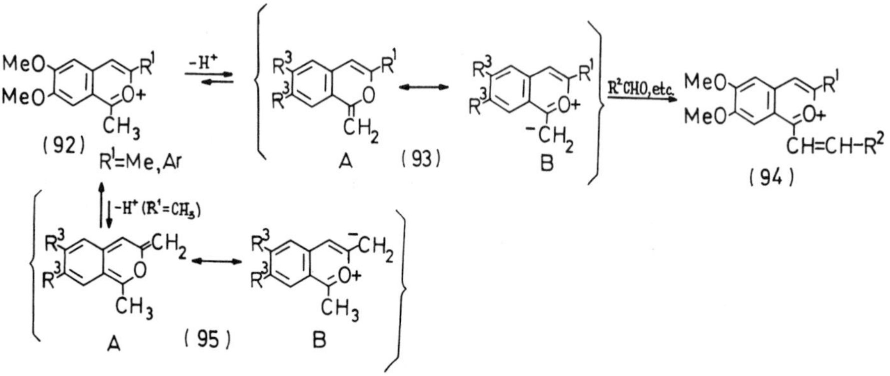

Among all these examples, the interaction of **92** with azomethines is the most interesting, and evidently the nitrogen-containing component acting as a base facilitates the primary step of deprotonation of the 1 (α)-methyl group followed by condensation.

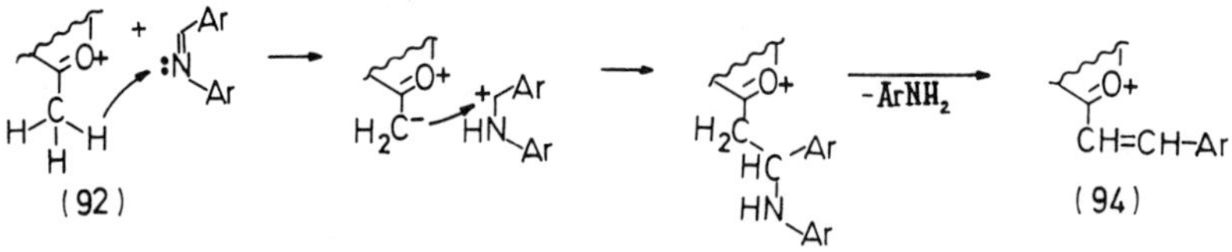

Interestingly, the 3 (α')-methyl group in benzo[c]pyrylium salts of type **92** (R^1 = Me) is never involved in the described condensation, even under drastic conditions, in contrast to the formation of distyrylpyrylium derivatives from α,α'-dimethylpyrylium salts (72BSF3173). The explanation, in accordance with the supposed mechanism, is a greater activation barrier for the formation of the deprotonated intermediate **95,** which occurs because of a perturbation of aromaticity in it, whereas the reactivity toward deprotonation is equal for both α- and α'-methyl groups in monocyclic pyrylium cations. The obvious energetic disadvantage of the *ortho*-quinonoid anhydrobase **95** is confirmed by quantum-chemical calculations by the modified intermediate neglect of differential overlap (MINDO/3) method, yielding a lower energy for the model $\alpha(1)$-methyleneisochromene **95** (R^3 = H) in comparison with $\alpha'(3)$-methyleneisochromene **93** (R^3 = H) (87UP1).

Intermediate formation of $\alpha(\alpha')$-methylenepyrans was also proposed for the isotopic-exchange reaction in the alkyl-substituted monocyclic pyrylium salts [82AHC(Suppl)]. These observations are in agreement with preliminary data on isotopic exchange in 1,3-dimethyl-benzo[c]pyrylium perchlorate **92** (87UP2), which reveal a very fast deuteration of the $\alpha(1)$-methyl group followed by deuteration of the $\alpha'(3)$-methyl group, whereas for symmetrically substituted monocyclic salts, the α,α'-deuteration is a simultaneous process [82AHC(Suppl)].

C. Reactions Beginning with Addition of Nucleophiles

1. *Survey*

The most interesting feature of different heterocyclic cations is their ability to be involved in recyclization reactions following the addition of nucleophiles. Among the whole variety of cations, the most distinguished in this respect are the pyrylium salts (79MI1; 87MI2). The conversions into substituted benzenes, pyridines, furans, etc. place this class of cations among the universal synthons [82AHC(Suppl)], and their recyclizations are characterized as "the amazing feature which makes organic chemistry such an attractive object for investigations" (79MI2).

The recyclization of monocyclic pyrylium salts **96** occurs in accordance with the addition of nucleophile-ring opening-ring closure (ANRORC) mechanism (78ACR462) which starts with the α-addition of the nucleophile[2] (Scheme 4).

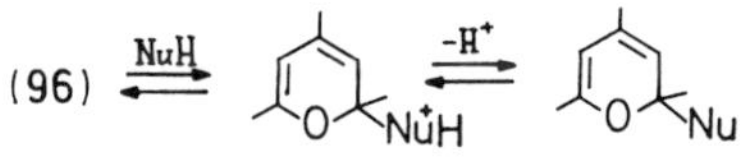

$$(96) \quad \xrightarrow{\text{NuH}} \quad (97a) \quad \xrightarrow{-\text{H}^+} \quad (97)$$

The benzo[c]pyrylium cation **30**, in contrast to the monocyclic analog **96,** has two nonequivalent α-positions as already mentioned (Section III,B). Evidently, pathways *a* or *b* (Scheme 5) are determined by the

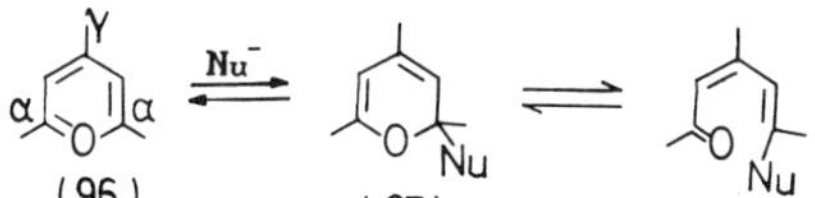

$$(96) \quad \xrightarrow{\text{Nu}^-} \quad (97) \quad \rightleftharpoons \quad (98)$$

SCHEME 4

[2] If the nucleophile is a neutral molecule of type HNR^1R^2, the protonated form **97a** is a precusor of adduct **97**. The conversion **97a→97** occurs because of excess NuH.

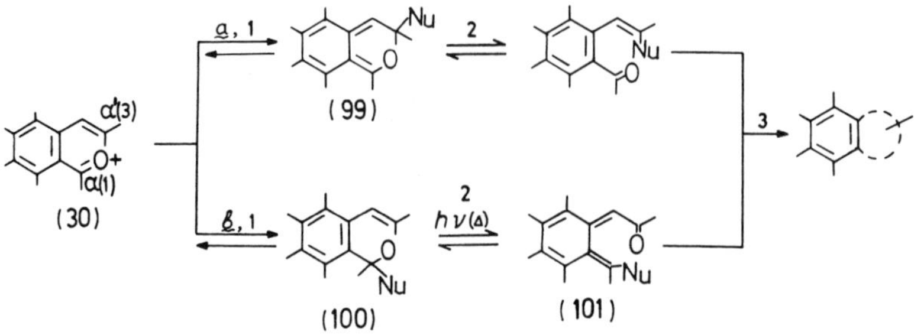

SCHEME 5

charge densities in positions $1(\alpha)$ and $3(\alpha')$ in **30** and by the thermodynamic stability of adducts **99** and **100**.

In analogy to the competing formation of anhydrobases **93** and **95**, the nucleophilic addition to position 3 (α') leads to the disturbance of aromaticity in the annelated benzenoid ring of adduct **99**. This process is therefore considered thermodynamically disadvantageous (85MI1), and until recently was believed to be impossible.

However, quantum-chemical calculations (87UP2) show that the difference in charge densities between positions 1 (α) and 3 (α') in 2-benzopyrylium cation **30** is not as great as reported earlier (70KGS1308) (cf. Section IV,C). From this data it is suggested that the attack on position 3 (α') might be possible, especially if the possibility of fast rearomatization of the annelated benzenoid ring exists, for instance the ring-opening in adduct **99** (pathway **a,2** in Scheme 5). At the same time, the addition of a nucleophile to position 1 (α) in **30** is preferred because, in this case, only the heterocyclic fragment of the cation loses its aromaticity. Moreover, according to existing opinion (85MI1), this process occurs more easily in comparison with the transformation of monocyclic cations **96,** which totally lose their aromaticity on nucleophilic addition. This is probably the reason for the lower tendency towards rearomatization of benzo[c]annelated adducts (cf. Section III,C,2 and 3) in comparison with the monocyclic analogs.

The further behavior of benzo[c]annelated adducts **100** depends on the structure of substituents in the initial cation **30**, nature of the nucleophile, thermodynamic parameters of the final products, and conditions of the experiment. The reaction may be stopped at the step of adduct **100** (*b,* 1 in Scheme 5) or may be continued with the formation of ring-opened intermediate **101** (*b,*2). However, the latter step has some specific features, in comparison with monocyclic pyrylium salts, as a consequence of the presence of the annelated benzenoid ring in benzo[c]pyrylium cations.

Thus, the thermally allowed electrocyclic ring-opening process occurs easily for monocyclic adducts **97** [82AHC(Suppl)], but in the case of benzo

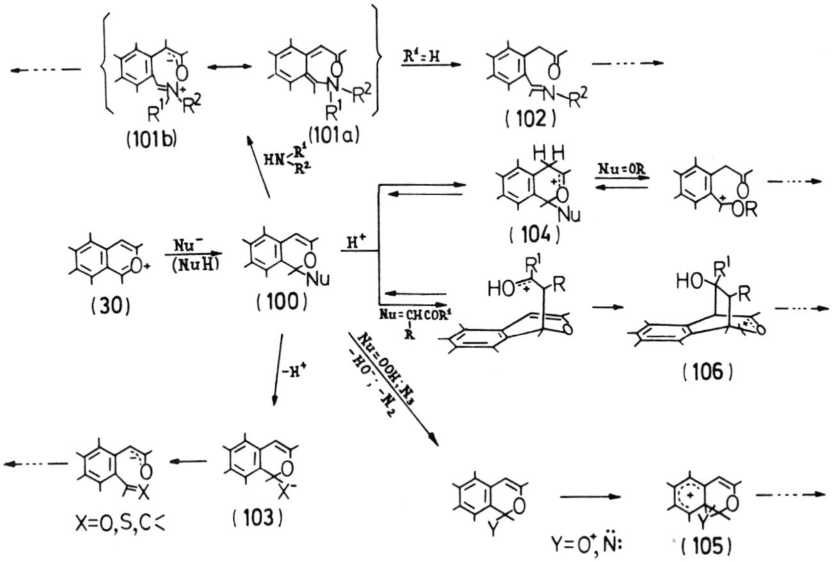

SCHEME 6

adducts **100** (Scheme 5) it is again accompanied by the disturbance of aromaticity in the annelated ring. Therefore, in the latter case, the direct electrocyclic ring opening occurs in irradiation and probably at elevated temperature. Under the usual conditions, this process may take place only in reactions with nitrogen nucleophiles where the *ortho*-quinonoid structure **101a** is stabilized by the resonance form **101b** (Scheme 6).

On the other hand, the annelated benzenoid ring gives other possibilities for the ring opening, and these pathways are determined by transformations in adduct **100**. Such transformations may include deprotonation of the nucleophilic fragment leading to **103**, protonation of the β-carbon atom in the vinyl fragment of the isochromene ring leading to **104,** or formation of *ortho*- (**105**) or *para*- (**106**) bridge-adducts[3]. In the former conversions, the sequence of the second (ring opening) and the third (ring closure) steps is interchanged by comparison with the ANRORC reaction.

2. Isolable Adducts and Their Reactions

The formation of isolable adducts in reactions of 2-benzopyrylium salts with nucleophiles depends on the structure of the nucleophile and conditions of the experiment. Thus, the stable adducts **108–113** were obtained as

---

[3] Formation of *para*-bridge-adducts of type **106** will be discussed in Section III,D,1 according to mechanism of the coordinated cycloaddition or *para*-cyclization (84UK1648).

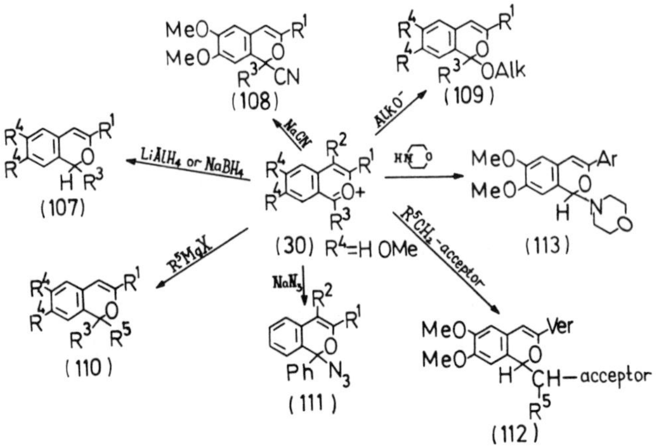

SCHEME 7

the result of interaction between 2-benzopyrylium salts **30** and sodium cyanide (87T409), alkoxide ions [66ACH(50)381], Grignard reagents (81KGS1177), sodium azide (80JHC847), some compounds possessing active methylene groups (90KGS315), and secondary amines (89TH2), respectively.

Interestingly, the addition of the hydride ion to benzo[c]pyrylium cations with the formation of adducts of type **107** occurs not only in reactions with lithium aluminum hydride (54JOC1533) or sodium borohydride (89KGS750), but also on short-time heating of the formate **30** (R^1 = ME, R^2 = Et, R^3 = Ver, R^4 = OMe) in formic acid [67ACH-(51)107] (Scheme 7).

It was not possible to isolate adducts of type **100** (Scheme 6), when the nucleophiles become attached by an α-heteroatom bound to a hydrogen atom as in the case of direct interaction of 2-benzopyrylium salts with the corresponding nucleophiles. However, this type of adduct may be synthesized by other reactions such as **114** (67JA5581) (cf. Section II,A).

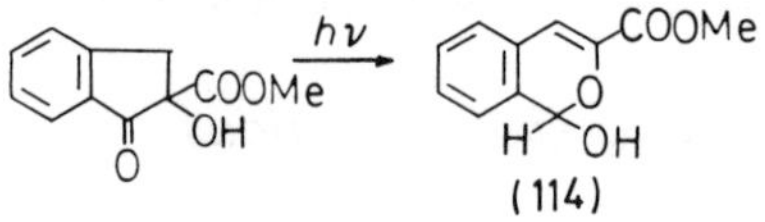

The transformations of adducts may be conventionally classified into two types. The reactions of the first duplicate transformations of the initial 2-benzopyrylium salts and are discussed in the corresponding sections, whereas the second type is different. The latter conversions of the isolable adducts are obviously determined by the annelated benzenoid ring; its presence causes resistance towards ring opening in comparison with monocyclic analogs [82ACH(Suppl)].

The result of transformations of adducts **107–113** depends on the nature of substituents at the geminally substituted center with sp³-hybridization, the structure of the reagent, and the experimental conditions. If the added nucleophile is a good leaving group, the action of strong acids in a nonnucleophilic medium leads to regeneration of the initial 2-benzopyrylium salts **30** [66ACH(50)381; 87T409]. That the protonation of such adducts

(108), (109), (113)   (30)

may occur at position 4 in the isochromene ring cannot be excluded. This course of protonation was established when the added nucleophile was a tightly bonded group such as the 1-methyl substituent in **110** (86UP3).

The protonated form **115** (R = H) was detected only by its ¹H-NMR spectral data; however, perchlorate **115** (R = COMe) was isolated, but it is also unstable and is readily converted into the β-acylated isochromene **116** on treatment with triethylamine. The latter compound may undergo acidic hydrolytic splitting to the initial adduct **110,** in accordance with transformations of β-diketones (70MI2). Thus, the carboxonium ion **115** (R = COMe) reacts by deprotonation without ring opening. If such a type of cation has a good leaving group at the geminal center, ring opening may take place [cf. Section III,C,4,b(i)].

However, the ring-opening process may occur even in the case of a tightly bonded group if the ring-opened intermediate is able to form a new ring under acidic conditions, for instance, as in a ring contraction in **117** leading to **118** [67ACH(51)107]. At the same time, the oxidation of **107** (Ar = Ver) is the main process in sulfuric acid which leads to the 2-

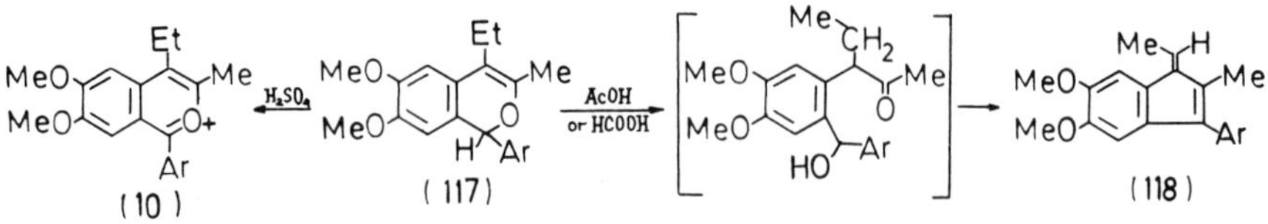

benzopyrylium salt **10** (54JOC1533), whereas isochromene **107** (R^1 = Ver, $R^2 = R^3$ = H, R^4 = OMe in Scheme 7) yields, regardless of the acid used, a complex mixture of unidentified compounds (89UP1).

The reaction of adducts with alkali hydroxides have been studied for cyanoisochromenes **108**, which are oxygen analogs of Reissert compounds described for the first time by Balaban and co-workers (87T409). In this case, sodium hydroxide, depending on its concentration in aqueous solutions and on the nature of R^3 in **108,** may act both as a base and as a

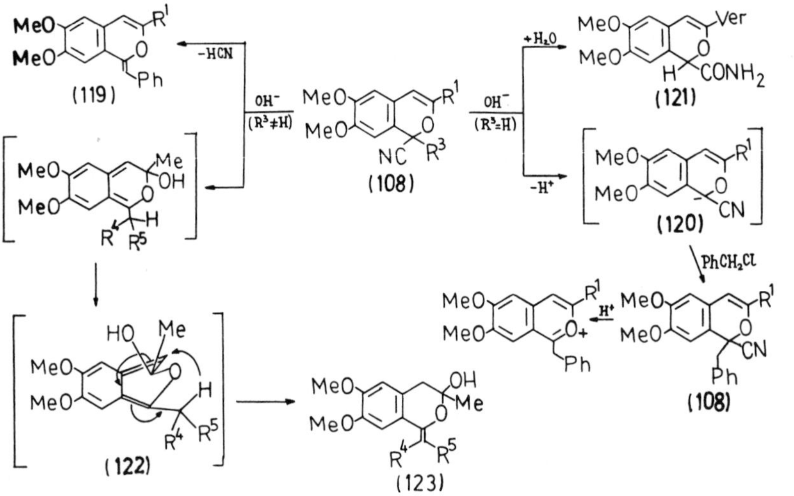

nucleophile (88T6217). The former course is adopted in the presence of an acidic proton (R^3 = H, CH₂Ph), which is split out with formation of an-hydrobase **119** or anion **120**. The latter is followed under conditions of the phase-transfer method of alkylation (87T409), affording 1-cyano-1-benzylisochromenes **108** (R^1 = Me, Ver; R^3 = Ch₂Ph). The application of these alkylation reactions may lead to various 1-cyano-1-alkyl(aryl)-substituted isochromenes, which are easily converted by perchloric acid into corresponding 2-benzopyrylium salts. At the same time, the usual hydrolysis of the cyano group in **108** (R^1 = Ver, R^3 = H) occurs in aqueous alkaline solutions, yielding amide **121**.

The nucleophilic addition of the hydroxide ion was described for cyanoisochromenes **108** with bulky substituents in position 1 (R^3 = Et,

CHMe$_2$). The steric hindrance for nucleophilic attack on position 1, in addition to approximately equal charge densities in positions 1 and 3, facilitates attack by the hydroxide ion to position 3 in **108,** with simultaneous elimination of a CN anion (a kind of tele-substitution (78JHC731). Then, a thermally allowed 1,5-hydrogen transfer in **122** leads to stable 1-alkylidene-3-hydroxy-isochromanes **123** (88T6217).

The direct substitution of the added nucleophile was described (76JA5581) for **114** with **124** as the final product; two authors postulated that the dissociation of the primary adduct takes place with the intermediate formation of cation **30.** Heating of the pseudobase **114** without nucleophiles leads to the acetal dimer **125.**

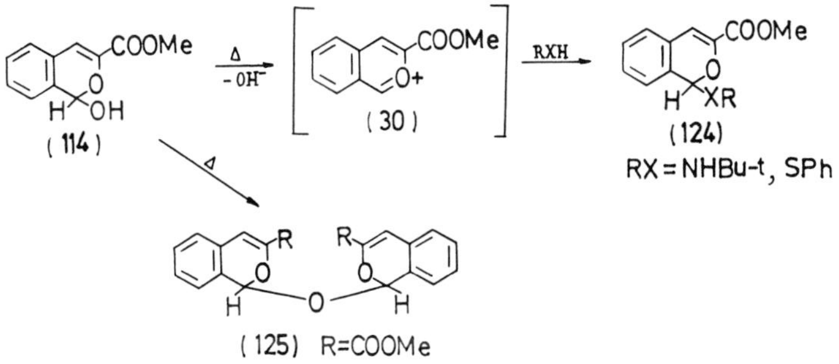

The photolysis of isochromenes of type **107** (76JA5581; 80JOC3524) or **109** (66JA4942) gives rise to electrocyclic ring-opening. The *ortho*-quinonoid intermediates of type **126** thus formed instantly undergo further conversion (80JOC3524).

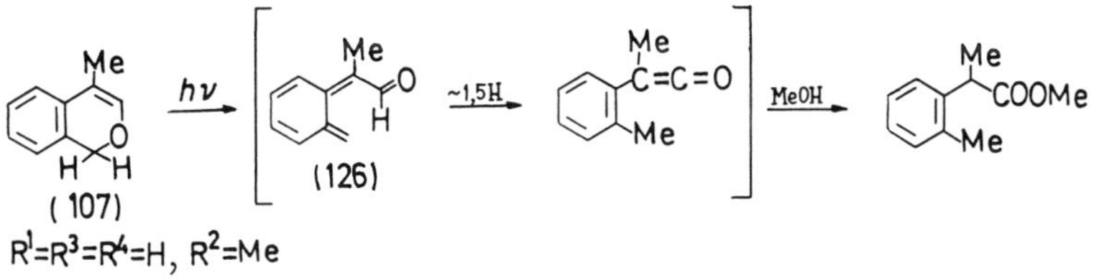

An analogous ring-opening occurs on photolysis of isochromenes **107** substituted with methoxy groups in the annelated benzenoid ring (87MI1). The thermolysis of azidoisochromenes **111** yields benzoxazepines **128** and anils **129** (80JHC847).

The formation of anils **129** occurs in accordance with the generally accepted mechanism for thermal rearrangements of aliphatic azides to imines which involves nitrenes **127** as the intermediate (74JA480). Nitrenes may undergo cyclization to aziridines **105** [cf. Section III,C,1 (Scheme 6) and Section III,C,4,b(ii)] which are converted into benzoxazepines **128**, the major products of this reaction. Isochromenes **108** react readily with dichlorocarbene to yield dichloropropane derivatives **130** (71BSF362; 89MI1), whose further behavior is determined by the substituents in the heterocyclic fragment of the molecule.

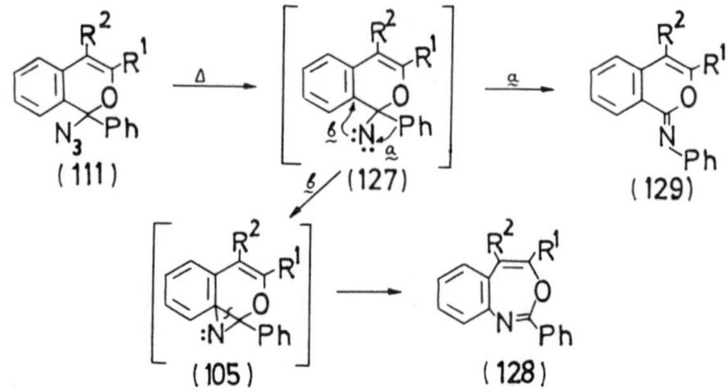

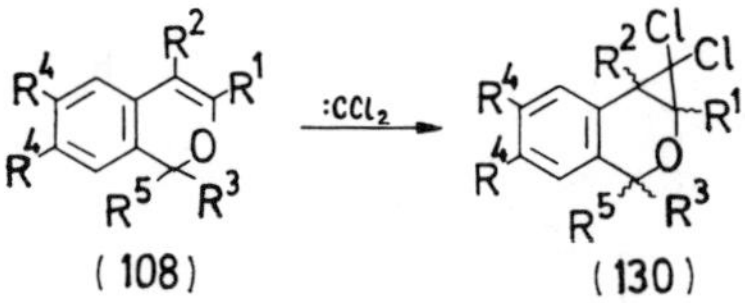

R^1=H, Alk, Ar; R^3=H,CN; R^4=H,OMe; R^5=Alk,Ar

3. *Reactions of Polycyclic Benzo[c]pyrylium Salts*

5-Oxoniachrysenes **67** are simultaneously 2-benzopyrylium and 1-benzopyrylium salts, but maintain a distinct chemical behavior. 5-Oxoniachrysene salts **67** are interesting mainly as precursors of benzo[c]phenanthridine derivatives **132**, as mentioned in Section II,C,6. However, in contrast to 2-benzopyrylium salts (cf. Section III,C,4,a), compounds **67** react with nitrogen nucleophiles to form stable adducts **131** or products of anomalous ring-opening, namely α-naphthols **66** and esters **70** (77KGS1176).

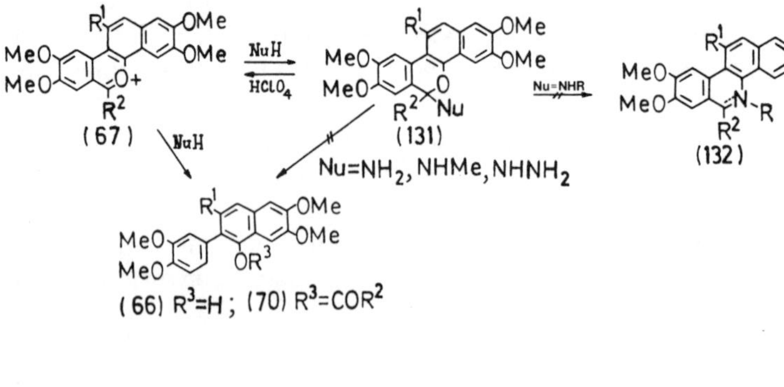

The ability of 5-oxoniachrysenes **67** to form stable adducts with ammonia, methyl amine, and hydrazine is not usual for 2-benzopyrylium (cf. Section III,C,2), but is often encountered for 1-benzopyrylium (chromylium) salts (51MI1). At the same time, no rupture of the C_6—C_{ring} bond was observed for 2-benzo- or 1-benzopyrylium salts under the conditions used. A remote similarity to C_6—C_{ring} bond rupture may be seen in reactions of 2-benzopyrylium salts with sodium azide (Section III,C,2) or with hydrogen peroxide (Section III,C,4,b,ii).

Since the rupture of the C_6—C_{ring} bond complies formally with the *ipso*-protonation of the *c* side-annelated veratryl ring, the occurrence of this process may be interpreted in terms of a specific addition of the nitrogen nucleophile. Thus, one of the hydrogen atoms of the added nucleophile is above the annelated ring plane, and the π-complex **133** is formed. The rehybridization of the C_6 atom at the geminal center thus formed disturbs, probably insignificantly, the total system of conjugation. In this case, the annelated ring plays the role of an internal base and accepts the proton whose mobility is increased because of the positive charge at the heteroatom of the added nucleophile. If deprotonation of the fragment of added nucleophile in **133** occurs under the action of excess nucleophile itself, the stable adducts **131** are formed, and they are easily converted into the initial salts **67** on treatment with 70% perchloric acid.

4. Reactions Followed by Formation of New Ring Systems

Since the reactions of 2-benzopyrylium salts with nucleophiles (in accordance with the ANRORC scheme) are seldom stopped at the ring-opening step (as a rule, formation of 1,5-dicarbonyl compounds takes place), it is expedient to examine at once the reactions completed by formation of a new ring. Such a process may occur with participation of both the added nucleophile and the α-alkyl substituent in the heterocyclic fragment of the cation. For the latter case, an intramolecular aldol-type condensation takes place where an excess of nucleophile, as a rule, plays the role of base. Along with these two general, but often competing pathways of recyclization, conversions were observed in which the formation of a new ring system occurs without participation of either the added nucleophile or the previously present alkyl substituents.

As reagents causing intramolecular recyclizations of 2-benzopyrylium salts, nucleophiles centered at nitrogen, oxygen, sulfur, or carbon atoms were used. As a fairly general rule, the ring opening of the heterocyclic moiety leads to an *ortho*-quinonoid structure, as shown in more detail in Section III,C,4,iv (Structure **201**).

a. *Reactions with Nitrogen Nucleophiles.*

i. *Ammonia.* The reaction of 2-benzopyrylium salts with ammonia is most important, practically, because it leads to the formation of isoquinolines. The latter compounds, especially those having methoxy groups in the annelated benzenoid ring, possess strong biological activity, and they are the main structural units for construction of different types of alkaloids (78MI1). The interaction between 2-benzopyrylium salts and ammonia [68DOK(181)345] is shown in Scheme 8 and is similar to the process known for monocyclic pyrylium salts [82AHC(Suppl)].

For monocyclic cations, the unique pathway of this reaction occurs regardless of conditions and is often used to identifiy pyrylium salts by converting them into pyridines. However, for 2-benzopyrylium salts, this reaction, under standard conditions (aqueous or alcoholic ammonia), is accompanied by side transformations whose extent depends on the nature of substituents in the cation.

Thus, Müller and co-workers [66ACH(50)381; 66ACH(50)387; 67ACH(52)261] found that reaction of salts **10** with ammonia in absolute

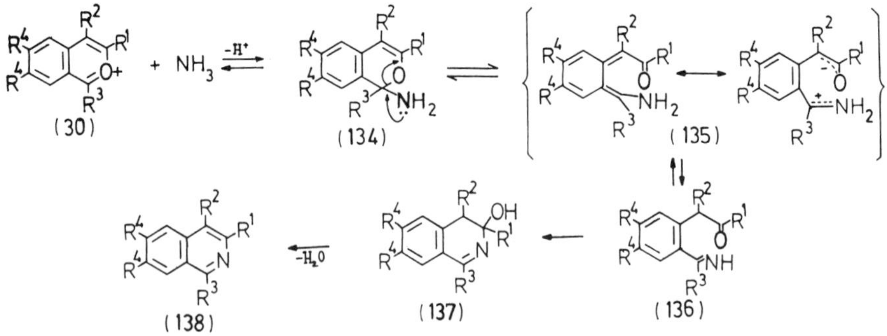

SCHEME 8

alcohol stopped with the formation of 3-hydroxy-3,4-dihydroisoquinolines **137** regardless of the temperature. These authors showed that conversion of **137** into isoquinolines **135** occurs on heating in acetic anhydride. On refluxing compound **137** in alkaline aqueous solutions, another recyclization was observed with β-naphthols **139** as the result, whereas heating of **137** with methyl iodide regenerated the initial 2-benzopyrylium cation **10**.

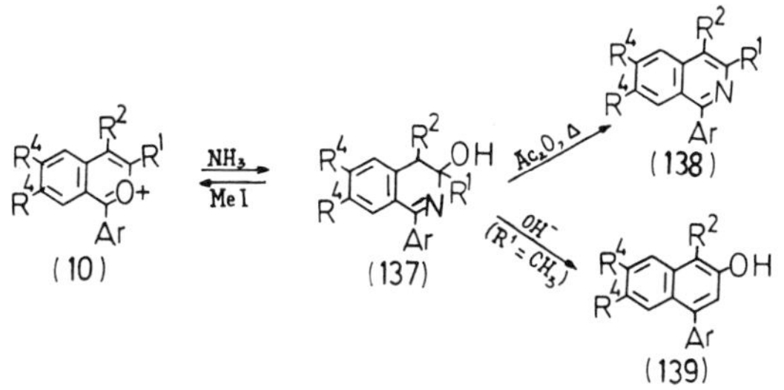

The interaction between ammonia and 1-R^3-3-aryl-substituted benzo[c]pyrylium salts **30**, as well as 3,4-diaryl-substituted indeno[e;1,2-]-benzo[c]pyrylium, and other substituted cations having a similar fragment is more complicated, especially in hydroxyl-containing solvents. In these cases, mixtures in different ratios are obtained that contain isoquinolines **138**, 3-hydroxy-3,4-dihydroisoquinolines **137**, OR-adducts **109**, anhydrobases **119**, diketones **29**, α-naphthylamines **140**, and α-naphthols **141** (88UP2).

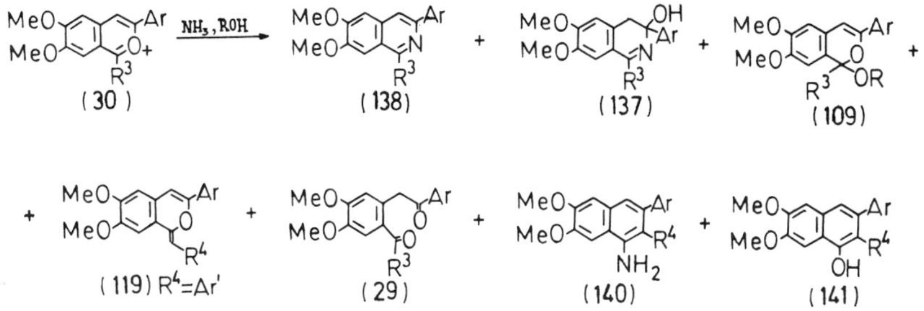

The easier nucleophilic addition to 2-benzopyrylium salts, compared with monocyclic analogs, leads to loss of selectivity by the cation towards nucleophiles present in the mixture when they possess approximately equal nucleophilicity (NH_3, H_2O, ROH). This forms stable adducts **109** (R = Alk), under these conditions, and labile pseudobases **109** (R = H) which undergo ring opening to diketones **29**. The benzo[c]annelation also plays its role in the easy formation of anhydrobases **119**, which are stabilized by two aryl rings.

Side transformations also take place after heterocyclic ring-opening.

The latter process probably occurs by a loosening action of a vacant electron pair of the nitrogen atom, as shown for intermediate **134** in Scheme 8. This electron pair stabilizes followed by the ring-opened intermediate **135**; deprotonation of the added nucleophile occurs after ring opening.

The reaction step of heterocyclization of iminocarbonyl compounds **136** is probably the rate-limiting stage, as for monocyclic pyrylium salts [82AHC(Suppl)]. Its rate is determined by the electrophilicity of the carbonyl group and by steric factors. In the case of a bulky 3-aryl group, especially with donor substituents, the rate of heterocyclization should be retarded, and because of this influence, the behavior of **136** becomes dependent on the nature of R^3. Thus, if R^3 = Me or Et, the possibility of intramolecular aldol-type condensation is real, and α-naphthylamines **140** are formed (82KGS552).

The easy implementation of this pathway can be explained by formation of the intermediate enamine ions of type **142**, which are stabilized due to conjugation with the aromatic ring. A similar stabilization is impossible for the corresponding anions of monocyclic ring-opened intermediates **98** [Section III,C,1, (Scheme 3)], and this is why the formation of anilines in the reaction of pyrylium salts with ammonia was seldom observed [82AHC(Suppl); 89RRC1425]. An excess of ammonia and elevated temperatures favors the formation of anion **142**, which is more nucleophilic than the iminic nitrogen atom in **136**. A retarding of heterocyclization may also lead to hydrolysis of the imino group in **136** and to the formation of 1,5-diketones **29**. If R^3 = CH_2R^4, this process may be completed by formation of α-naphthols **141**, and their increased yields in this case are due to 1,5-diketones **29** formed from pseudobases **109**.

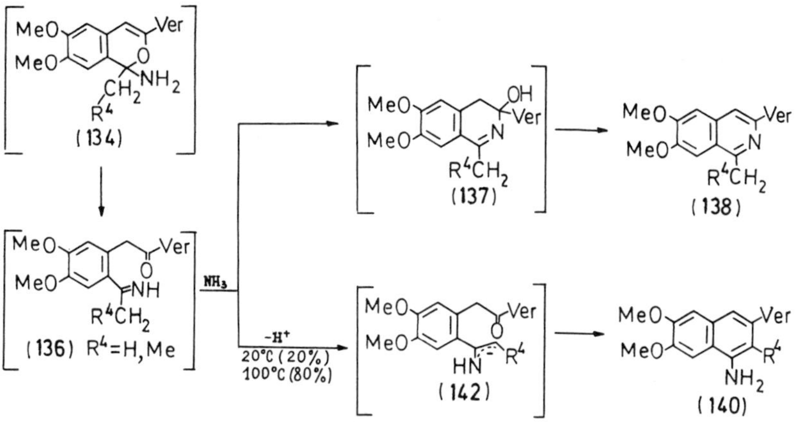

Evidently, the stability of 3-hydroxy-3,4-dihydroisoquinolines **137**, formed as the result of heterocyclization, is also determined by the annelated benzenoid ring. The lower tendency toward aromatization for these compounds, compared to monocyclic analogs, leads to the ability of **137** to react as a cyclic azomethine. The addition of a molecule of nucleophile to the C=N bond causes opening of the isoquinoline ring and formation of a new ring system (for instance, α-naphthols **141** in alkaline aqueous solutions). Such conversions occur even under conditions of the recyclization reaction of 2-benzopyrylium salts, namely, on heating **137** in alcoholic ammonia; a mixture of isoquinoline **138** and α-naphthylamine **140** results (88MI1).

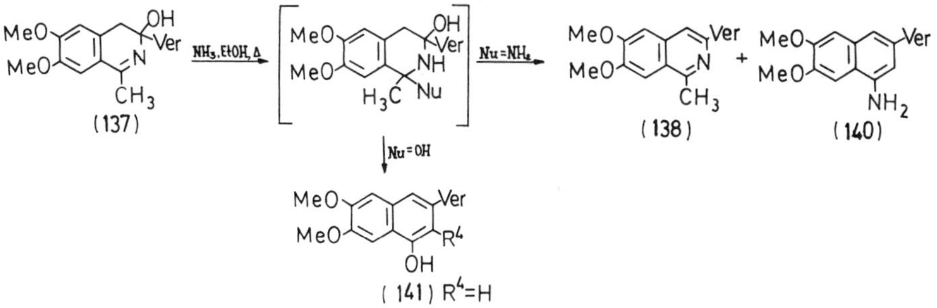

Among the different methods for controlling competing transformations in the interaction between 2-benzopyrylium salts and ammonia (such as varying temperature in the different steps of reaction and using different solvents), the most effective is the application of acid catalysis, and the best experimental condition is heating the mixture of 2-benzopyrylium salt and ammonium acetate in acetic acid. In this case, the high selectivity of the whole process is determined by the weak nucleophilicity and basicity of the acetate ion, the absence of excess free ammonia, the suppression of deprotonation processes, and the activation of the carbonyl group in the ring-opened intermediate **144** for nucleophilic attack because of its protonation, is the only pathway to the intermediate of 3-hydroxy-3,4-dihydroisoquinolines **137**.

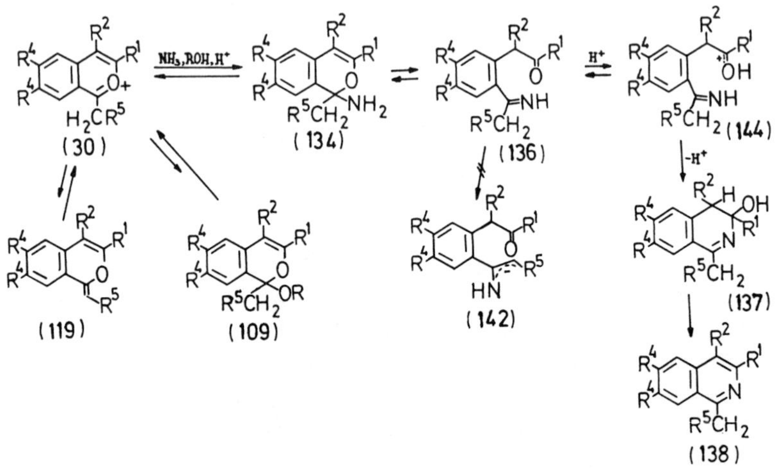

In some cases, another version of acid catalysis was used: aqueous or alcoholic ammonia was added to a suspension of 2-benzopyrylium salt in alcohol, and after a few seconds, an excess of acetic acid was added to the mixture which was then heated to boiling. After chilling, isoquinolines **138** were obtained in good yield as free bases or salts. If the carbonyl group in intermediates **136** is electrophilic enough, and the hydrogen atom in position 4 is sufficiently mobile, the conversion of 2-benzopyrylium salts to isoquinolines occurs easily without acidic catalysis, for instance, in the case when this intermediate is **145** (75KGS25).

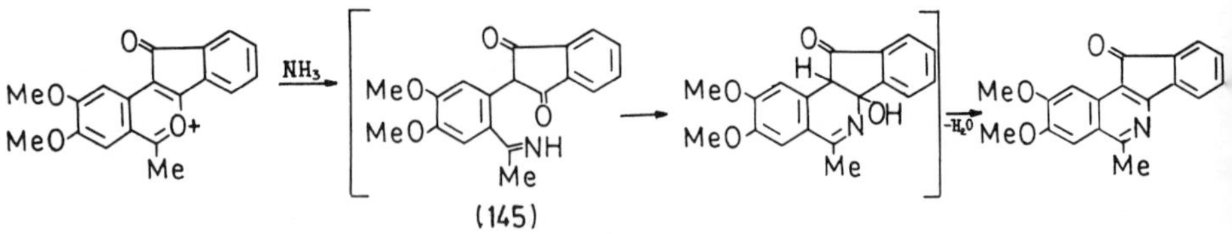

Interestingly, the recyclization of **146**, having hydroxyl groups in the 3-aryl substituent, proceeds readily also without acidic catalysis affording isoquinolines **138** (70KGS1013). Probably, deprotonation of **146** is the primary process in this case, and a quinonoid compound **147** thus formed adds a molecule of ammonia similarly to other quinonoid forms (cf. Section III,E,2) affording **148**, which is converted into isoquinoline **138**.

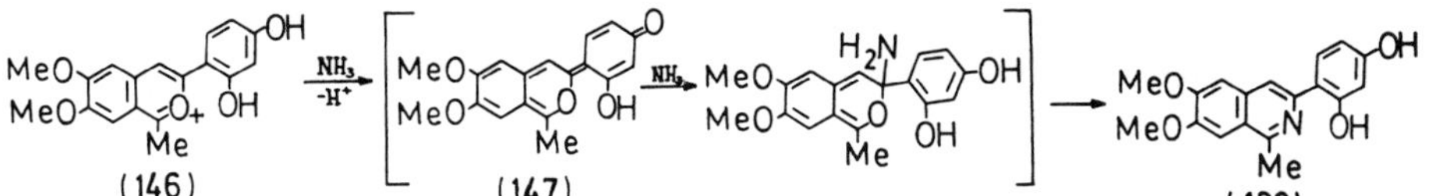

One can conclude that the choice of methods for O → N exchange in 2-benzopyrylium salts independent of their structure, gives the possibility of converting all known types of 2-benzopyrylium cations into isoquinolines. This approach to isoquinolines has obvious advantages, especially for the synthesis of 3- and 4- (or 3,4-) substituted isoquinolines, which are difficult to synthesize using classical methods (80T1279). In this connection, one should highlight syntheses of 3-arylisoquinolines, which are the main structural units in the construction of tetracyclic isoquinoline alkaloids (78MI1; 80T1279).

ii. *Primary amines.* The formation of isoquinolinium salts **152** in reactions of 2-benzopyrylium salts **30** with the primary amines obviously proceeds in a manner similar to the interaction with ammonia (Section III,C,4,ai) as shown in Scheme 9. However, the number of side transformations is smaller in this case than for the reaction with ammonia. Thus, no competing addition of other nucleophiles or formation of anhydrobases of type **119** occurs; 3-hydroxyisoquinolinium derivatives **151** were also not isolated.

At the same time other processes compete with the formation of isoquinolinium salts **152,** and the result is evidently determined by the characteristic properties of the ring-opened intermediate **150** (Scheme 9). The tendency towards formation of α-naphthylamines of type **153** is especially strong. This fact can be explained, on one hand, by higher steric interfer-

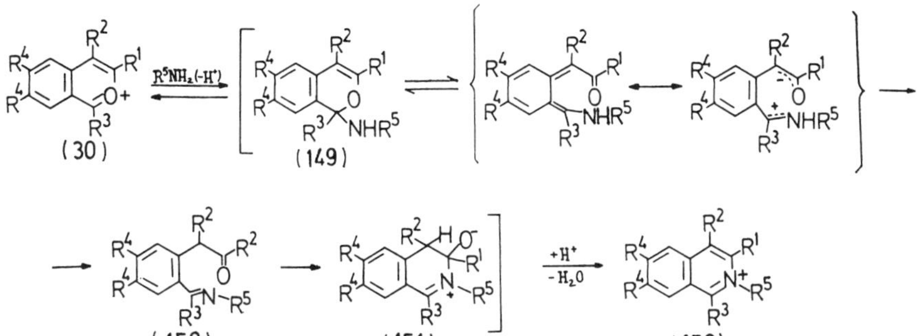

SCHEME 9

ence in the heterocyclization step because of the presence of a substituent at the nitrogen atom in imine **150,** unlike the unsubstituted imine **136** (Scheme 8). On the other hand, the higher basicity of aliphatic amines is also a factor. The latter property plays its role already in the reaction of methylamine with salts **30** (R^1 = H,Me), devoid of bulky substituent in position 3 (70KGS1308; 85KPS815).

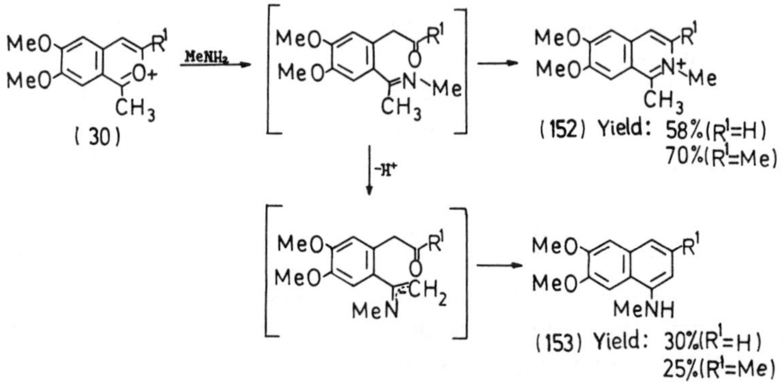

With the presence of aryl or aralkyl substituents in position 3 (and 4), the additional steric factors lower the rate of heterocyclization. As a result of these steric effects in **150,** the carbocyclization to an aniline derivative often becomes the only process, especially with methylamine (70TH1; 71KGS1437; 88MI1). This conversion occurs even on treatment of 1 mol of salt **30** with 2 mol of methylamine, which is enough only for the ring opening in intermediate **149.**

In the reaction with a less basic amine such as aniline, mixtures of naphthylamines **153** and isoquinolinium salts **152** were obtained (71KGS1437; 88UP2). The alternative pathway of formation of naphthylamines **153** from isoquinolinium salts **152** by "isomeric recyclization" (81T3425) is completely excluded, since model experiments with isoquinolinium salts showed no reaction under the conditions of synthesis of naphthylamines **153.**

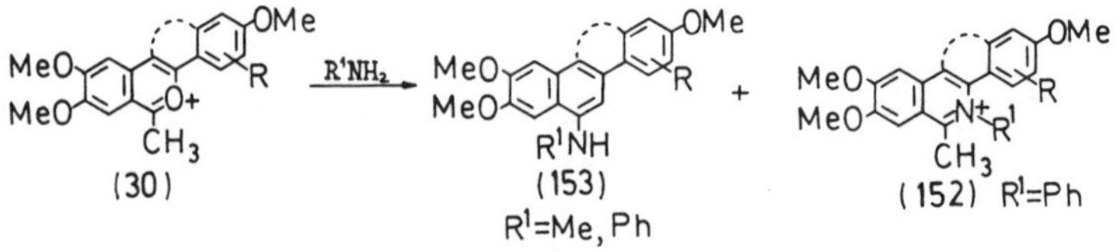

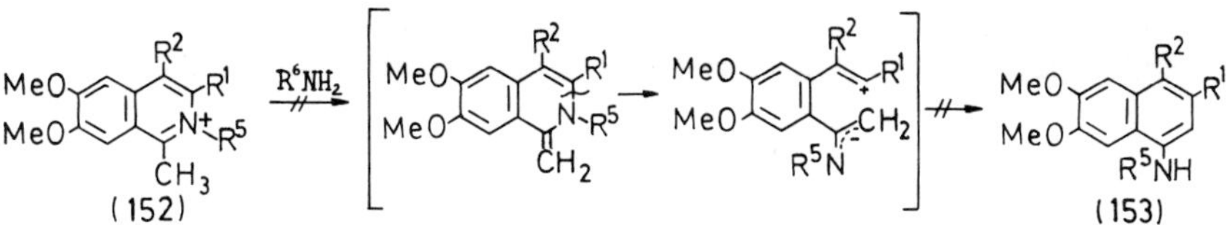

As in the reactions with ammonia (preceding section), primary amines do not lead to intramolecular condensations involving the 3-methyl group of 1-aryl-substituted salts **154** (88UP2).

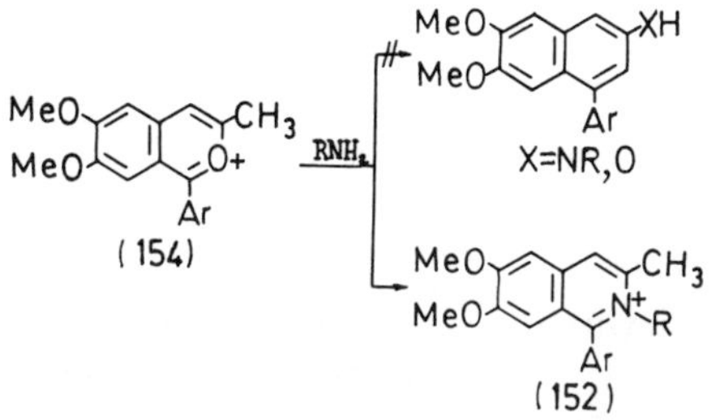

1,3-Diphenylbenzo[c]pyrylium perchlorate, in reaction with primary amines, forms isoquinolinium salts **152** in medium yields (71CB2984).

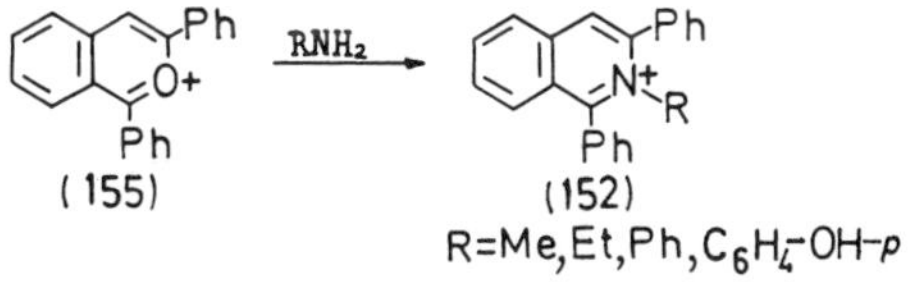

R=Me,Et,Ph,C₆H₄-OH-p

At the same time, 1,3-diarylbenzo[c]pyrylium salts **156** with electron-donor groups in the aryl substituents and in the annelated benzenoid ring give only diketones **158** under the same conditions (70TH1). This fact may be explained by a low rate of heterocyclization, and as a consequence, the hydrolysis of intermediate **157** leads to **158.**

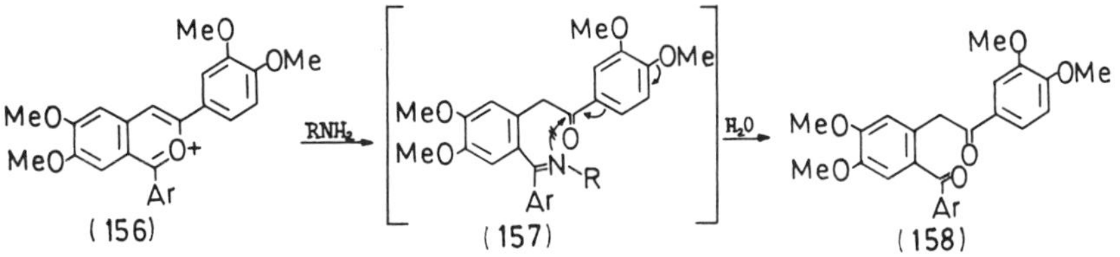

If the 2-benzopyrylium cation has no substituent in position 1, the interaction of the salt **159** with amines having a primary α-carbon atom or with aniline leads to the formation of isoquinolinium salts **160** in high yield (81KGS313; 89KPS75). However, the use of amines with secondary α-carbon atoms, e.g. alanine, does not yield the heterocyclization products, probably because of steric interference (89KPS75).

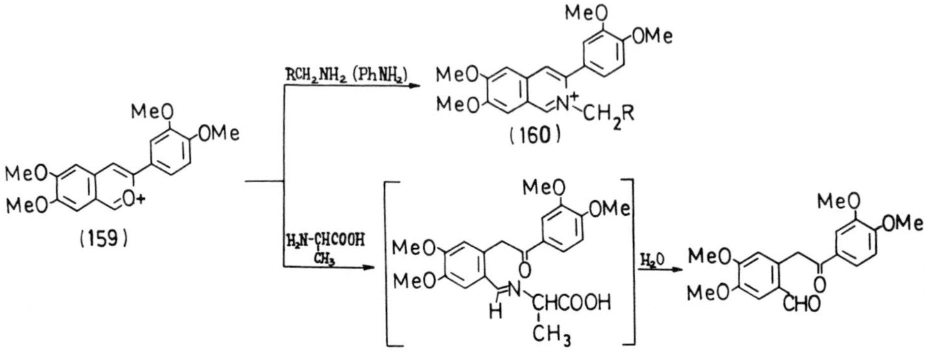

The effect of the structure of the alkyl substituent in primary amines on the course of reaction with these amines was observed also for monocyclic pyrylium salts [66T(57)9; 87TL3143; 89RRC1425]. The interaction of benzo[*c*]pyrylium-4-oxide **15** with aniline follows a different recyclization pathway, which is interpreted as shown here (64JA3814).

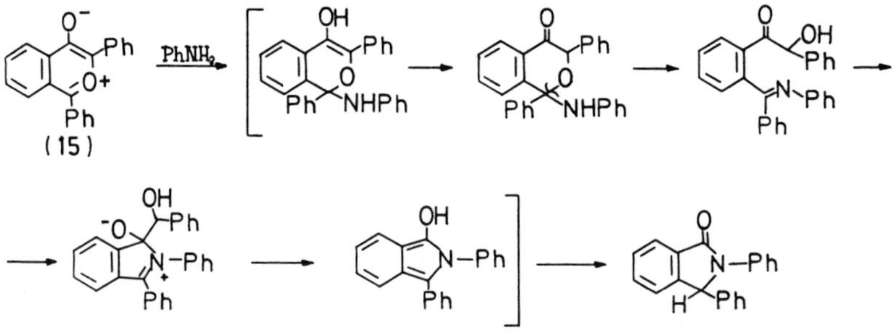

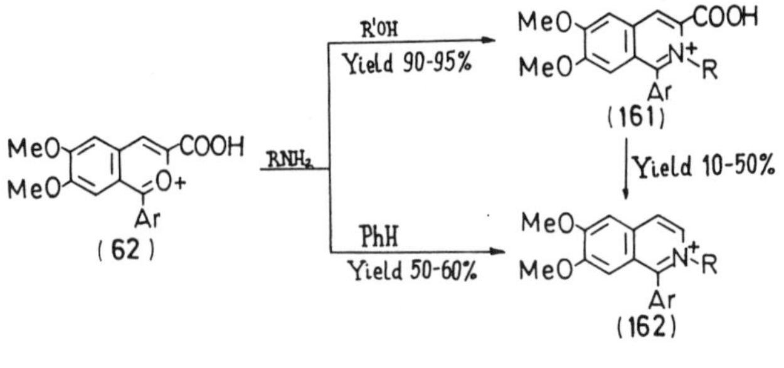

A specific behavior in reactions with primary amines was described for 1-aryl-3-carboxy-benzo[*c*]pyrylium salts **62** (89KGS454). While their monocyclic analogs undergo decarboxylation with heteroatom exchange, independent of the nature of the solvent (82JOC492; 82JOC498; 84KGS1428), salts **62** form 3-carboxy-subsituted isoquinolinium salts **161** in alcohol or chloroform and decarboxylated derivatives **162** in benzene.

Since the decarboxylation of **161** to **162** proceeds in a poor yield, it was suggested that formation of **162** in benzene occurs directly from 2-benzopyrylium salts **62** through primary nucleophilic attack by amine in position 3. In this case, an enamine fragment of pyruvic acid appears in the ring-opened intermediate **163**, which undergoes easy decarboxylation (82TL459). The vinylic carbanion **164**, formed by the loss of carbon dioxide, captures a proton by intra- or intermolecular process, then heterocyclization takes place.

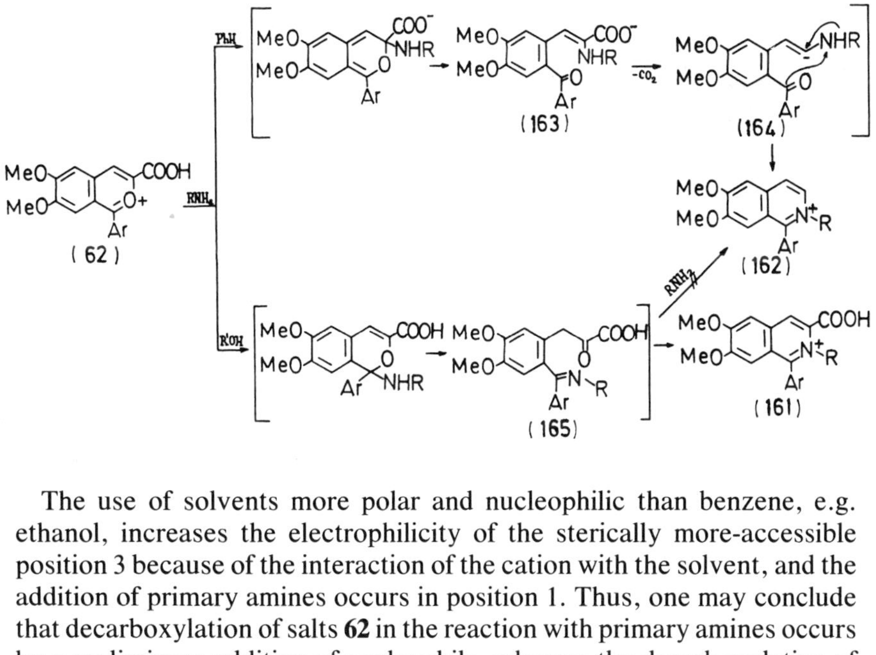

The use of solvents more polar and nucleophilic than benzene, e.g. ethanol, increases the electrophilicity of the sterically more-accessible position 3 because of the interaction of the cation with the solvent, and the addition of primary amines occurs in position 1. Thus, one may conclude that decarboxylation of salts **62** in the reaction with primary amines occurs by a preliminary addition of nucleophile, whereas the decarboxylation of isoquinolinium salts **161** to **162** proceeds by a ylide mechanism as is shown here. As known, the latter pathway requires the use of stronger bases and more elevated temperatures (79MI3).

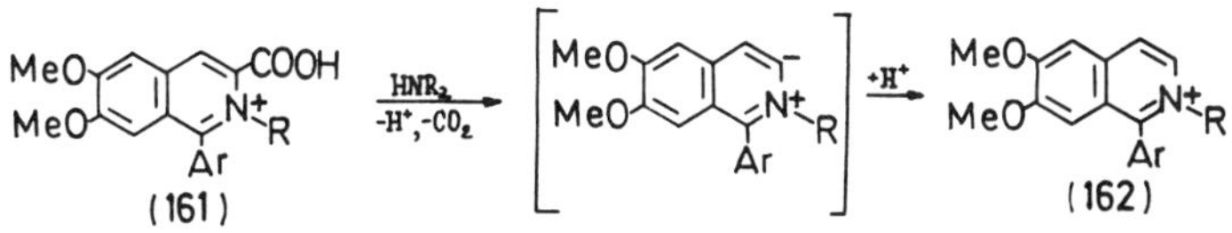

The application of acetic acid catalysis in reaction of 2-benzopyrylium salts with primary amines[4], in contrast to the reaction with ammonia, does not lead to a simple result. Thus, if in **30** R^3 is not Alk, excluding the alternative formation of α-naphthylamines of type **153,** the use of acetic acid catalysis leads to isoquinolinium salts **152** in high yields (89KPS75), whereas without acetic acid, diketones **166** were the only products of interaction between 2-benzopyrylium salts **30** and primary amines.

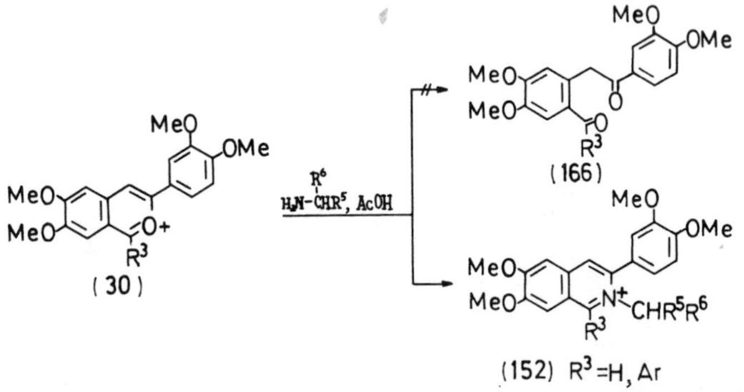

In **30** (R^3 = Alk), given that there is possibility of a competing formation of α-naphthylamines **153,** the application of acetic acid catalysis, in some cases, leads to an adverse effect by increasing the yields of carbocyclization products **153.** Such a marked influence of acid catalysis may be explained by the greater susceptibility of a substituted imino group in the ring-opened intermediate **150** toward exhibiting basic, but not nucleophilic, properties, in comparison with ammonium intermediate **136** (Scheme 8). This may lead to the formation of more nucleophilic enamines

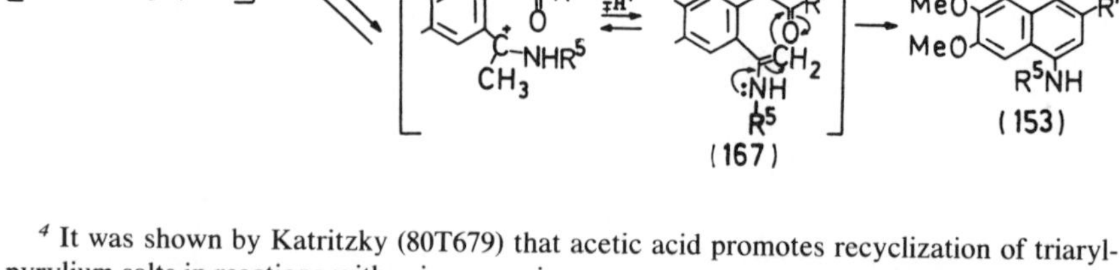

---

[4] It was shown by Katritzky (80T679) that acetic acid promotes recyclization of triarylpyrylium salts in reactions with primary amines.

167 and then to carbocyclization as the result. The preceding scheme provides, for the first time, an explanation for the different course of reactions with ammonia or primary amines and for the role of acetic acid as catalyst.

The formation of α-naphthylamines **140** and **153** in reactions of 2-benzopyrylium salts with ammonia and primary amines is not always an undesirable process. The synthetic application of α-naphthylamines is connected with the chemistry of benzo[c]phenanthridine alkaloids **132** (78MI1; 81H474).

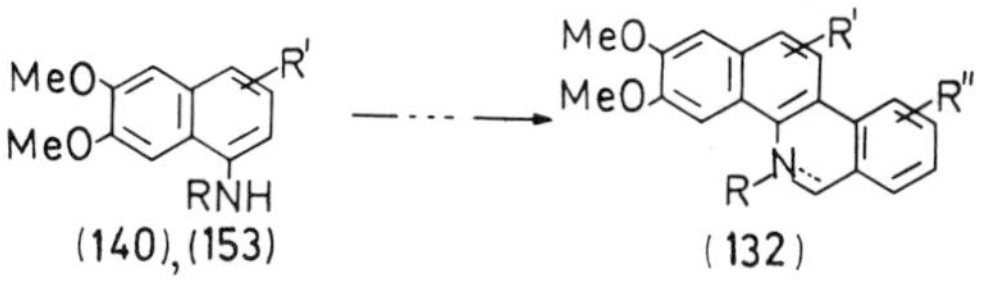

This approach to the synthesis of such alkaloids is considered a good prospect (78MI1), but its development was hampered by the absence of suitable syntheses of the initial amines (78MI2; 81H221). At the same time, by using easily available 3-aryl-isoquinolinium salts **168** unsubstituted in position 1^5, the synthesis of dibenzo[a,g]quinolizinium derivatives **169** was carried out, including the alkaloid dehydronorcoralidine **170** (89KPS75).

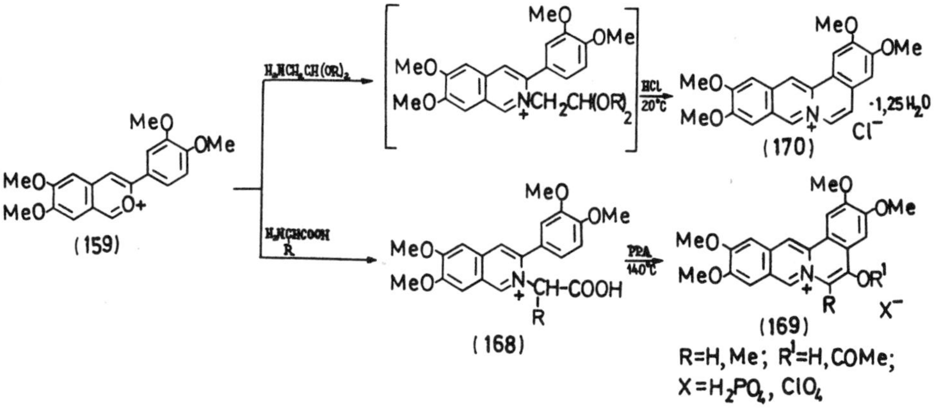

Another approach to the synthesis of dibenzo[a,g]quinolizinium derivatives includes interaction of 3-(2-carboxymethylaryl)-benzo[c]pyrylium

---

^{5} The formation of these salts by quaternization of the corresponding isoquinolines proceeds in a poor yields, probably because of a steric influence of the bulky 3-aryl substituent (69IJC527).

salts **56** with ammonia. In this case, the heterocyclization step is accompanied by intramolecular N-acylation affording **171** (77KGS1481; 82KGS1381).

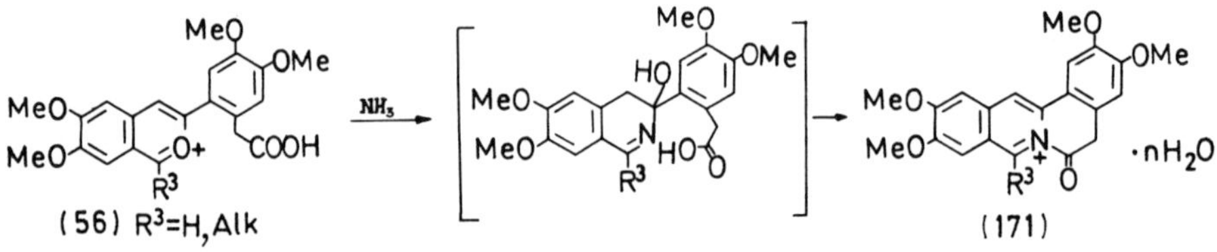

The interaction of primary diamines with 2-benzopyrylium salts that do not contain functional groups in the cation leads to bisisoquinolinium salts **172** (70KGS1308; 81KGS1608).

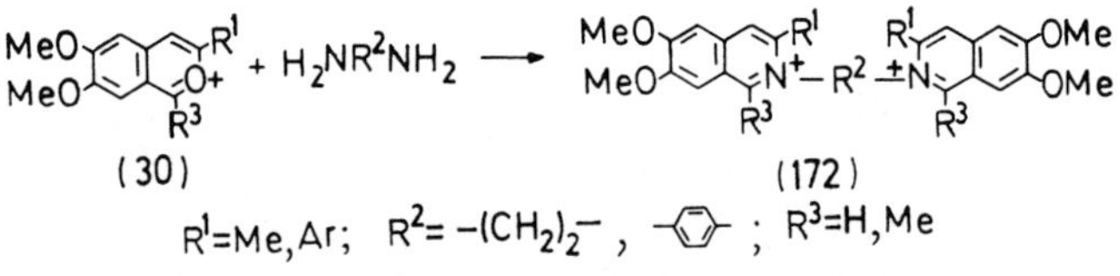

R^1=Me,Ar; R^2= -(CH$_2$)$_2$- , -◯- ; R^3=H,Me

The analog of the alkaloid dauricine **173** is a representative of another group of bisisoquinolinium salts, and this type of compound is easily obtained from the corresponding 2-benzopyrylium salt **43** (77KPS714).

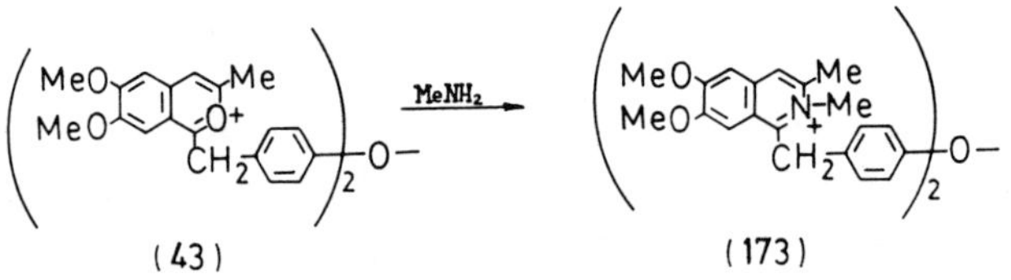

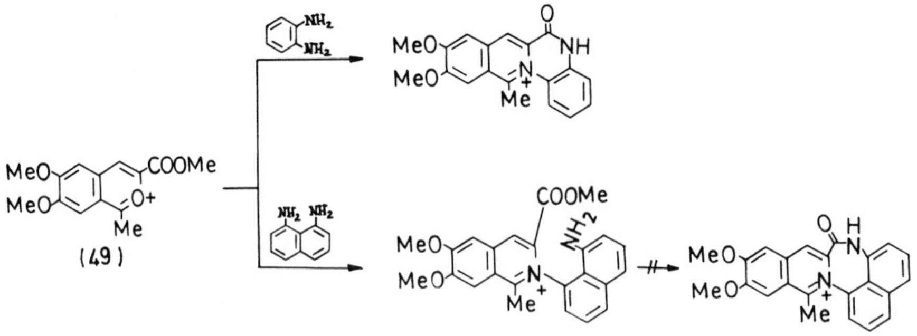

The presence of a functional substituent such as carbalkoxy in position 3 of benzo[*c*]pyrylium salts **49**, or alkoxy in compound **52**, or acylamino in compound **50** determines the formation of mono- or bis-condensed systems. The result depends both on the nature of the amine used (74KGS342) and on experimental conditions (74KGS1695).

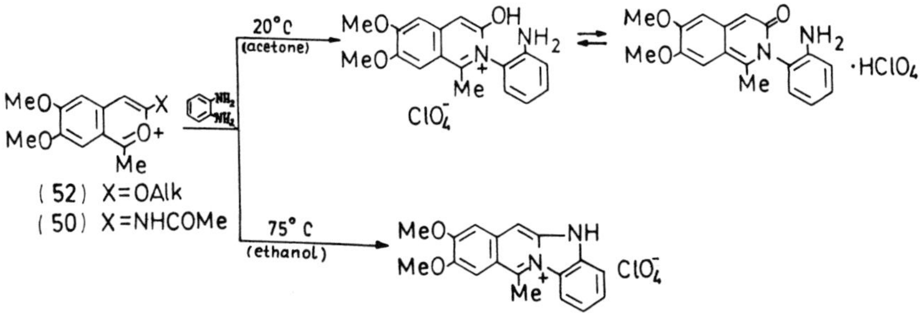

iii. *Hydrazine and hydroxylamine derivatives*. 1,3-Dimethylbenzo-[c]pyrylium salts **35** react with hydrazine, as in the interaction with ammonia or primary amines (Section III,C,4,a i,ii), and hydrazine acts as a 1,1-dinucleophile in this reaction, yielding isoquinolinium derivative **174** (R = H) (70KGS1308).

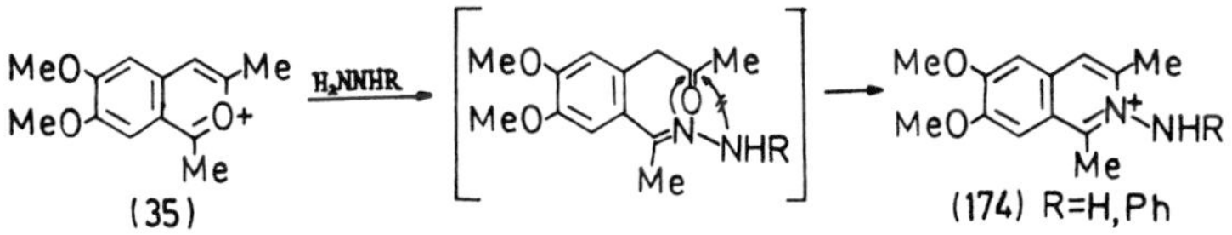

For 1-aryl-substituted benzo[c]pyrylium salts **10,** hydrazine acts as a 1,2-dinucleophile with the formation of benzodiazepines **175** (69MIP1; 74CB3883).

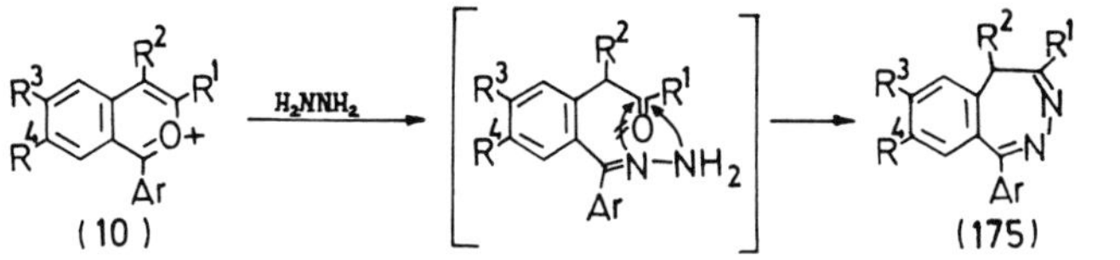

Similar differences leading to pyridinium salts or diazepines have been observed in the reaction of monocyclic pyrylium salts with hydrazine [82AHC(Suppl)]. Diazepine **175** (R^1 = Me, R^2 = Et, Ar = Ver) is known as the tranquillizer drug "Tophizopame" (Gideon Richter, Hungary).

The reaction of a 4-cyano-substituted salt with hydrazine gives rise to the product of a different transformation of 2-benzopyrylium cation (86KGS999), leading to a pyrazole ring. The authors (86KGS999) suggested that the cyano group participates in the construction of the new six-membered heterocyclic ring in **177,** assuming an intermediate **176** is formed. In the view of the present reviewers, this intermediate **176** could arise by an attack of hydrazine in position 3, which involves several steps.

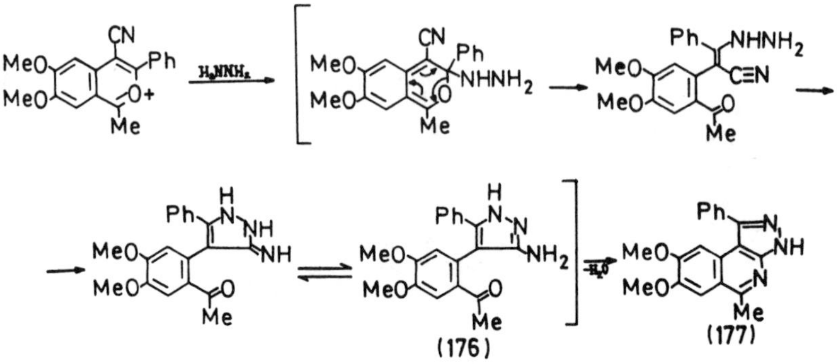

Different results, depending on the structure of the initial 2-benzopyrylium salt, were obtained also in reactions with phenylhydrazine. Thus, the 1,3-dimethyl-substituted cation **35** is converted into an isoquinolinium salt **174** (R = Ph) (70KGS1308), whereas the 1,3-diphenyl derivative **178** forms an indole **180** via the bishydrazone **179** by the Fischer reaction (71CB2984).

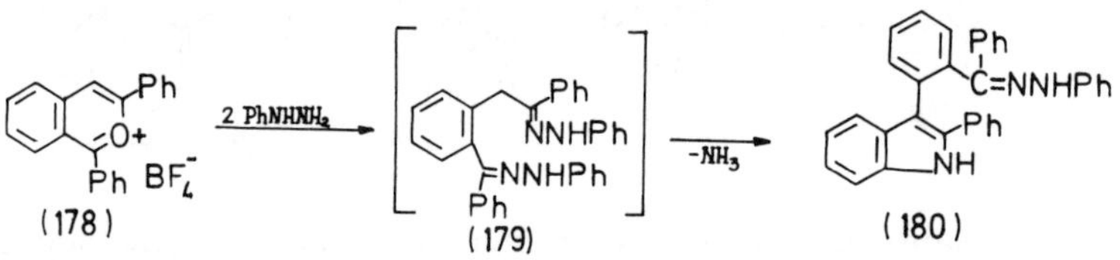

The only known example of a reaction of 2-benzopyrylium salts with hydroxylamine is a normal O → N exchange and formation of isoquinoline *N*-oxide **181** (70KGS1308).

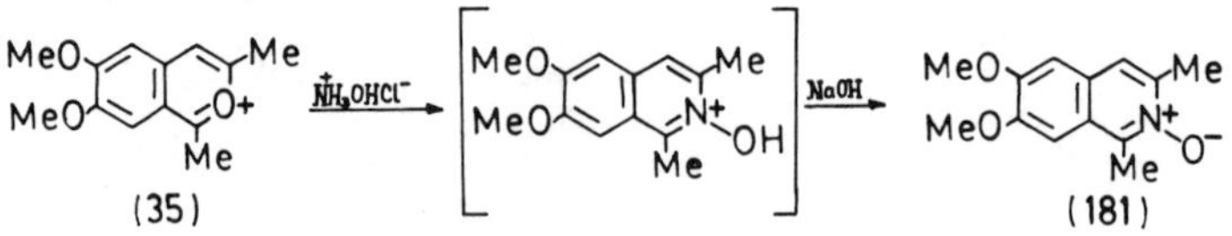

iv. *Secondary amines.* Whereas reactions of 2-benzopyrylium salts with ammonia and primary amines result mainly in heterocycles, the interaction of these salts with secondary amines accompanied by recyclization leads to formation of carbocyclic compounds. As described in Section III,C,4,a i,ii, the formation of naphthylamine derivatives in reactions of 2-benzopyrylium salts with ammonia or primary amines occurs only when α(1)-alkyl substituents are present. The construction of a new carbocycle in reactions with secondary amines may also involve the α'(3)-

alkyl substituent (81KGS1608). The regioselectivity of α- vs. α'-substituents depends on several factors such as the structure of the cation, the nature of amine and solvent, temperature, and concentration, as well as on the correlation of these conditions. Thus, this reaction is hardly predictable in its outcome, and it obviously needs additional investigation to be systematized in its mechanism and results.

However, this reaction has only one pathway if the 2-benzopyrylium cation **30** has a single substituent able to take part in the aldol condensation (71KGS1437; 81KGS1608).

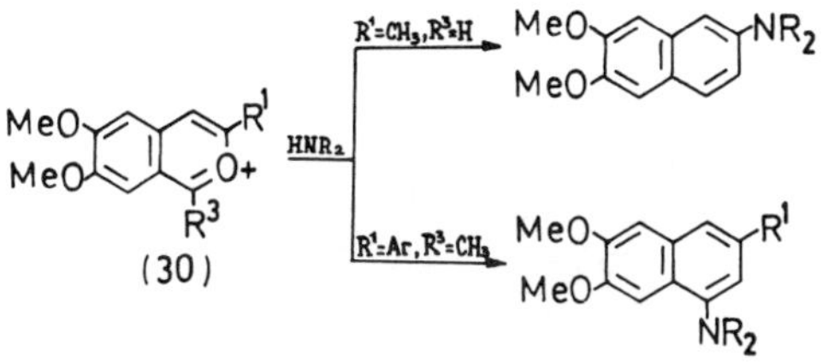

One pathway only was observed also for 1,3-dimethyl-substituted salt **35,** although either of the two methyl groups might be involved in the formation of the carbocycle: α-naphthylamines **182** were the only products independent of the nature of amine used (70KGS1308). This means the nucleophile preferentially attacks position 1.

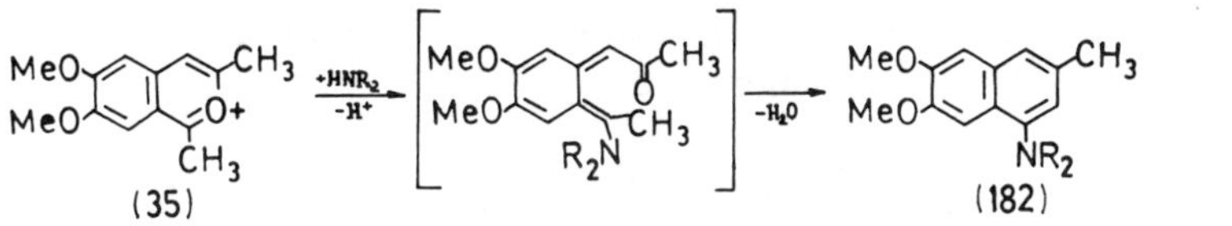

For benzo[c]pyrylium salts having different alkyl substituents in positions $\alpha(1)$ and $\alpha'(3)$, the reaction with secondary amines loses its predictable regioselectivity. Thus, reaction with dimethylamine, independent of the solvent, gives rise to β-naphthylamine **185,** from 1-ethyl-3-methyl-substituted salt **183** (81KGS1608), and α-naphthylamine **186** from 1-benzyl derivative **184** (88UP1). In the former case, the nucleophilic attack occurs in position 3, whereas in the latter case, it takes place in position 1.

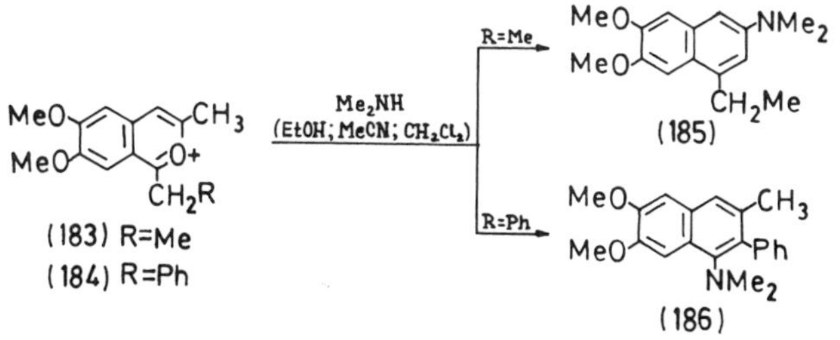

Interaction of the same 1-benzyl-3-methyl-substituted salt **184** with morpholine leads to α-naphthylamine **187** in ethanol and to the mixture of α- (**187**) and β- (**188**) naphthyl-amines in a ratio depending on the temperature if morpholine itself was used as a solvent (88MI1; 88UP1).

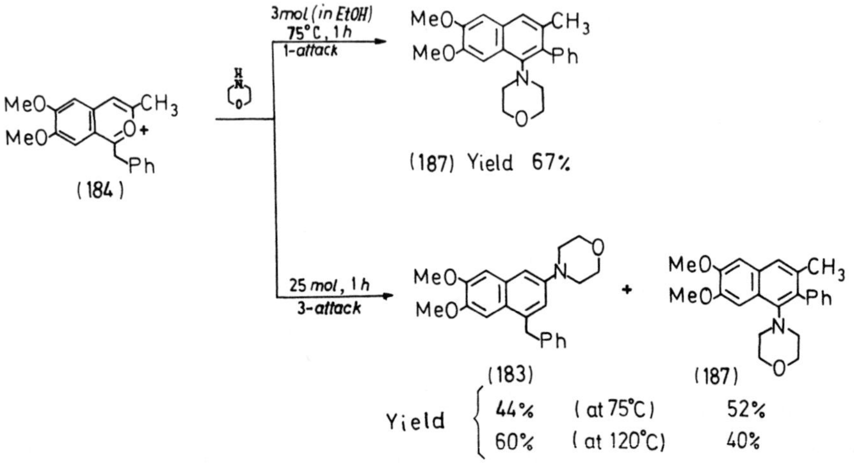

Similar to the reaction of 2-benzopyrylium salts with ammonia (Section III,C,4,a,i), an excess of competing nucleophiles in reactions with secondary amines is undesirable. Thus, the interactions of 1-ethyl-3-methyl-substituted salt **183** with 40% aqueous solution of dimethylamine leads to α-naphthol **189** as the major product (79TH1).

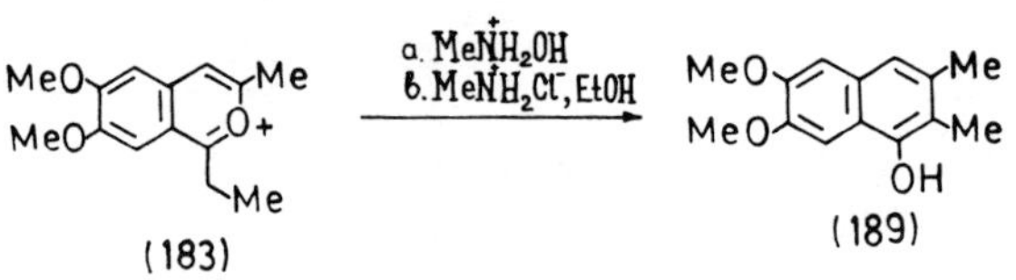

The formation of α-naphthols of type **189** from 1-alkyl-substituted salts **30,** on heating with dimethylamine hydrochloride in ethanol, occurs by another mechanism and will be explained in Section III,C,4,b,i. The interaction of 1-aryl-3-carboxy-substituted salts **62** with secondary amines in benzene is initiated probably as in the reaction of these salts with primary amines, and the attack by the secondary amine on position 3 is the primary step of this reaction. However, since protonation of the intermediate anion **190**, a masked acyl anion, becomes difficult, an interaction of this anion with the carbonyl group of the benzophenone fragment occurs (86KGS125). The enamines **191** thus formed are usually hydrolyzed on purification, yielding five-membered cyclic acyloins **192**.

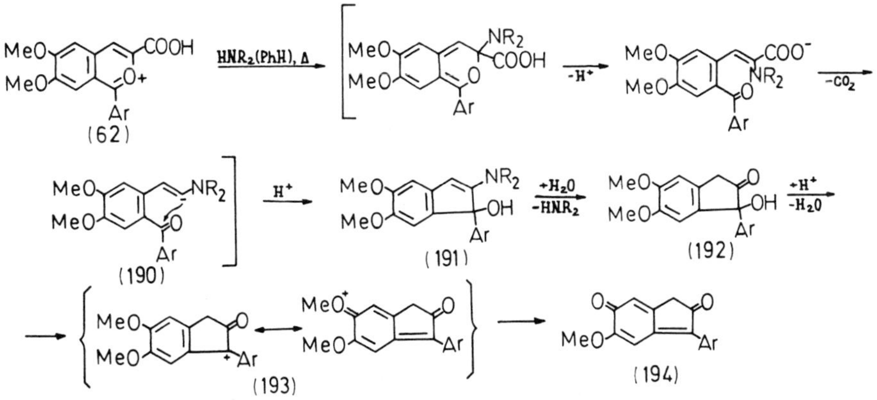

(62)

(190)

(191)

(192)

(193)

(194)

On acidic hydrolysis of enamines **191**, the acyloin intermediates **192** are converted on protonation into quinonemethides **194** via the intermediate formation of destabilized α-acyl carbenium ions **193**. Obviously, the higher conformational stiffness in ions **193** compared to their acyclic analogs (83CC7; 83TL4911) is due to the methylene bridge in the five-membered ring, leading to a considerable increase in the contribution of the annelated benzenoid ring to resonance stabilization and consequently to increased stability. Thus, cation **193** (Ar = Ver), obtained on treatment of acyloin **192** in acetic acid with perchloric acid, can be kept in a crystalline state for a long time.

Interestingly, α-carboxy-substituted monocyclic pyrylium salts **195**, probably because of a lower tendency towards nucleophilic addition in comparison with 2-benzopyrylium salts (Section III,C,1), react via two pathways. One is similar to reactions of 2-benzopyrylium salts (*a*), and the other resembles the behavior of isoquinolinium salts (*b*) (Section III, C,4,a,ii). In this case, monocyclic five-membered acyloins **198** were obtained in 15–40% yield. Obviously, their formation occurs by initial addition of amine (pathway *a*). The difficultly obtainable α-unsubstituted py-

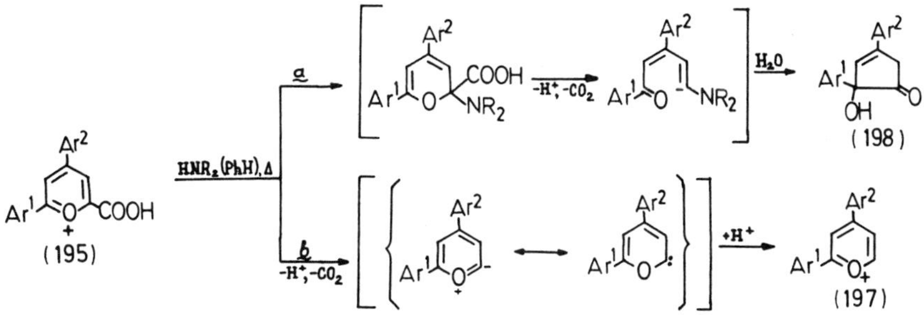

(195)

(198)

(197)

rylium salts **197** were isolated on acidification of the solution, obtained after separation of acyloins **198** (89KGS1564).

The formation of these salts assumes the intermediate formation of carbenic ylids **196,** which were detected earlier in reactions of salts **195** with carbonyl compounds [77DOK(236)1364]. As noted in Section III,C,i, reactions of 2-benzopyrylium salts with nitrogen nucleophiles occur by direct electrocyclic ring-opening. Good support for this suggestion was described in 1989 by Verin and Kuznetsov (89MI2, 89TH2).

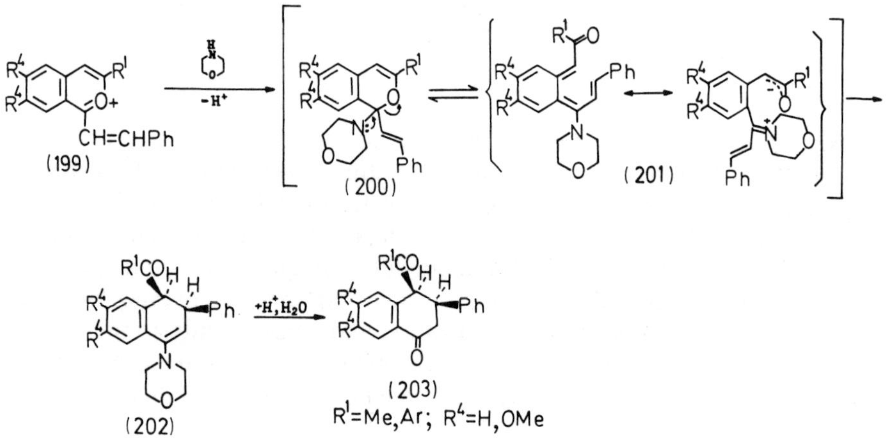

A substituent containing a double bond at position 1 was put into 2-benzopyrylium cation **199** with the goal of obtaining a hexatriene fragment in the ring-opened intermediate **201**. The analogs of **201,** as known (72MI2), form carbocycles with a high degree of stereoselectivity by a disrotatory process, in accordance with steric reasons intermediate **201** should have a trans-cis-trans configuration, therefore, its cyclization must lead to a cis-isomer. Indeed, the ^{1}H-NMR data for enamines **202** confirm their existance as cis-isomers, with $J_{1,2} = 6$ Hz in accordance with known data (78JOC286, 78JOC2852; 80JOC5067). On heating with aqueous alkali, the cis-isomer **202** are converted into trans-isomers, whereas heating with acidic aqueous solutions leads to formation of tetralones **203** by hydrolysis of the enamine group.

b. *Reactions with Oxygen Nucleophiles.* Depending on the nature of oxygen nucleophiles, their interaction with 2-benzopyrylium salts leads to products of carbocyclization, as in the reaction with secondary amines (Section III,C,4,a,iv), or the process is accompanied by oxidation.

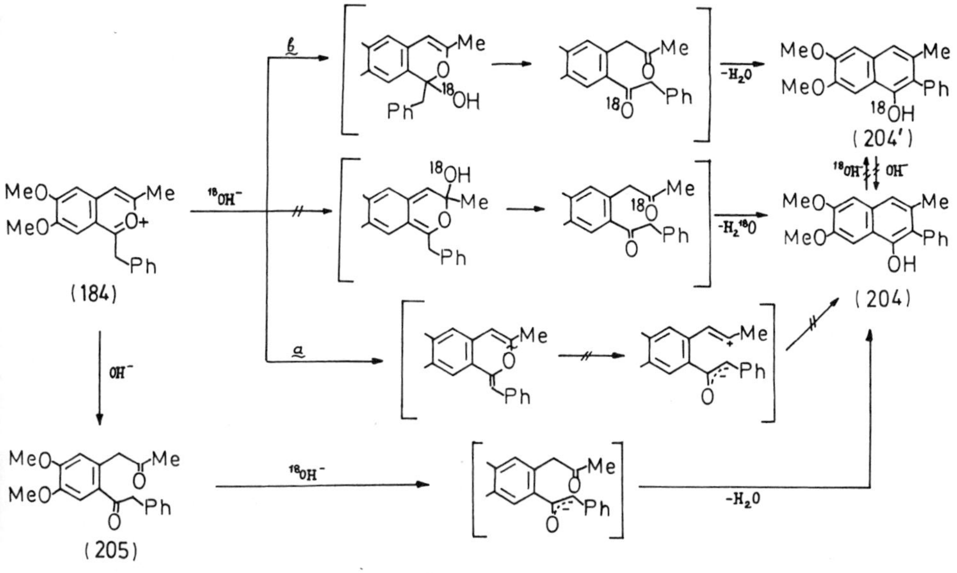

SCHEME 10

i. *Formation of α- or β-naphthols.* The transformations of 2-benzopyrylium salts **184** into α-naphthols of type **204** (as well as transformation of monocyclic pyrylium salts to phenols, pyridinium salts to anilines, etc.) may be viewed as isomeric rearrangements if the onium atom and the resulting exofunctional group have the same heteroatom (81T3425).

Since such reactions as a rule proceed under the action of nucleophilic reagents whose heteroatoms have the same nature as the onium hetero-atom, the problem is whether these recyclizations are really isomeric, i.e., if the onium atom becomes the heterofunction (pathway *a*, Scheme 10), or if they are pseudoisomeric, i.e., the heterofunction has a heteroatom that belonged earlier to the reagent (pathway *b*, ANRORC mechanism).

It should be stressed that isomeric recyclizations were found to be predominant for reactions of primary amines with isoquinolinium salts **152** (81T3425; 82KGS291), which are isoelectronic analogs of 2-benzopyrylium salts. Moreover, this transformation assumes a rupture of the N—C$_3$ bond in **152** after its deprotonation.

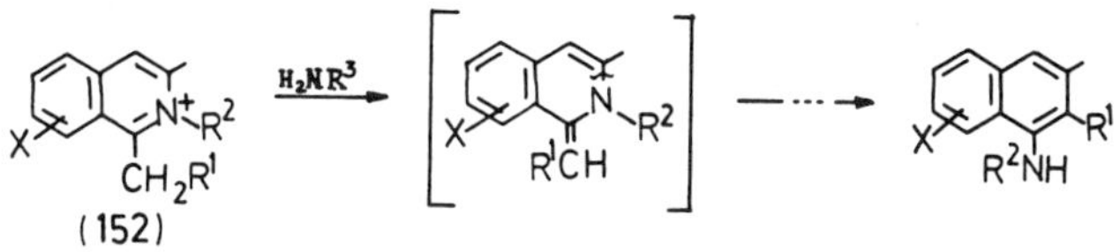

Mass spectrometric determination of the ^{18}O isotope in the product obtained by recyclization of **184** with oxygen-labelled sodium hydroxide showed that the resultant compound was naphthol **204'**. That oxygen isotope exchange does not take place for α-naphthols **204'** or **204** under these conditions was checked. The transformation of diketone **205** into the unlabelled naphthol **204** under the described conditions signifies the inclusion of the ^{18}O-isotope label on the primary recyclization step of the salt **184**. Moreover, the ^{18}OH group is added to position 1 in **184**, since the addition to position 3 could lead to the unlabelled α-naphthol **204**. On the basis of these data, one can conclude that this conversion, under the described conditions, is not a real isomeric recyclization, and that it occurs by electrophile–nucleophile interaction in accordance with pathway *b* as shown in Scheme 10. In other words, the reaction is analogous to the transformation of monocyclic pyrylium salts according to the ANRORC mechanism [82AHC(Suppl)]. However, in the case of 2-benzopyrylium salts, deprotonation of the added hydroxyl group probably precedes the ring opening (79AHC2). The experimental data described here leads to the conclusion that the rate of enolate formation from the diketone **205** (Scheme 10) by basic catalysis is higher than the oxygen exchange rate in the carbonyl group, whereas the opposite data are shown for aliphatic ketones (66JA1916). This fact stresses again the influence of the annelated benzenoid ring in 2-benzopyrylium cations.

The clarification of the recyclization mechanism of 2-benzopyrylium salts into naphthols under the action of alkali permits some generalizations (low temperature at the beginning of reaction followed by heating, and the use of cosolvents that are unable to form stable adducts). This leads to an increase in the yield of this reaction, considered earlier (70MI1) as lacking preparative interest, similar to the formation of phenols from monocyclic pyrylium salts. Consider the example shown in Scheme 11[6].

At the same time, it was impossible to suppress the deprotonation completely if the 2-benzopyrylium cation had a benzyl group in position 1 (76KGS745; 77KGS1176); a stable anhydrobase results. As expected, the

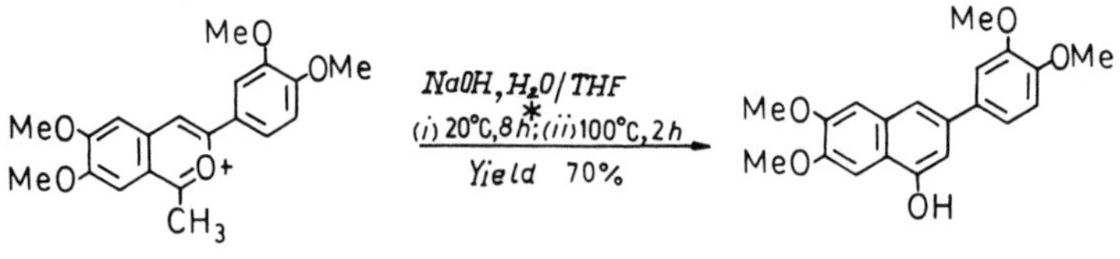

SCHEME 11

---

[6] The use of elevated temperature at the beginning of this reaction leads to the formation of chrysenes, as described in Section III,F,2,b.

formation of the unconjugated enolates **207'** or **207**, on recyclization of the salt **30** or diketone **206**, respectively, proceeds at a lower rate than the isotope exchange in a benzyl carbonyl group. As a result, the oxygen-labelled β-naphthol **208'** is formed in both cases (88UP1). These data do not lead to a conclusion about the position of nucleophilic attack if the 3-methyl group participates in recyclization, as in **30** or **206**.

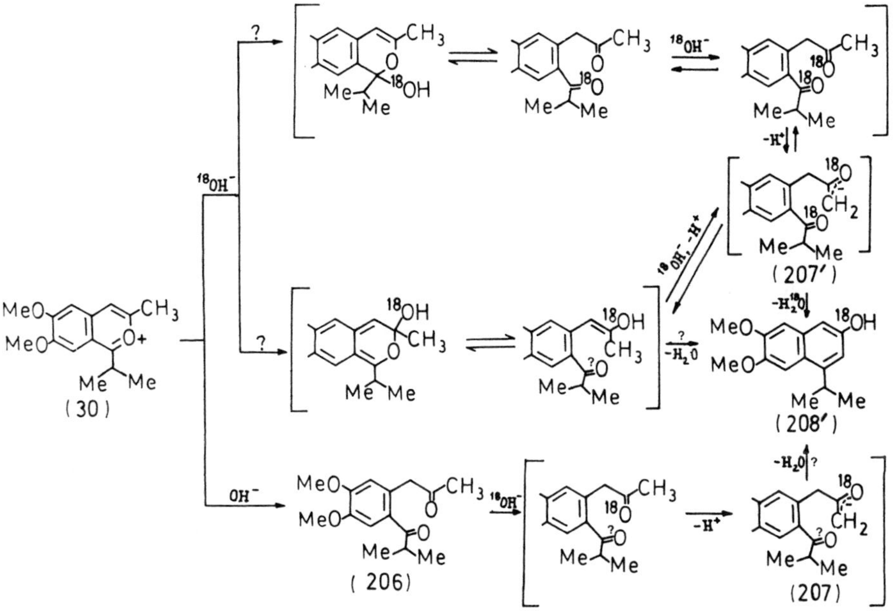

The formation of β-naphthols **209** occurs readily on treatment of salts of type **154** with dilute alkaline aqueous solutions, but this reaction competes with other processes when concentrated alkali hydroxide solutions are used (cf. Section III,F,2,c).

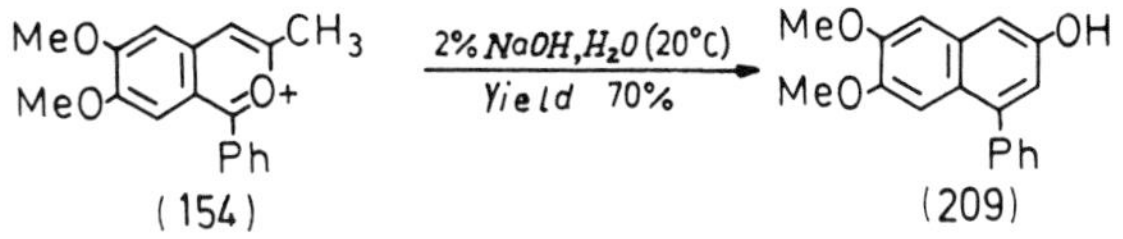

Except for 1-methyl-substituted cations (cf. Section III,F,2,b), these complications may be avoided if the recyclization of 2-benzopyrylium salts is carried out in acidic nucleophilic media, i.e., on heating the salt in aqueous alcohol or in water with or without acidification (82KGS1184). Probably, in this case, the reaction occurs by two simultaneous processes. One of them proceeds by direct ring opening of the heterocycle in pseudo-base **210**. However, the main process includes protonation of the nucleophilic vinyl fragment in pseudobases **210** (R^2 = H), and especially in the corresponding ethers (R^2 = Alk).

Contrary to the oxonium intermediate **115** (Section III,C,2), the intermediate **211** having the alkoxy leaving group readily opens its heterocyclic ring under these conditions. Diketones **205** (R^3 is not H) thus formed are then cyclized into α-naphthols **204** (or β-naphthols of type **209**). If anhydrobases **212** were formed, they would have to take part in the analogous conversions described previously. Obviously, the same mechanism may be applied to the scheme of formation of α-naphthols **204** from 2-benzopyrylium salts under the action of dimethylamine hydrochloride in ethanol (77KGS996) (cf. Section III,C,4,a,iv).

Interestingly, on briefly heating the salt **184** in an aqueous suspension, the intermediate α-tetralone **213** was isolated, and it was easily converted into the α-naphthol **204** (82KGS1184).

Tetralone **213** may be considered as the specific analog of 3-hydroxy-3,4-dihydroisoquinolines **137** (Section III,C,4,a,i). Its isolation is more evidence of the lower tendency toward aromatization in recyclization products of benzo[c]annelated pyrylium salts in comparison with monocyclic cations.

Contrary to the recyclizations in alkaline medium, the transformations of either the salt **184** and the diketone **205** in acidic oxygen-labelled aqueous solutions lead only to the labelled α-naphthol **204'**. These facts are evidence that the rate of isotopic exchange in the carbonyl group in this case is higher than the enolization rate under basic conditions. The conversions **204** $\rightleftarrows$ **204'** under acidic conditions also do not occur.

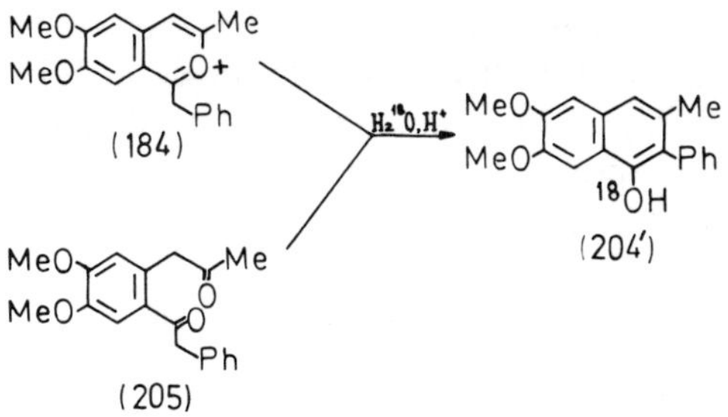

The conversion under acidic conditions of 2-benzopyrylium salts into naphthols is not observed for isoquinolinium or monocyclic pyrylium salts, which become converted into phenols only in alkaline medium. Moreover, the ring opening of the heterocycle in **211,** determined by its protonation, may be considered as the specific type of 1,4-addition of R^2OH to the heterodiene fragment of the 2-benzopyrylium cation. This is a new version of the corresponding step in the ANRORC scheme.

Probably, the same mechanism is involved in the conversion of 2-benzopyrylium formates **10** into indenes **118** on heating in formic acid (Section III,C,2), or in the formation of indanes **214** on heating the fluoroborate **178** in hydrochloric acid with phosphorus or in glacial acetic acid with phosphonium iodide (71CB2984). The primary step of this conversion includes the addition of a hydride anion; then the intermediate isochromene of type **107** becomes protonated and takes part in further transformations.

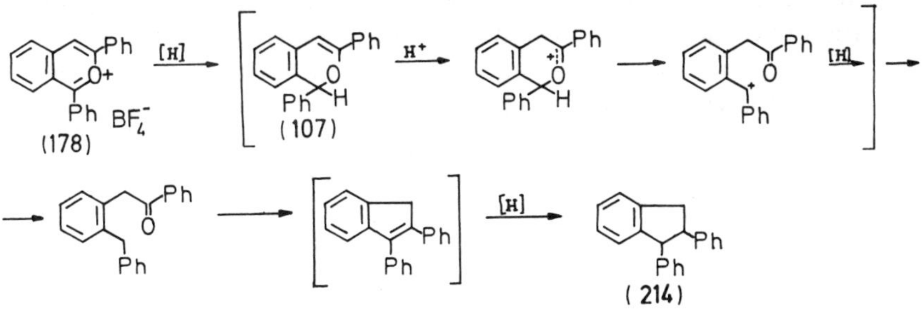

The formation of α- or β-naphthols also occurs in the interaction of aqueous alkaline solutions with cyanide adducts of 2-benzopyrylium salts, i.e., cyanoisochromenes **108** (88T6217), and the yields of naphthols obtained this way are higher than on recyclization of the 2-benzopyrylium salts themselves.

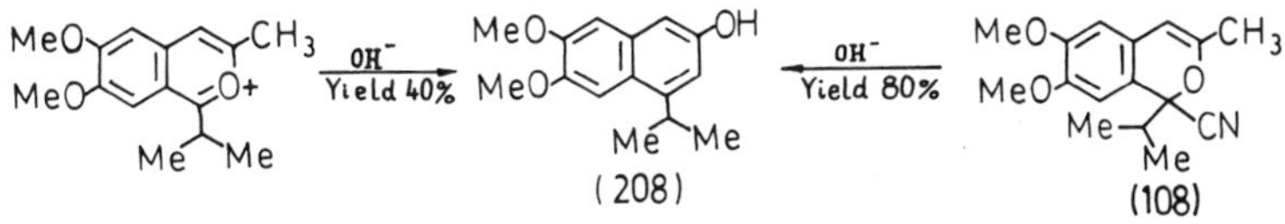

Supposedly cyanoisochromene **108** reacts by tele-substitution, with intermediate formation of 1-alkylidene-3-hydroxyisochromanes **123** (Section III,C,2). Indeed, on heating with aqueous alkaline solutions, these intermediates are quantitatively converted into the corresponding α- (**204**) or β-naphthols (**208**) (88T6217).

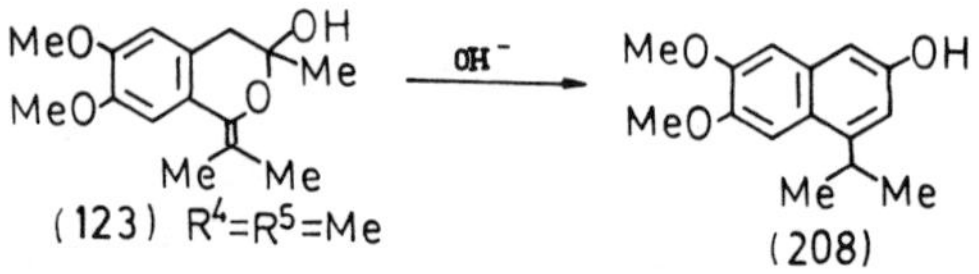

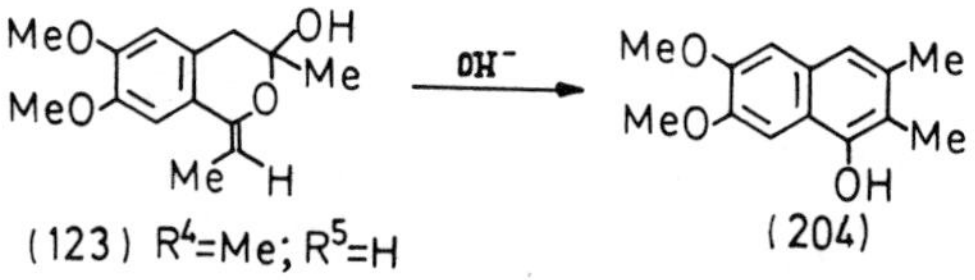

ii. *Oxidation reactions.* The oxidation of 2-benzopyrylium salts was one of the first reactions investigated in this class of cations (50JA1118; 54JOC472). Several papers by Hungarian groups were devoted to oxidations of 2-benzopyrylium salts with hydrogen peroxide [54JOC472; 64ACH(41)451] and chromic anhydride [67ACH(52)261], resulting in benzo[*b*]furans (coumarones), in the former case, and anthrone or anthraquinone in the latter. The formation of coumarones has some analogy with the oxidation of monocyclic pyrylium salts to 2-acylfurans (60CB599).

It was suggested that for hydrogen peroxide, the reaction begins with the addition of the HOO⁻ anion, followed by formation of the cationic intermediate **216**. According to terminology, such a mechanism, which implies the presence of intermediate **105,** may be called *ortho*-bridging or *ortho*-cyclization (84UK1648), as it was described for the reaction of 2-benzopyrylium salts with sodium azide (Section III,C,2).

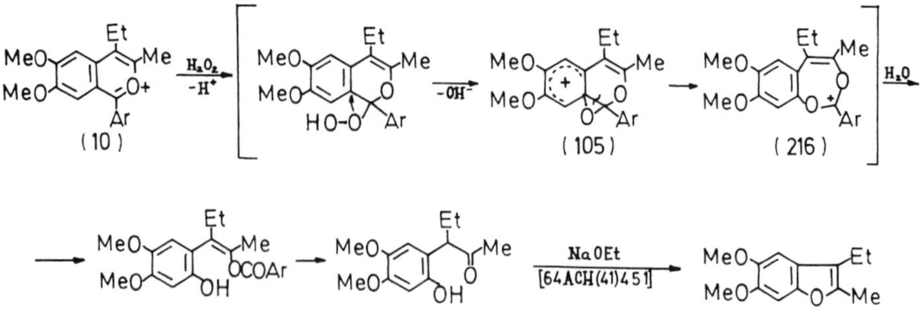

Indeed, as shown by Verin and Kuznetsov (89UP1), the oxidation of **159** with hydrogen peroxide in ethanol gives rise to cyclic orthoester **217**, and the formation of this compound corroborates the mechanism suggested earlier for this reaction.

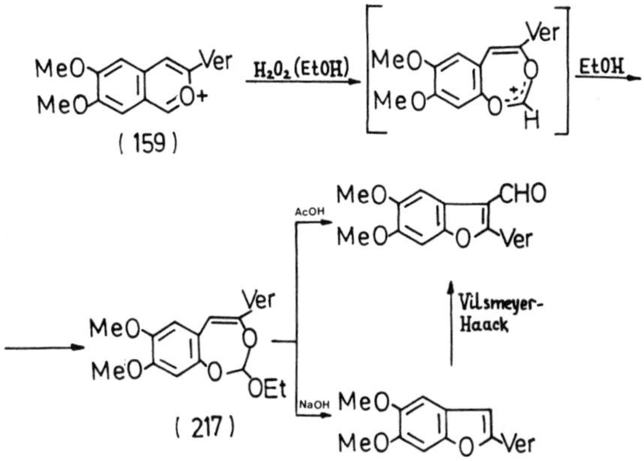

The oxidation of **10** with chromic anhydride in aqueous acetic acid leads to anthrone **218** and to anthraquinone **219** with excess oxidant. According to the authors' opinion (54JOC1533), the oxidation process was supposed to include the primary formation of diketone **9**; it was this compound, but not the initial salt **10**, which was oxidized to anthrone **218**.

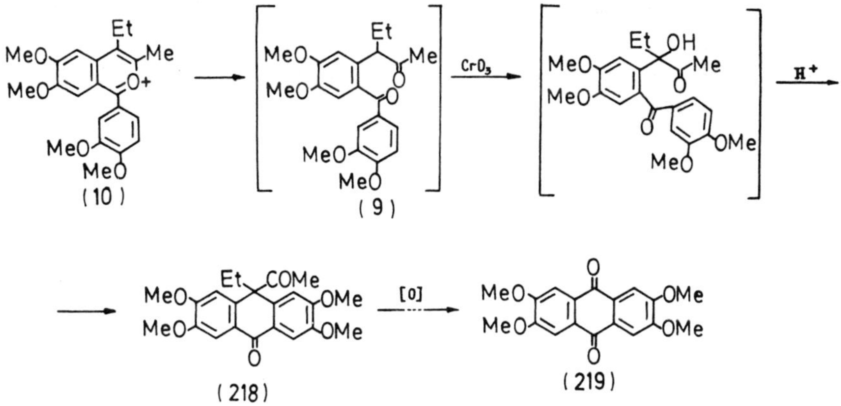

However, special experiments showed that diketones of type **220** are not oxidized with chromic anhydride under the described conditions. Such an oxidation occurs via formation of bridged intermediates **221** (89UP1).

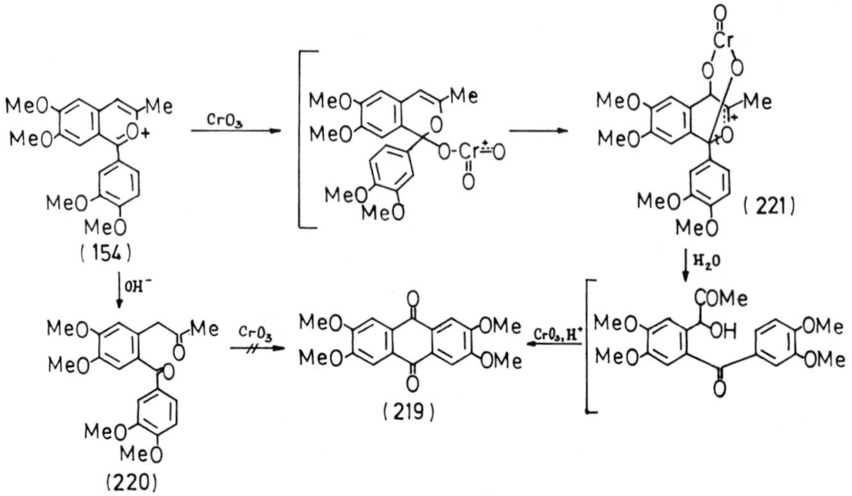

The only known reaction of a 2-benzopyrylium salt with ozone leads to a destructive oxidation of the cation (51JOC1064).

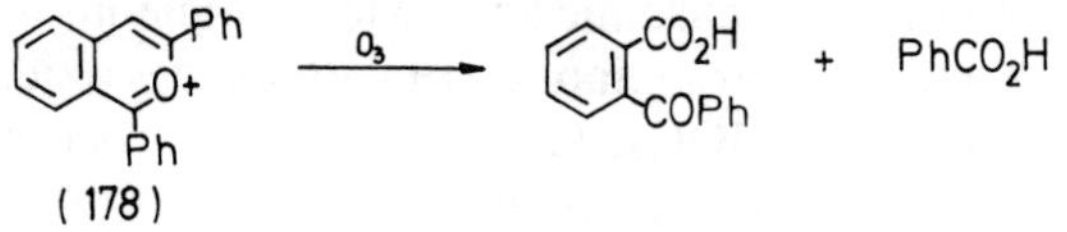

c. *Reactions with Sulfur Nucleophiles.* The recyclization of 2-benzopyrylium salts to the corresponding thio-analogs does not occur as easily as was described for monocyclic pyrylium salts [82AHC(Suppl)]. Usually the dominant process is the dimerization reaction in the interaction of 2-benzopyrylium salts with alkali sulfides (Section III,F,2). All attempts to suppress this undesirable conversion failed.

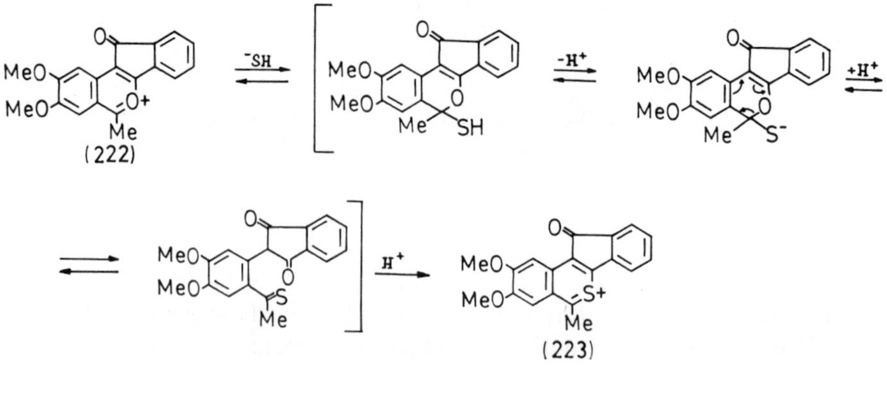

Only a single example of the formation of a 2-benzothiopyrylium salt **223** from its oxygen analog **222** is known (76KGS858; 77TH1). The 2-benzopyrylium salt **222**, having an electron-withdrawing substituent in the cation, takes part in the recyclization without dimerization.

d. *Reactions with Compounds Possessing Active Methylene Groups.* Reactions of 2-benzopyrylium salts with compounds possessing active methylene groups occur in the presence of specially added bases in accordance with the ANRORC scheme. This idea is supported by isolation of the adduct **112** (R^3 = H, R^1 = Ver, R^2 = MeCO, C$\equiv$X = CO_2Et), a protonated form of the ring-opened intermediate **224** (R^1 = Ver, R^2 = (C$\equiv$X) = CN), and by the conversion of these compounds under the action of excess base into the corresponding naphthalene derivatives **225** (90KGS315).

R^1=Ar; R^2=CO_2Et, CN;
R^3=H,Alk,Ar;
X=OH,NH_2, Me

Evidently, deprotonation of adduct **112** precedes the ring opening. Unlike monocyclic pyrylium salts, only one pathway may be followed for the adducts of 2-benzopyrylium salts with the nucleophile possessing two carbon atoms, both of which might participate in the construction of the new skeleton. That the added component does not react as a 1,1-nucleophile may be explained by the difficult formation of intermediate **226**, since it should be a fixed *ortho*-quinonoid form.

For monocyclic pyrylium salts, the realization of an alternative pathway is equally possible [82ACH(Suppl)]. But for 2-benzopyrylium salts, even in the reaction with nitromethane, which is able to act only as a 1,1-nucleophile, the formation of naphthalene **228** was not more than 5%, whereas adduct **227** was isolated quantitatively (90KGS315).

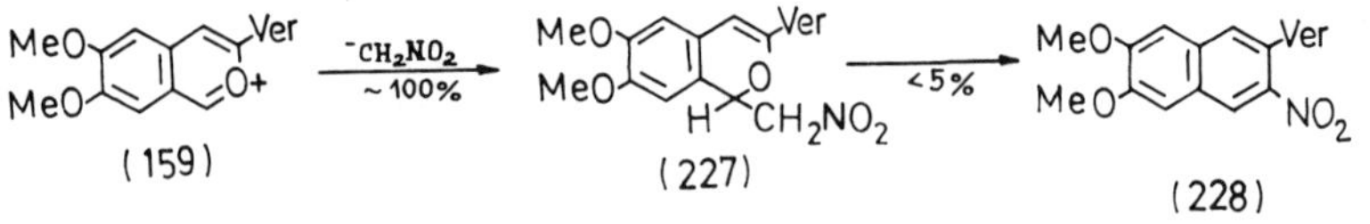

The formation of acylnaphthalenes **225** occurs under mild conditions (catalysis by sodium *tert*-butoxide or with the use of the phase-transfer method and TEBA as catalyst), and it affords high yields (60–90%). In this connection, it is surprising to recall the opinion about the incapability of 2-benzopyrylium salts to take part in recyclization reactions with compounds possessing active methylene groups, with secondary amines, or with sulfur or phosphorus nucleophiles (71CB2984).

The only competing reaction in interactions of 2-benzopyrylium salts with active methylene compounds is the formation of naphthalene **229,** which is constructed with participation of the 1-methyl group of the benzo[c]pyrylium cation (90KGS315).

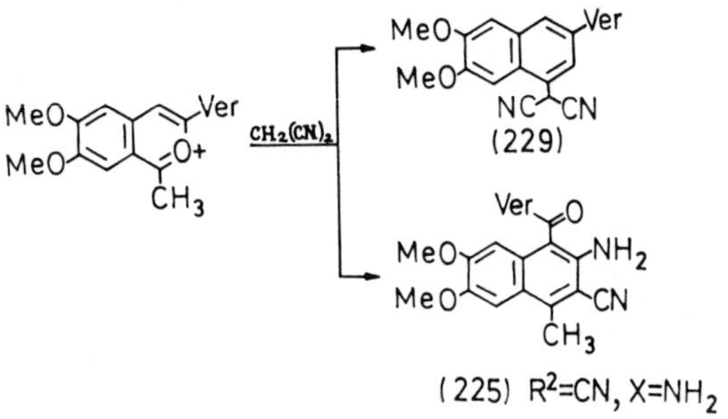

It was also shown (90KGS315) that adduct **230,** which exists in the enolic form, according to ^{1}H-NMR data, is converted into the corresponding acylnaphthalene **225,** both in alkaline solutions and on thermolysis; the enol **230** affords the deacylated naphthalene **232** in acidic solutions.

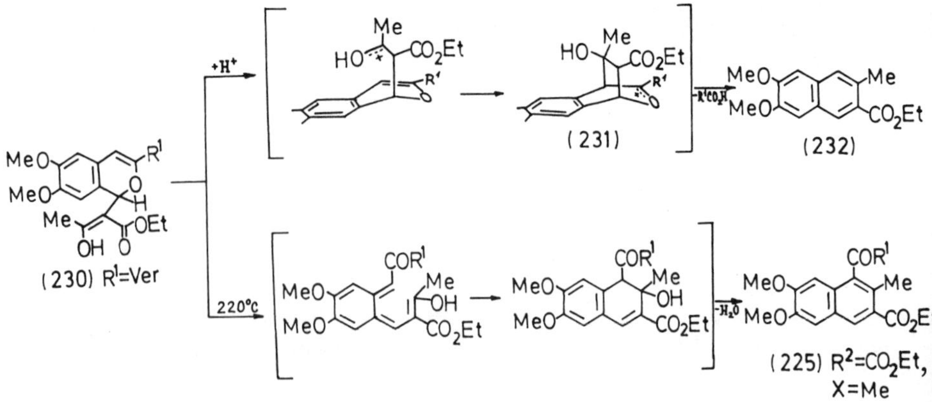

The latter reaction assumes a sequence of steps that differ from the ANRORC scheme. Thus, first the formation of bridged ring intermediate **231** takes place, and then the ring opening occurs. Interestingly, the step following aromatization is the result of splitting a molecule of carboxylic acid and not of water, as observed under basic catalysis (cf. Section III,F,2,c).

On thermolysis, one cannot exclude the provisional disturbance of aromaticity by a reverse electronic process followed by an electrocyclic ring-closure to acylnaphthalene **225** as described for the reaction of 1-styryl-substituted 2-benzopyrylium salts **199** with secondary amines (Section III,C,4,a,iv).

It should be noted that such a disturbance of aromaticity, on thermolysis, was detected for isochromanone **233** (84T4383), which may be considered as a reduced form of 3-hydroxy-benzo[c]pyrylium salt **52** (R = R′ = H).

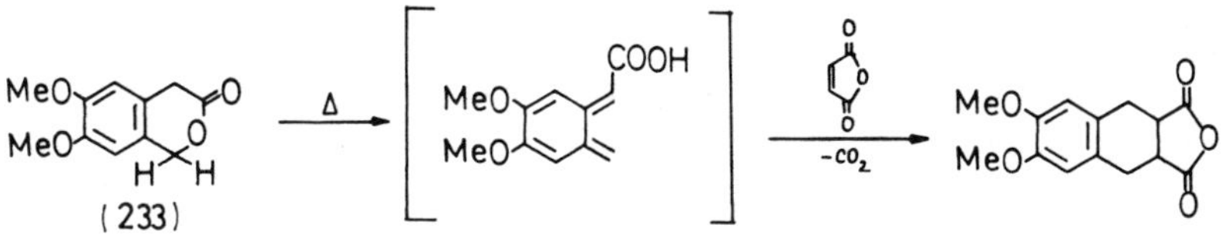

D. Cycloaddition Reactions

1. Reactions with Ethyl Vinyl Ether

Owing to the annelated benzenoid ring the double bonds in the heterocyclic ring of 2-benzopyrylium are considerably fixed, and this fact, correlated with the high electronegativity of the oxonium atom, causes these cations to react as heterodienes. Thus, interaction of 2-benzopyrylium

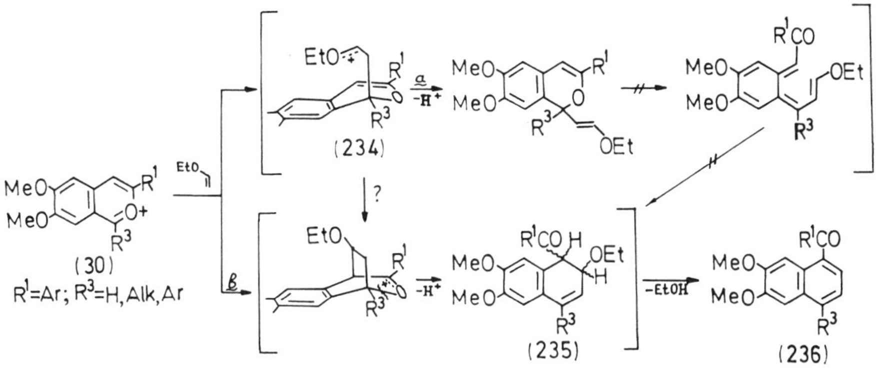

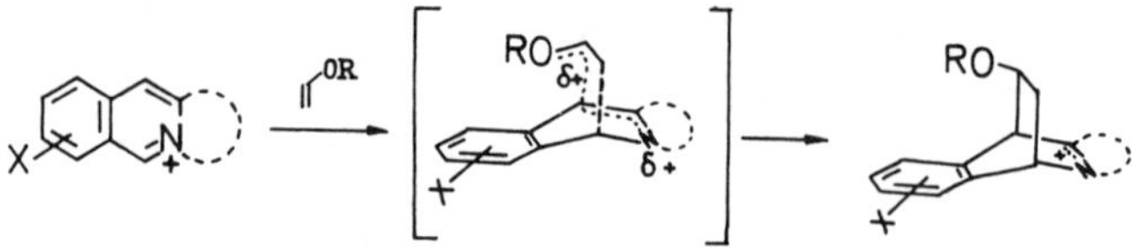

SCHEME 12

salts **30** with ethyl vinyl ether gives rise to 1-acylnaphthalenes **236** (90KGS315). Under the mild reaction conditions (temperatures below 100°C; as a rule, no added base), it is unlikely that ethoxy groups would be able to assist in the electrocyclic ring opening of the adduct **234** after deprotonation (pathway *a*) in accordance with the ANRORC scheme.

However, the evident analogy with the well-known Bradsher–Falk reaction (Scheme 12) (74AHC289; 79JOC727; 82JOC5021; 84CC761), which is considered as a coordinated[7] [4 + 2] cycloaddition with inverse electron demand (coordinated *para*-bridging), allows for confident classification of the reaction with such types of interactions described here. At the same time, the good leaving group of **235** does not allow the isolation of this intermediate in order to infer, from its stereochemistry, the extent of coordinated addition, i.e., the existence of the cation **234** as a real intermediate.

The greater electronegativity of the onium oxygen atom gives the possibility, in the case of 2-benzopyrylium salts, of overcoming some limitations which are characteristic for the Bradsher–Falk reaction, i.e., the compulsory absence of a substituent in the α-position of the heterocation and the presence of an electron-withdrawing substituent at the nitrogen atom in isoquinolinium salts. Thus, the reaction of 2-benzopyrylium salts with ethyl vinyl ether is the first example of a heterodiene synthesis (83T2869; 84TL721; 86CRV781) where the heterodiene includes the oxonium atom. This reaction has not been reported for monocyclic pyrylium salts, probably because these cations do not possess to this extent the fixation of double bonds.

2. Reactions with Azomethines

Monocyclic pyrylium salts react with azomethines according to a postulated four-centered mechanism yielding products of exchange between the oxonium atom and the amine fragment of the Schiff base (73ZOR1079; 82KGS465).

---

[7] "Coordinated" implies concerted, but not necessarily synchronous.

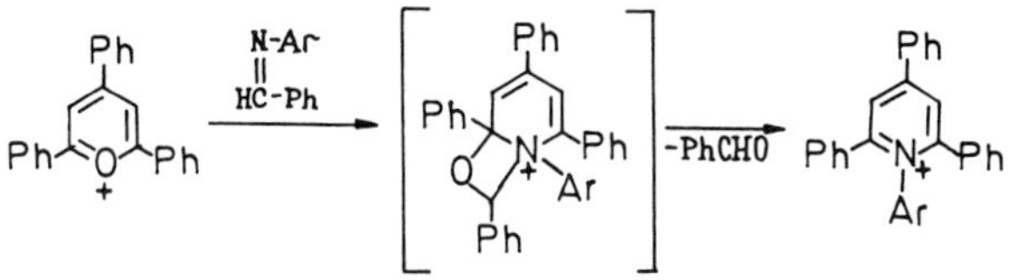

Benzo[*c*]pyrylium salts having a methyl group in position 1 form, in the reaction with the same azomethines, 1-styryl-substituted benzo[*c*]pyrylium salts **94** (Section III,B).

As shown by Verin *et al.* [90KGS(ip2)], the reaction of azomethines with 2-benzopyrylium salts **30** having substituents other than a methyl group in position 1 results quantitatively in 3,4-dihydroisoquinolinium salts **238**. Interestingly, isoquinolinium salts of type **152** and their benzenoid analogs do not react with such weak dienophiles as azomethines.

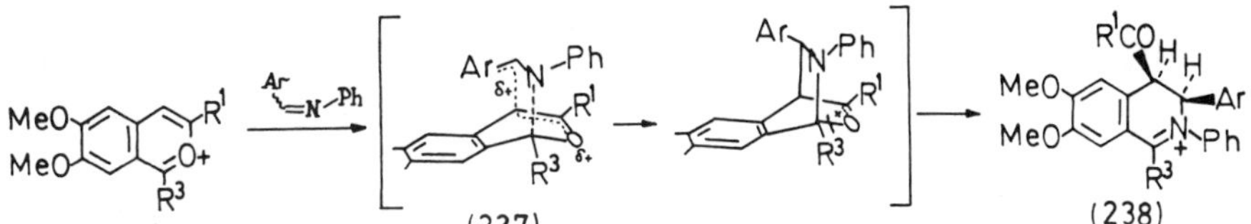

The formation of isoquinolinium salts **238** as trans-isomers is in good agreement with the concept of maximal charge separation in **237** by Coulomb interaction, which was accepted as the explanation of stereoselectivity in the Bradsher–Falk reaction (78JOC822).

In the reaction of monocyclic pyrylium salts with azomethines, however, an analogous *para*-bridged intermediate was excluded on the basis of experimental data. It is necessary to stress that the desired conversion is similar to the condensation of substituted homophthalic anhydrides with the same azomethines (77JOC1111; 78JOC286; 79JOC409; 80S847):

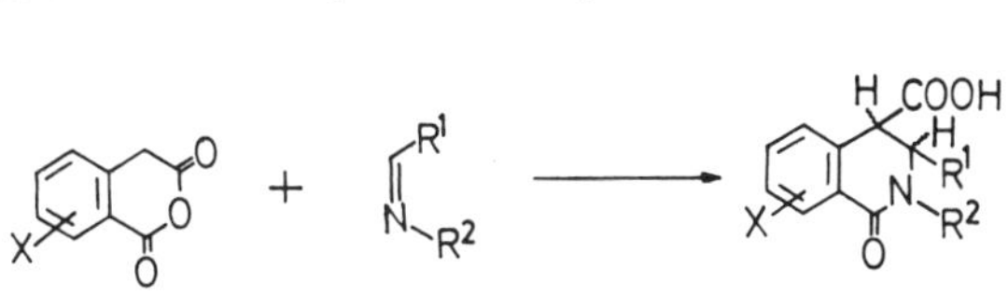

2-Benzopyrylium salts are more suitable than homophthalic anhydrides for the synthesis of dihydroisoquinolinium salts of type **237,** by the method described previously, because of the easier availability and greater variety of the former compounds.

E. Deprotonation

When 2-benzopyrylium cations have a substituent with a fairly mobile hydrogen atom (α-alkyl group or heteroatom group in any other position of the cation), deprotonation of such a substituent occurs even under mild conditions by an acid-base interaction as the primary step (Section III,A). Although deprotonation in both cases leads to compounds whose structures can be depicted by two resonance formulas, either with charge separation (betaines) or without (anhydrobases), on discussing products of C-deprotonation, the term (and the corresponding formula) "anhydrobase" is more often used, whereas products of O-deprotonation are called "betaines."

1. *Formation of Anhydrobases and Betaines*

Examples are known for deprotonation of hydroxyl groups attached either to the 2-benzopyrylium cation itself or to its subsituents. In both cases, the stability of the betaine thus formed is determined by the position of the deprotonated group. Thus, unstable betaine **15,** obtained as described in Section II, A, may be formally considered as products of deprotonation of the nonisolated 4-hydroxybenzo[*c*]pyrylium salts (84CC702; 86NKK1622). A valence-isomeric structure **14** is possible, and photochemical processes were discussed in Section II,A.

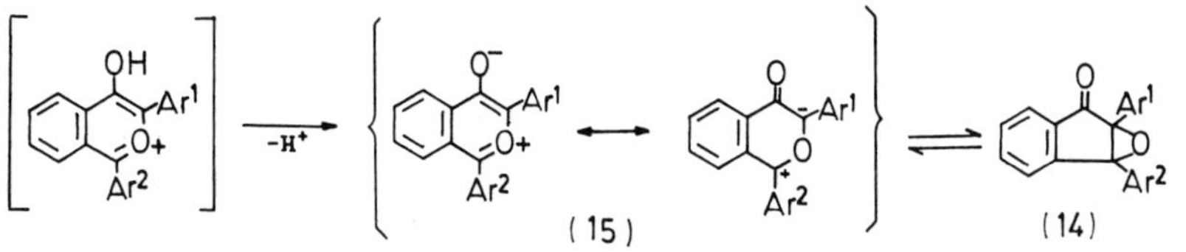

Deprotonation of 3-hydroxy-substituted benzo[*c*]pyrylium salts **239** leads to betaines **240a** having an *ortho*-quinonoid structure **240b** as a resonance form (84TL3659).

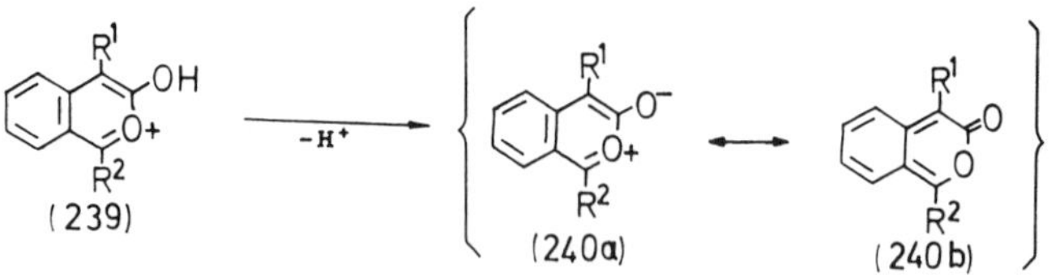

The stability of compounds thus formed is determined by the nature of R^1 and R^2. If the hydroxyl group is situated at the annelated benzenoid ring,

deprotonation and addition of nucleophiles are competing pathways in
alkaline solutions, and the chemoselectivity is determined by the structure
of the initial salt **30**.

Thus, treatment of 6-hydroxy-substituted salts **30** (R^4 = H) with aqueous or alcoholic alkaline solutions results in stable betaines **241** (vinylo-
gous pyrones) having a *para*-quinonoid resonance structure, whereas
7-hydroxy-substituted salts **30** (R^5 = H) give rise to diketones **244**
[66ACH(50)381; 75ACH(85)79].

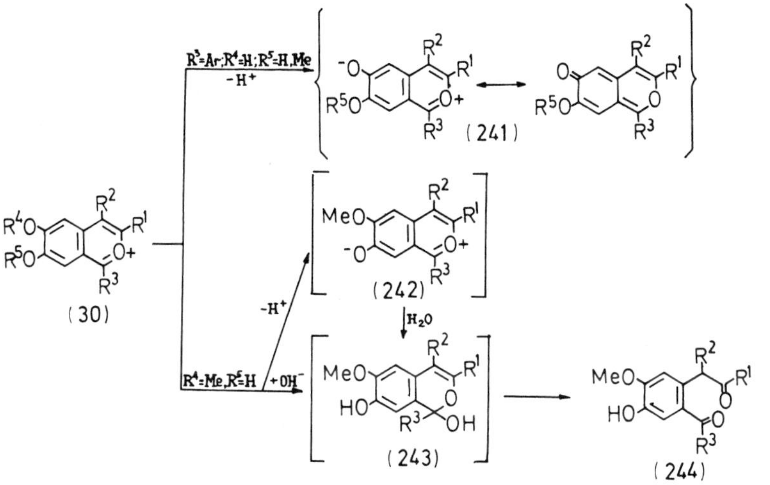

The existence of a deprotonated form **242** as intermediate, together with
pseudobase **243**, was assumed in this case, but the betaine was so short-
lived that even its electronic absorbtion spectrum could not be measured.
Deprotonation of the 6-hydroxyl group was preferred for salt **10** (R^1 = H)
having another hydroxyl group in the 1-aryl substituent [75ACH(85)79],
and betaine **245** was formed in spite of the possibility of alternative deprotonation, which would have conserved aromaticity in the annelated benzenoid ring, as in **246** (R^1 = H).

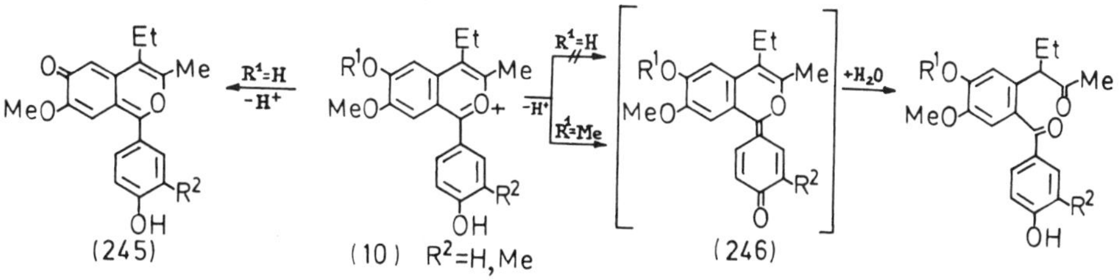

At the same time, if R^1 = Me in salt **10,** the deprotonated form **246** was detected by UV spectra [75ACH(85)79], and this unstable betaine was rapidly converted into the corresponding diketone. The explanation for the instability of betaine **246** was based on the suggestion of steric hindrance for coplanarity of the aromatic ring attached to C_1 and the heterocyclic ring. However, data obtained by Shcerbakova and co-workers (87RRC417) contradict such a conclusion. Thus, betaines **248** have been isolated in the crystalline state; on the other hand, steric shielding by two *t*-butyl groups may also contribute in this case to stabilization.

(247)   (248)

The stability of betaines of type **246** and **248** is thus mainly determined by the environment of the deprotonated hydroxyl group, and their further transformations may be sterically retarded (cf. Section III,E,2). At the same time, on treatment of salt **249** with triethylamine in anhydrous solvents, betain **250** was detected by electronic absorption spectra as a short-lived compound. Its instability is obviously explained by the disturbance of aromaticity in the annelated benzenoid ring, resulting in an *ortho*-quinonoid fragment (87RRC417).

(249)   (250)

Interestingly, such a perturbation does not destabilize a molecule with a *para*-quinonoid fragment if this involves deprotonation of a hydroxyl group at the annelated ring itself, as in **241.** At the same time, the tendency to avoid a disturbance of aromaticity in the annelated benzenoid ring results in deprotonation of the methylene group instead of the hydroxyl in 6,7-dimethoxy-substituted salts **251** (87KGS1032).

The stability of anhydrobases of type **252** is determined by the nature of substituents at the double bond: if R is H or Alk, it was impossible to isolate anhydrobases (cf. Section III,F,2); if R is Ar, anhydrobases **252** are stable (82KGS1184), obviously because of conjugation of the resulting double bond with two aryl rings.

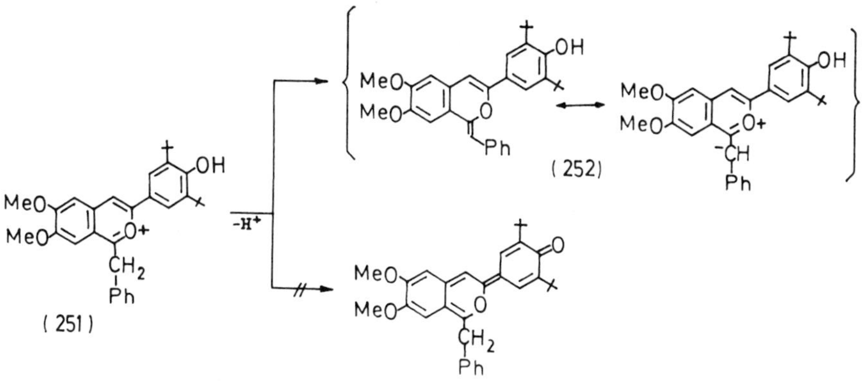

(251)   −H+

(252)

2. Reactions of Anhydrobases and Betaines

Benzo[c]pyrylium-3-oxides **240** having an *ortho*-quinonoid resonance structure act as dienes in reactions with dienophiles (84TL3659).

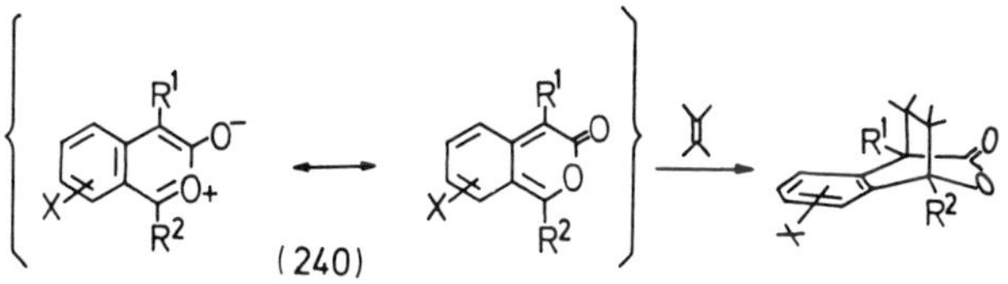

(240)

Benzo[c]pyrylium-4-oxides of type **15** act as 1,3-dipoles in cycloadditions (84KGS167).

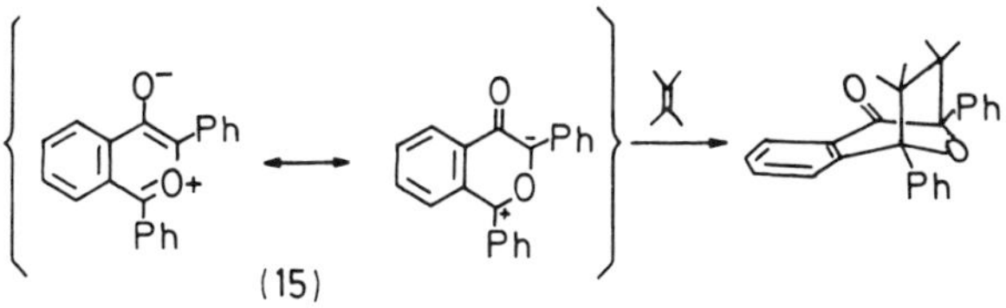

(15)

Interestingly, betaine **15** possesses high reactivity towards cycloaddition. Thus, even in conditions of its formation from epoxyindanone **14** in the absence of foreign reagents, betaine **15** captures the carbonyl group of the initial indanone as a dipolarophile (62JA1315; 64JA3814), forming a dimer or a mixture of dimers.

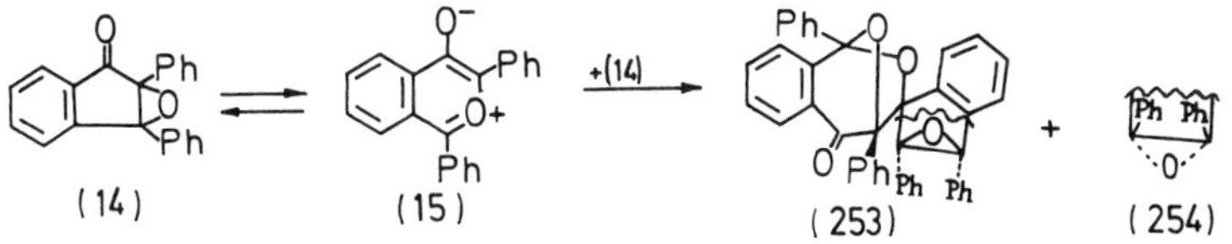

(14)   (15)   +(14)   (253)   +   (254)

The stereochemistry of the resultant adducts depends on the reaction conditions: on thermolysis of **15,** two steroisomers **253** and **254** are formed, whereas photolysis leads only to the isomer **253.** In the absence of external dipolarophiles, betaines of type **15** are dimerized with formation of the symmetrical dimers (Section III,F,1).

Betaines of type **241** react slowly with ammonia, yielding isoquinolines **255,** whereas the interaction with aqueous-alkaline solutions in the presence of alkylation reagents leads to β-naphthol derivatives **256** [68ACH(57)181].

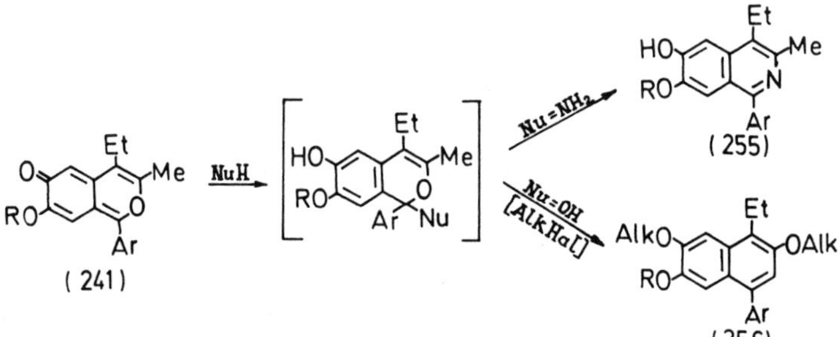

The mechanism of these reactions probably includes 1,6-addition (cf. Section III,C,4,a,i).

The deeply colored betaines **248** are stable in concentrated aqueous solutions of alkali hydroxides, but in organic solvents in the presence of water, even traces of bases cause their ring-opening conversion into the corresponding colorless diketones **257,** probably also via a 1,6-addition mechanism.

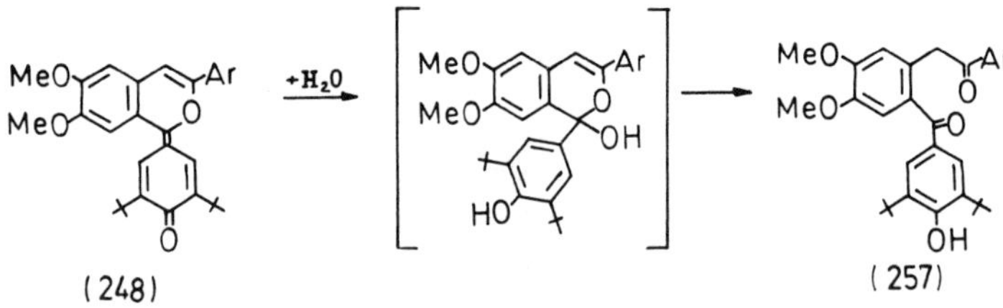

In contrast to **248,** anhydrobases **212,** formed on C-deprotonation, do not react with alkalies. On the other hand, in acidic nucleophilic medium, according to the duration of heating, these compounds are converted into diketones **205** and then into α-naphthols **204** (82KGS1184).

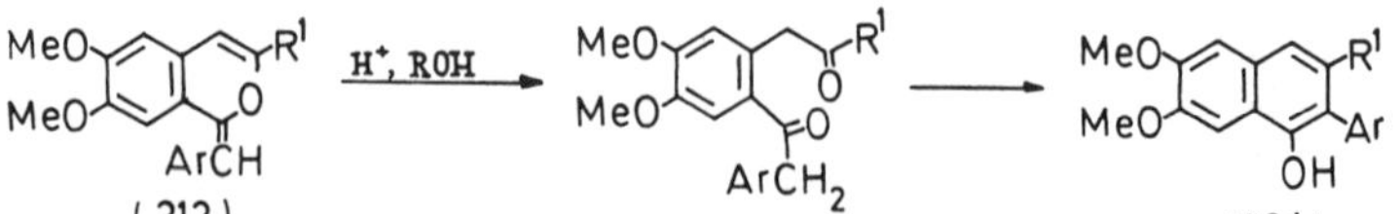

Due to an excess of electron density at the exo methine group in **212**, these anhydrobases are readily oxidized to isocoumarines **258** (78KGS1320), and depending on the nature of the cationic substrate, compounds **212** may form a charge-transfer complex **259** with the initial salt **184** or may act as one-electron donors in reaction with the 1-benzopyrylium salt **260** (77KGS1138; 77TH1).

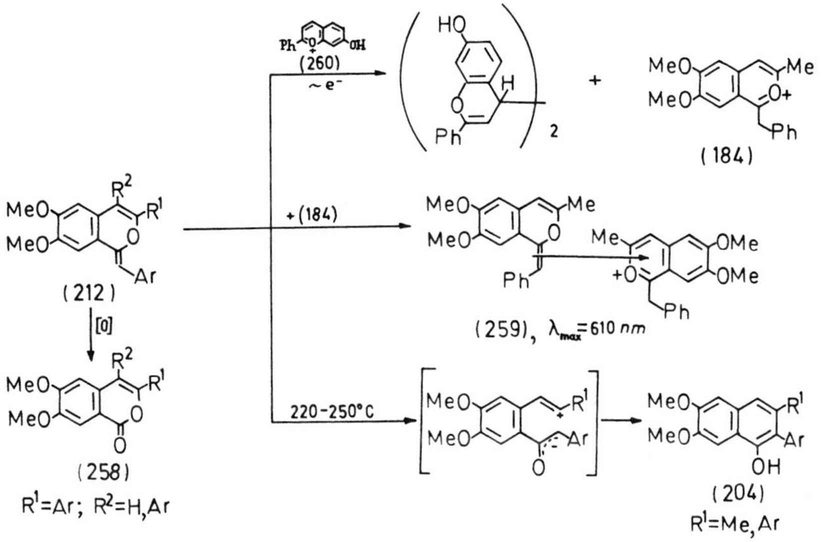

Heating anhydrobases **212** without any external reagent at elevated temperature gives rise to α-naphthols **104** (86KGS276). Therefore, a real isomeric recyclization of 2-benzopyrylium salts (cf. Section III,C,4,a,ii) may occur and is carried out in two steps, with the last one requiring severe conditions.

F. DIMERIZATION REACTIONS

1. Formation of Symmetrical Dimers

A classical type of dimerization of heterocyclic compounds is the formation of a twin molecule consisting of the same fragments. Such a reaction is known for 2-benzopyrylium salts. Thus, the reduction of the 2-benzopyrylium salt **261** by zinc dust leads to the intermediate radical **262,** which is dimerized *in situ* to the bisisochromene **263** (76KGS999) (cf. Section IV,B).

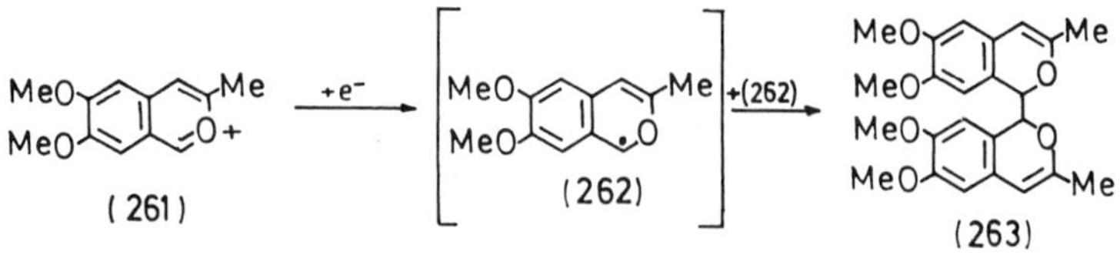

All attempts to convert dimer **263** into a dimeric 2-benzopyrylium salt, on treatment with triphenylmethyl or acetyl perchlorate, lead only to the rupture of the newly formed C—C bond and to the regeneration of the initial monomeric salt **261,** unlike the behavior of dimers of monocyclic pyrylium cations [73DOK(212)370]. Dimerization may be considered a typical reaction for benzo[*c*]pyrylium-4-oxides of type **19,** which react in dimerizations as 1,3-dipoles by analogy with their behavior in cycloadditions (Section III,E,2).

Interestingly, the stereochemical result of dimerization of the unsubstituted oxide **19** depends on the substituent in adduct **264,** from which this betaine is generated (Scheme 13) [76ACS(B0)619].

The stereochemistry of dimer **265** (R^1 = Ph, R^2 = Me) obtained from the corresponding betaine (64TL1847) was not investigated.

2. Formation of Nonsymmetrical Dimers

a. *Introduction.* As discovered, 2-benzopyrylium salts take part in a series of dimerization reactions yielding nonsymmetrical dimers, contrary to the symmetrical ones obtained from 2-benzopyrylium derivatives by the classical type of dimerization described earlier. The essence of the newly found types of dimerization includes the formation of a thermodynamically unstable intermediate generated from the 2-benzopyrylium cation with participation of a nucleophile or a base. Such an intermediate assumes nucleophilic properties towards the initial 2-benzopyrylium cation, leading to interaction between these two components, which results in various nonsymmetrical dimers.

The wide range of conditions for such a process, the different types of

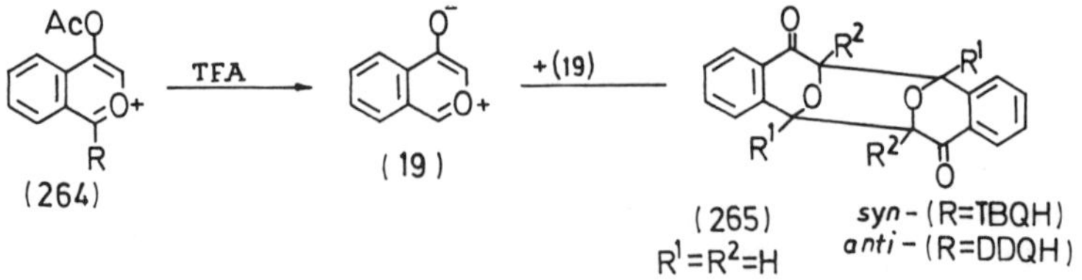

SCHEME 13. DPQ, 2,3-dichloro-5,6-dicyano-1,4-benzoquinone; TBQ, tetrachloro-1,2-benzoquinone.

dimerization, and the multiplicity of further conversions of dimers thus formed into chrysenes, anthracenes, naphthalenes, etc., constitute unique and characteristic reactions of 2-benzopyrylium cations. No other hetero-cycle is known to afford such a variety of dimeric products.

b. *α-1′ Dimerization with Formation of Chrysenes.* It was found that 1-methyl-substituted benzo[c]pyrylium salts with a vacant β-position in the heterocyclic moiety give rise to acyl-chrysenes **270** on heating in aqueous alcoholic alkaline solutions (83KGS274; 84KGS1472; 85KGS-910).

1,5-Diketones **29** are not intermediates in the formation of chrysenes because under the reaction conditions they form only α-naphthols **204**, which are products of intramolecular cyclization. The real intermediate for chrysene formation is the anhydrobase **267**. The key role of this compound is supported by the isolation of the dimeric pseudobase **269** (85KGS910), as well as by trapping of the latter compound with the "proton sponge," which possesses pronounced basic properties (68CC723).

Obviously, the formation of dimer **269** is due to the negative charge at the exo methine group of **267**, which is not screened by substituents, and because of this fact, the α-methine carbon atom can add to position 1 of another unreacted 2-benzopyrylium cation. Such an interaction may be called *α-1′* dimerization; it leads to dimeric salt **268** and then to pseudo-base **269**.

A similar type of dimerization is known for quinolinium (82KGS291) and 1-benzopyrylium (69TL2047) salts, but dimers thus formed are unable to undergo further intramolecular transformations. Contrary to these results, dimer **269,** which may be considered as a C-adduct between the initial benzopyrylium cation (fragment B) and a compound possessing an active methylene group (fragment A), is converted into acetyl-chrysene **270** on heating in alkaline solutions (85KGS910). On brief heating, a hydroxy-dihydro-derivative **271** may be isolated, and this compound is readily dehydrated to the aromatic chrysene system on further heating or on recrystallization from acetic acid.

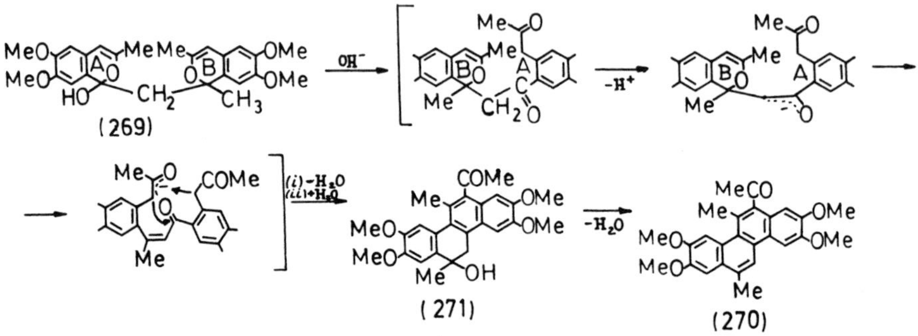

Therefore, most probably, the formation of chrysenes **246** from 2-benzopyrylium salts occurs via the process of α-1' dimerization. At the same time, one cannot exclude the action of anhydrobase **267** as dieno-phile, which leads to its addition to positions 1 and 4 of the initial 2-benzopyrylium cation by analogy with [4 + 2] cycloadditions in reaction with ethyl vinyl ether (Scheme 14) (cf. Section III,D,1).

Probably, the experimental conditions (temperature and solvent) determine the conversions of 2-benzopyrylium salts **266** into the chrysene derivatives **270** via α-1' dimerization or [4 + 2] cycloaddition (Scheme 14) (89TH2).

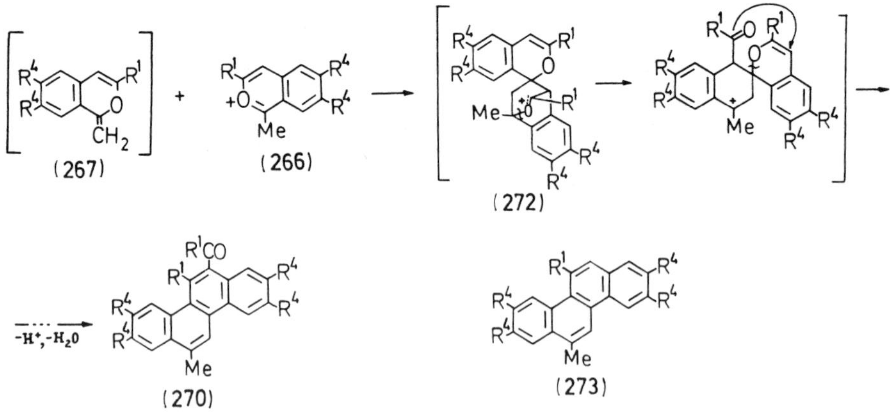

SCHEME 14

The formation of acyl-chrysenes 270 from 1-methyl-substituted ben-zo[c]pyrylium salts 266 occurs not only on heating in alkaline solutions, but also in acidic nucleophilic media (cf. Section III,C,4,b,i). However, in the latter case, together with acylchrysenes 270, their deacylated analogs 273 are formed (85KGS910). Under these conditions, the same results were obtained for conversions of diketones 29, therefore it is difficult to conclude which compound (anhydrobase 267 or diketone 29) is the inter-mediate in the formation of chrysenes 270 and 273 from salts 266 in acidic nucleophilic medium. It was not possible to trap or detect the dimeric pseudobase 269 under these conditions. However, the latter compound, under the described conditions or on heating in acetic acid, forms the mixture of the same products (270 and 273).

The pathway to chrysenes, in this case, was assumed to include the formation of dimeric salts 268. Then the intramolecular interfragment interaction takes place with the formation of the C_{4B}—C_{1A} bond, which leads to the already familiar intermediate 272 (Scheme 14).

However, on treatment of the dimeric pseudobase 269 in acetic acid with 70% perchloric acid without heating, the initial monomeric salt 266 is regenerated. Obviously, at first the dimeric salt 268 is formed, but then it undergoes a decomposition into the initial 2-benzopyrylium salt in analogy with the bisisochromene 263 (cf. Section III,F,1). On rapidly heating the initial mixture, the interactions result in chrysene 273 in 30% yield along with regeneration of the 2-benzopyrylium cation (85KGS910).

The formation of deacylated derivatives 273 in acidic media may be explained by ipso-protonation of acyl-chrysenes 270. On the other hand, they may also be explained by reactions at preceding stages by analogy with transformations known for monocyclic pyrylium salts (68TL4379; 69JOC2736; 72TL4439) (cf. Section III,C,4,d and III,F,2,c). Chrysenes

270 are also formed from some adducts of 2-benzopyrylium salts under the action of concentrated alkaline solutions (88T6217). Since dissociation of adducts **100** with formation of the initial cation is hardly probable under these conditions, the dimeric salt **268** is formed probably by an S_N2 mechanism, and the anhydrobase **267** generated by elimination of NuH is one of the components, whereas the other one is the initial adduct **100**.

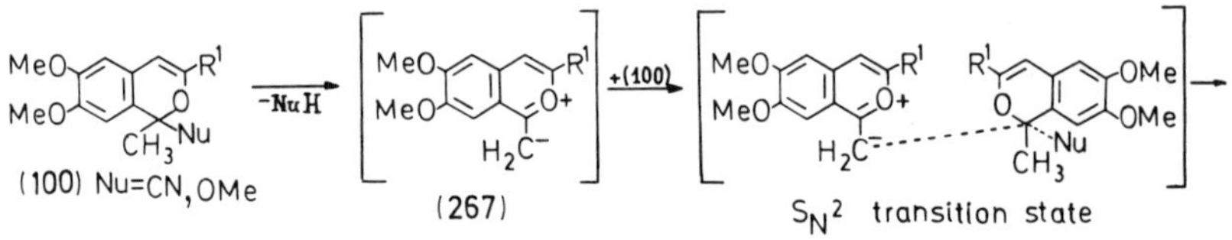

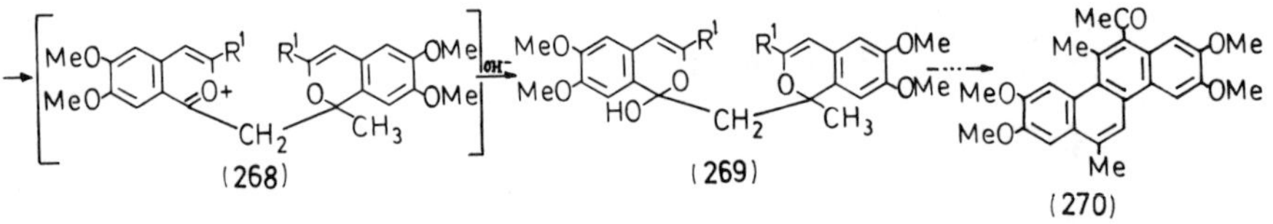

Interestingly, yields of acyl-chrysenes **270** thus obtained are higher than on their formation from 2-benzopyrylium salts (85KGS910). Moreover, salt **266** (R^1 = H), unsubstituted in position 3, forms a complex mixture on heating with alkaline solutions from which the corresponding acyl-chrysene was detected in low yield only by thin-layer chromatography, whereas it was isolated in a yield of 60% from cyanoisochromene **100** (R^1 = H, Nu = CN) under the same conditions (88UP1).

Acyl-chrysene **270** is formed on thermolysis of methoxyisochromene **109** in 50% yield (86UP3). By analogy with behavior of the adducts between 2-benzopyrylium salts and acetoacetic ester **235** (cf. Section III,C,4,d), one cannot exclude the electrocyclic ring-opening of the heterocyclic fragment in **109**. The *ortho*-quinonoid compound **274** thus formed is a diene, and the vinyl fragment in the unreacted adduct **109** plays the role of an external dienophile.

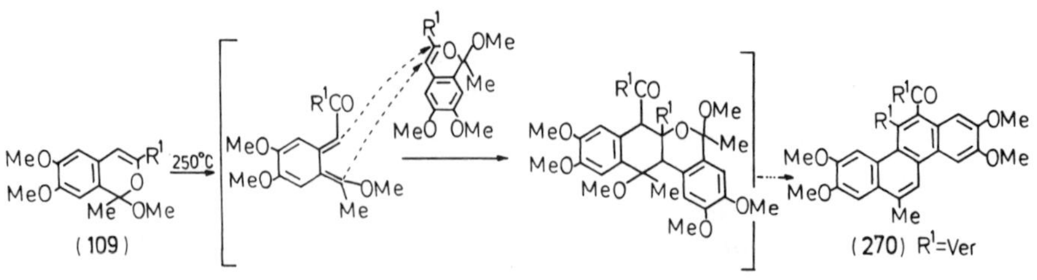

c. *4-1' Type Dimerization.* Whereas α-1' dimerization involves a participation of the exocyclic double bond of anhydrobase **267**, in 4-1' dimerizations, the reactive center is the endo fragment of cyclic vinyl ether in adducts of type **100**. However, for the success of the latter process, the adduct **100** must not be too stable (Nu should not be OAlk, CN, Alk; cf. Section III,C,2), and at the same time, it must not be too labile towards ring opening (Nu should not be NH_2R, NHR_2; cf. Section III,C,4,a). These restrictions are satisfied on carrying out the reaction in DMF, pyridine, alkaline solutions of specific concentrations, or in the presence of the thiolate ion, and in some cases, in the presence of morpholine or enamines.

The reaction of 4-1' dimerization is similar to the primary step of recyclization of 2-benzopyrylium salts in acidic nucleophilic media (cf. Section III,C,4,b,i), but the reactive electrophile is the initial cation in this case, and not a proton. Probably for this reason, the 4-1' dimerization of 2-benzopyrylium perchlorates is not observed in acidic nucleophilic media, in contrast to α-1' dimerization (cf. Section III,F,2,b). At the same time, the scope of 4-1' dimerizations is less restricted in terms of structural requirements for 2-benzopyrylium salts in comparison with a α-1' dimerization. Thus, in the latter case, the presence of a methyl group in position 1 of benzo[c]pyrylium cation is compulsory, whereas for 4-1' dimerizations, the nature of the substituent in this position may be different, leading to a variety of 4-1' dimers and, as a consequence, to a wide variety of their transformations.

Thus, dimer **276** is obtained from salt **30** at room temperature in a diphase system of 10% aqueous solution of sodium hydroxide and ether. On heating with alkali in isopropyl alcohol, these dimers are converted quantitatively into benz[a]anthracenes **278** (88KGS1185), and likewise the formation of dihydro-derivatives of chrysenes **271** (85KGS910) and **277** precedes the final result. In analogy with the formation of chrysenes, on treatment of acetylanthracene **278** with strong acids, an *ipso*-substitution of the acetyl group occurs, leading to a deacylated anthracene **279.**

(276) R^1=Me; R^3=Ar

(277)   (278) Ar=Ph,An   (279)

The formation of anthracene **278** (Ar = Ph) directly from the corresponding 2-benzopyrylium salt **30** occurs in low yield, and the final result is dependent both on the nature of the alcohol used and on the concentration of the alkaline solution. Thus, on treatment of this salt **30** with 2% aqueous alkaline solution, β-naphthol **209** was the only product, whereas with 50% concentration of alkaline solutions, anthracene **278** was formed, but in a yield of less than 20%. A somewhat better yield (40%) of **278** (Ar = Ph) was reached on carrying out the conversion of 2-benzopyrylium salt in isopropyl alcohol containing sodium isopropoxide. However, one can conclude that independent of conditions, the formation of anthracene **278** from 2-benzopyrylium salt in alkaline media does not occur via the intermediate diketone **29** (R^1 = Me, R^3 = Ph). This is because the latter compound gives rise only to β-naphthol **209** under the reaction conditions.

A specific feature of these conversions of dimers **276** (R^3 = H) is produced by the formyl group. Thus, treatment of **281** with a mixture of aqueous alkaline solution and dioxane at room temperature leads to cyclic semiacetals **284.** The precursor compounds towards **284** are the ring-opened intermediates **282,** which can be also isolated on treatment of dimer **281** with aqueous sodium acetate [90KGS(ip3)]. Heating of semiacetal **284** (as well as **282** and dimer **281**) in concentrated alkaline solutions results in the naphthalene derivative **285,** which is obviously formed via ring opening to the hydroxyaldehyde **283,** followed by elimination of water.

Acid **286** is another product of this reaction, and its formation may be explained by a hydride ion transfer between aldehydic and ketonic groups in the hydroxyaldehyde **283,** followed by elimination of a molecule of

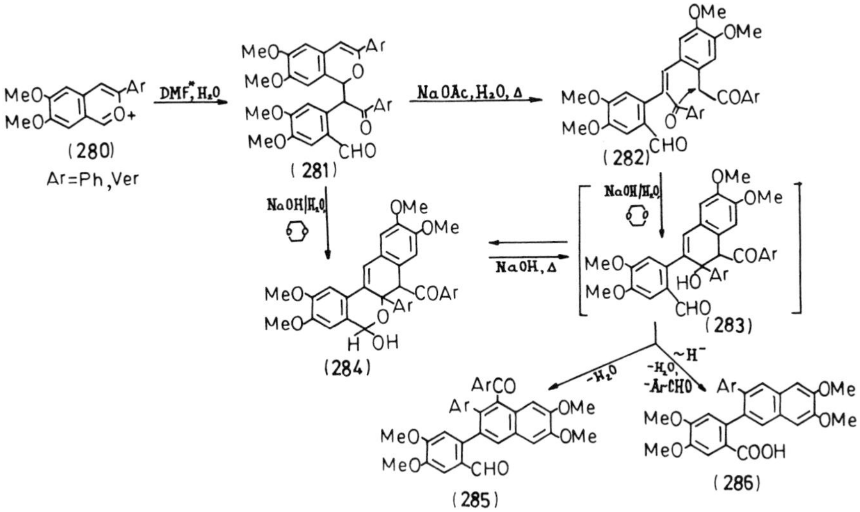

SCHEME 15. On treatment of 2-benzopyrylium salts **280** with DMF, the formation of dimers **281** is accompanied by disproportionation to a small degree, as described in Section III, F,2,d.

aromatic aldehyde. Such an unusual 1,7-analog of the Cannizzaro rearrangement is favored obviously by a conformational flexibility of the molecule **283**, allowing close proximity. Such a process does not occur in the case of aromatic naphthalenes **285** (Scheme 15).

Contrary to α-1′ dimers **269**, the 4-1′ dimers **276** are more susceptible toward regenerating the starting 2-benzopyrylium salts in acidic media. However, under special conditions, one can carry out some transformations of dimers **276** determined by their interfragment interactions. Thus, thermolysis of dimer **281** with catalytic amounts of acetic acid results in the naphthalene derivative **287**. This compound, together with another naphthalene derivative **285**, is also formed from semiacetal **284** on heating in acetic acid [90KGS(ip3)].

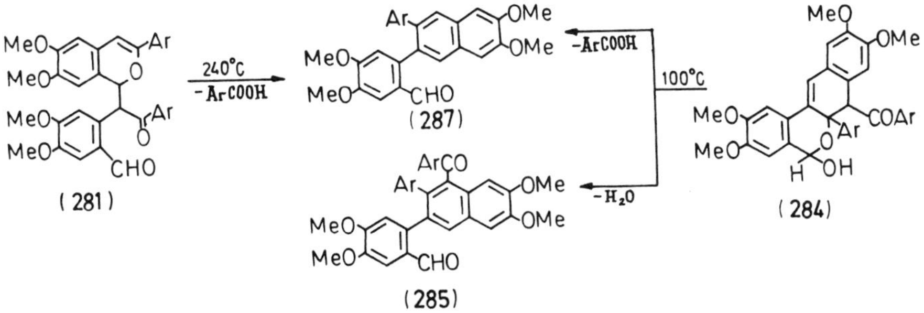

The aromatization step with elimination of a molecule of carboxylic acid occurs in the case of dimer **276** at room temperature in formic acid.

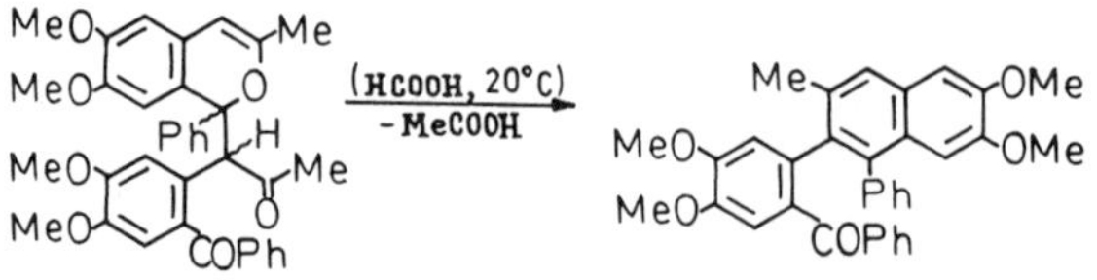

Probably the formation of naphthalenes **287** and **288** from dimers **276** occurs in the presence of acids, in accordance with the mechanism proposed for acid-catalyzed transformations of adducts formed from 2-benzopyrylium salts and acetoacetic ester **235** (cf. Section III,C,4,d).

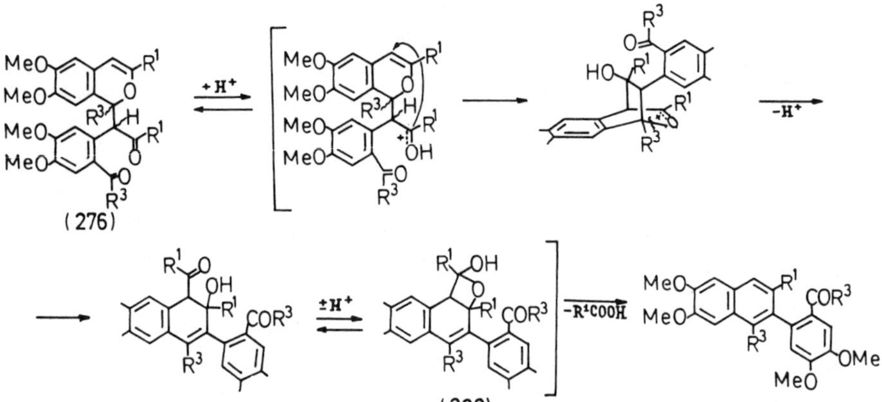

The preferred elimination of a molecule of carboxylic acid instead of water in this case may be explained by the intermediate formation of a bicyclic oxetane **289**. Along with conversions described here, which have some analogy with what is known from α-1′ dimerizations and reactions of monomeric C-adducts, a new transformation has been found in the case of 4-1′ dimerization, namely disproportionation, as shown in the following paragraphs.

d. *Disproportionation of 2-Benzopyrylium Cations via Intermediary 4-1′ Dimerization.* On reacting 1-unsubstituted 3-aryl-benzo[c]pyrylium salts with some nucleophiles, a new type of disproportionation was found [90KGS(ip1)]. Its peculiarity in relation to similar reactions for heterocyclic cations (79MI1), includes dimerization of 2-benzopyrylium cations followed by disproportionation of the dimer. The extent of this process, and consequently its results, depends on the reaction conditions used and on the nature of the added nucleophile. The latter governs the stability of cationic intermediates **291** formed by the ring opening of the "lower" ring in the cyclic carboxonium ion **290**. If a stabilization of **291** is possible (Nu = NR$_2$), then mainly hydride ion transfer takes place; if Nu = OH, the cation **291** mainly undergoes deprotonation to dimer **281** (Scheme 16).

A corroboration of the key role of the intermediate cation **291** is the observation of a sharp yield increase of disproportionation products **297** and **298,** compared to their direct formation from salt **280** on treatment of dimer **281** with catalytic amounts of triethylammonium perchlorate, which acts as the protonating agent generating intermediate **293**. The formation of "unsaturated" dimers **296** is possible not only by the direct hydride transfer, followed by deprotonation of dimeric salt **294** (pathway *a*, Scheme 16), but also by equivalent 1,5-hydrogen transfer in the *ortho*-quinonoid intermediate **295** formed by deprotonation of the acidic methine group in intermediate **291** (pathway *b*).

The cleavage of unsaturated dimers **296** is determined by hydration of the double bond, followed by rupture of the C—C bond by analogy with

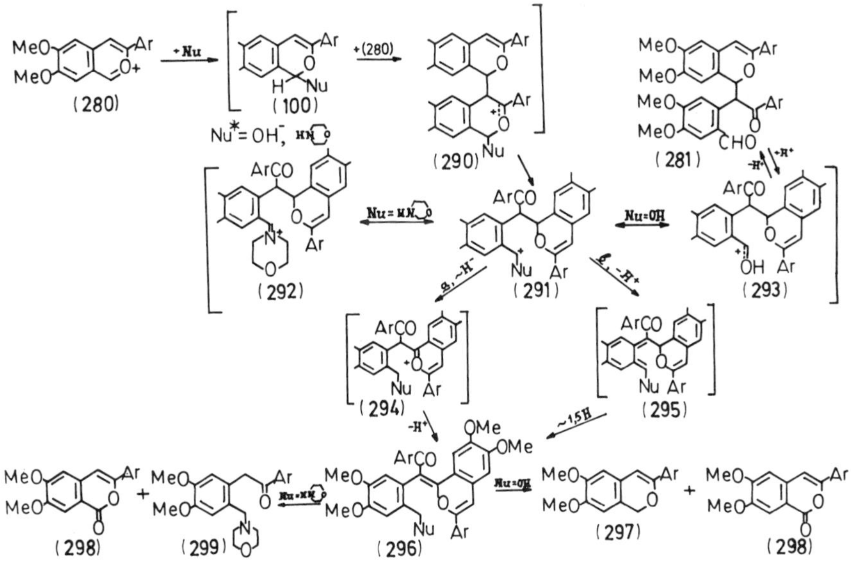

SCHEME 16. In addition to these nucleophiles, *N*-cyclohexenomorpholine was used, which reacted as nitrogen nucleophiles with formation of the corresponding adduct, whose ring opening is hindered.

the behavior of β-diketones (50MI1). Model experiments support this conclusion (cf. Section III,B). Since disproportionation of monomeric heteroaromatic cations is similar with the Cannizzaro reaction, the behavior of ketoaldehydes **300** (which are the ring-opened isomers of pseudobases of the corresponding 2-benzopyrylium salts) was checked in analogous conditions (89KGS1039). The previously unknown version of intramolecular Cannizzaro rearrangement, namely, the 1,5-rearrangement (cf. Section III,F,2,c) was found to take place in this case, but not in the intermolecular Cannizzaro reaction.

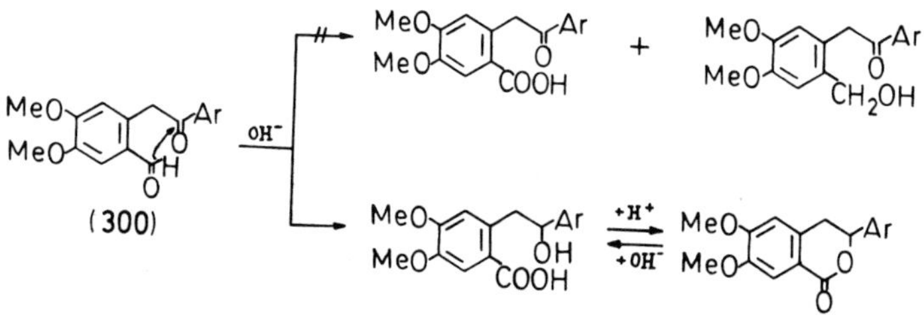

e. *Conclusion.* The unusual ease of dimerization of 2-benzopyrylium cations is determined, as one might suppose, by a possible stacking aggregation formed analogously by the planar pyridinium salts in solution [85JCS(P2)1895] and by the skew *N*-phenylpyridinium chloride hydrate (78RRC1065), which presents pairwise aggregates in crystals. Probably the mutual attraction of 2-benzopyrylium cations is favored by the partial charge localization in the hetero fragment, and the annelated benzenoid ring in this case preserves some π-excess. Such a separation of the π-deficient and π-excess fragments belonging to the common conjugated system probably overcomes the repulsions by the formation of the remote π-complexes as is shown here.

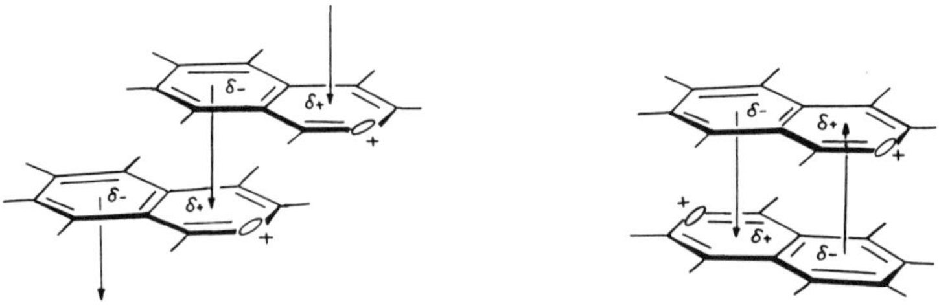

IV. Physical Properties of Benzo[c]pyrylium Salts

A. SPECTRAL PROPERTIES

1. Optical Spectroscopy

a. *Electronic Absorbtion Spectra.* A systematic study of a large series of 1-aryl-substituted benzo[c]pyrylium salts was published by Vajda and Ruff [64ACH(40)225]; the absorbtion bands in UV absorbtion spectra were investigated for substituents in positions 3, 4, 6, and 7, as well as in 1-aryl ring. The observed maxima have been divided into five groups.

Band group I appears as a few peaks with molar extinction coefficients about 10,000 at wavelengths lower than 235 nm, but no general rules can be made for the substituent effects on these bands.

Band group II can be found at 230–280 nm; intensities (E_{max}) vary from 20,000 to 30,000. Donor substituents shift the maximum to higher wavelengths.

Band group III can be recognized only in the spectra of unsubstituted, mono- and di-substituted compounds at 280–300 nm (E_{max} between 6000–12,000). In other compounds, the band seems to coalesce with the last band of group II.

Band group IV appears at 330 nm if no auxochromic substituent is attached to the molecule. 6-Methoxy-substitution displaces the maximum to 318 nm, while substitution in position 6 displaces the band bathochromically. This band is also characteristic for 1,3-diaryl-substituted salts, and it appears at ~320–325 nm in acidic alcoholic solutions (87RRC417).

Band group V is recorded at about 400 nm in the spectra both of 1-aryl-[64ACH(40)225] and 1,3-diaryl-substituted benzo[c]pyrylium salts (87RRC417). The batho- and hypochromic effects are discussed in detail in the case of multiple substitution in 1-aryl-benzo[c]pyrylium cations.

An additional band at ~490 nm was noted for 1,3-diaryl-substituted benzo [c]pyrylium salts (87RRC417), and the position of this band, as well as of the band group V, gives a good correlation with the polarity of solvents for these compounds.

b. *Infrared Spectra.* Vibrational spectra of monocyclic pyrylium salts have been investigated fairly extensively; the data obtained for these heterocyclic cations were discussed in comparison with other six-membered monocyclic aromatic systems with which they formed a regular series [82AHC(Suppl)]. The same approach was used by Vajda and Ruff

[64ACH(40)217] for investigating IR spectra of different 1-aryl-substituted benzo[*c*]pyrylium salts **10** in comparison with the corresponding iso-quinolines. Dorofeenko *et al.* (66ZOR1492; 70TH1) analyzed IR spectra of various 1,3,6,7-substituted benzo[*c*]pyrylium salts **30,** and some common features have been revealed for the whole series.

(10) $R^1, R^2 = H, Alk;$
$R^3 - R^6 = H, OH, OMe$

(30) $R^1 = H, Me, Ar;$
$R^2 = H, Alk; R^3 = R^4 = H, OMe$

Since compounds **10** and **30** are rather complicated, the authors could not attempt to determine the normal frequencies of the molecule, but had to restrict the analysis of spectra to the determination of characteristic frequencies. Thus, six frequencies have been found that can be used to identify 2-benzopyrylium cations. Band I appears between 1650–1610 cm^{-1}, and the pyrylium ring is responsible for this vibration [8a band according to Wilson's notation (34MI1)]. The position of this band and its intensity are dependent on the nature and position of substituents in the cation, and these changes are similar to data of monocyclic pyrylium salts [82AHC(Suppl)].

Band II appears in the region 1550–1530 cm^{-1}, and it is analogous to vibration 8b of the pyrylium cation, with a similar frequency range. Inter-estingly, methoxy groups in positions 6 and 7 enhance the intensity of this band until it becomes the most intense one of the spectrum. Comparison of bands I and II with similar type bands of isoquinolines with analogous structures [64ACH(40)217] shows results similar to those for pyridine and pyrylium salts (62T1083).

The next two characteristic frequencies are bands III (1480–1460 cm^{-1}) and IV (1430–1410 cm^{-1}), which are assigned to the normal pyrylium ring vibrations 19a and 19b, respectively. Although bands V and VI (670 and 580 cm^{-1}) can always be found in the spectra of 2-benzopyrylium salts, they are not assigned to definite vibrations, in contrast to monocyclic pyrylium salts (62T1083). As for stretching and bending vibrations of the ring hydrogen atoms on the 2-benzopyrylium ring, only one stretching vibration could be identified at 3120 cm^{-1}, which appears in the spectrum of 1-phenyl-benzo[*c*]pyrylium perchlorate [64ACH(40)217]. In addition to the bands described here, the out-of-plane bending vibrations of the hydro-gen atoms on the hetero ring appear at 1615–1600 and 1525–1500 cm^{-1}, as in aromatic compounds.

Bands at 1280–1260 cm^{-1} and 1040–1010 cm^{-1} are responsible for vibrations of the methoxy groups in spectra of the methoxy-substituted salts, and frequencies of the perchlorate anion were found at 1090 and 623 cm^{-1}.

2. NMR Spectra

So far, only a few papers have appeared on NMR spectra of 2-benzopyrylium salts. The ^{1}H-NMR spectra of eleven 1-R^1-3-R^2-6,7-dimethoxy-benzo[c]pyrylium salts were interpreted with the use of a complete neglect of differential overlap (CNDO/2) method in standard geometry, and a fair correlation between charge densities and chemical shifts has been obtained (90MI3). The experimental δ values for this series are presented here.

The ^{13}C-NMR spectra of 1-arylbenzo[c]pyrylium salts have been investigated in comparison with those of 1-arylnaphthalenes, N-methylisoquinolinium, and 1-arylisoquinoline derivatives [75ZN(B)943]. The correlation obtained is analogous to relationships found for a series of pyrylium and pyridinium salts, and pyridine derivatives [82AHC(Suppl)]. The ^{13}C shifts of 1-aryl-benzo[c]pyrylium salts are also a valuable source of information for pK_a values, aromatic character, and conjugative effects of these compounds (760MR324).

An unusual assumption has been made for correlating the ^{13}C-NMR chemical shifts of some alkoxy-substituted 1-aryl-benzo[c]pyrylium salts with all-valence charge densities of the various atoms calculated by the Del Bene-Jaffe method [78ACH(96)373]. It has been supposed that the sp^2-hybridized C-1 atom exits in the geometry of sp^3 hybridization, and only in this case is a good linear correlation between the charge densities and shifts obtained. For 1,3-dimethyl-6,7-dimethoxybenzo[c]pyrylium cation, both planar and pyramidal valence orientations of the methyl group at the C-1 atom gave a good correlation between the charge densities and the ^{13}C chemical shifts.

3. *Electron Spin Resonance Spectra*

Although some radicals and cation radicals are postulated for chemical and electrochemical transformations of 2-benzopyrylium cations (Sections III,F,1 and IV,B), attempts to record their electron spin resonance (ESR) spectra failed, obviously because of a low stability of these radicals. However, the structural combination of hydroxyaryl and 2-benzopyrylium fragments favors the formation of radical cations 301–303, and their ESR spectra were recorded on oxidation of the corresponding 2-benzopyrylium salts with lead tetraacetate (87RRC417).

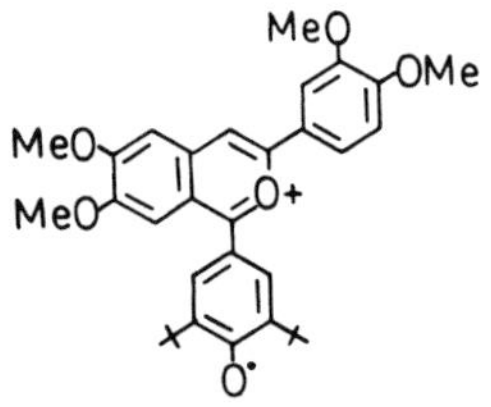

(301) 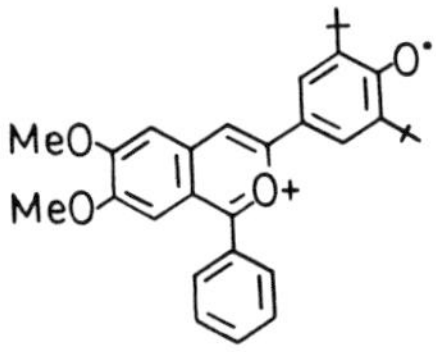   (302) 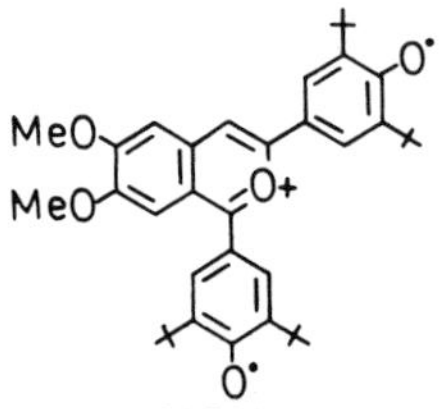   (303)

Similar to aroxyl-benzo[*b*]pyrylium systems (78ZOR1643), there is no detectable delocalization of the unpaired electron on the benzo[*c*]pyrylium ring in 301–303 (87RRC417). The stability of the radical cation is determined in this case by the position of the aroxyl substituent in the cation, and the most stable is 301. The formation of diradical species 303 together with the monoradical cation, was indicated on oxidation of the corresponding 2-benzopyrylium salt.

B. ELECTROCHEMICAL PROPERTIES

Vajda and Ruff [64ACH(40)225] were the first to determine the pK_a of 2-benzopyrylium ions, namely different hydroxy-substituted 1-aryl-benzo[*c*]pyrylium perchlorates of type 10.

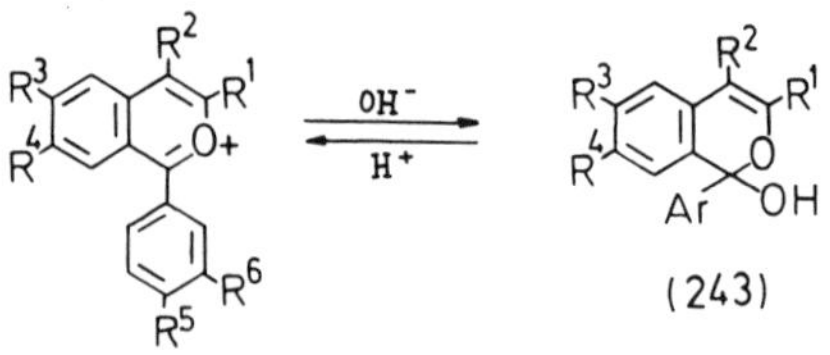

(10)   (243)

The pK_a values obtained by polarographic study were influenced by kinetic factors (recombination at the electrode) and thus did not represent the thermodynamic values [64ACH(40)225]. In the same paper, the UV spectral data on 1-arylbenzo[c]pyrylium salts were reported to be dependent on pH values. From the pH dependence of the spectra, pK_a values of the salts were computed, and although the general trend of polarographic and spectrophotometric pK_a values was similar, the numerical values were markedly different. The correlation of the $E_{1/2}$ values with spectrophotometric pK_a values also showed a marked break, therefore, it seems possible that the stability of the bases 243 is one of the parameters which influence the polarographic behavior and possibly the reduction mechanism as well.

The spectrophotometric pK_a values were found in the range of 0.60–5.40, depending on the nature of substituents in the cation 10. Thus, alkoxy substituents R^3, R^4, and R^5 enhance the stability of the salt, and a 6-methoxy group exerts the greatest influence since, in the presence of this group, the pK_a value is always greater than 3.

As was reported later [75ACH(85)79], all compounds 10 show well-defined one-electron reduction waves in solutions of strong acids, and detailed polarographic measurements have been carried out for hydroxy-substituted 1-arylbenzo[c]pyrylium salts in order to study their deprotonation and further transformations at various pH values. The data obtained are in good agreement with earlier results on electronic absorption spectra [64ACH(40)225] and chemical transformations [65M369; 68ACH(57)181].

Interestingly, the reduction mechanism of 2-benzopyrylium salts 10 at the dropping mercury electrode involves the disproportionation of a radical intermediate 304 in acid to 10 and 107 (72MI3).

the structures 10, 304, and 107

At the same time, dimerization of one-electron reduction product was indicated by oscillopolarographic studies in acid for 1-R-3,4-diphenyl-6,7-dimethoxybenzo[c]pyrylium perchlorates (R = Me, Et, Ph) (76MI1). As was noted in Section III,F,1, the benzo[c]pyrylium cation 261, unsubstituted in position 1, undergoes 1,1'-dimerization via the postulated intermediate radical, and the bisisochromene 263 thus formed is readily converted into the initial monomeric salt, but not into the corresponding biscation on treatment with chemical electron acceptors (76KGS999).

In order to compare the chemical and electrochemical behavior of 1,1'-bisisochromene 263, its electro-oxidation was studied (80KGS1421).

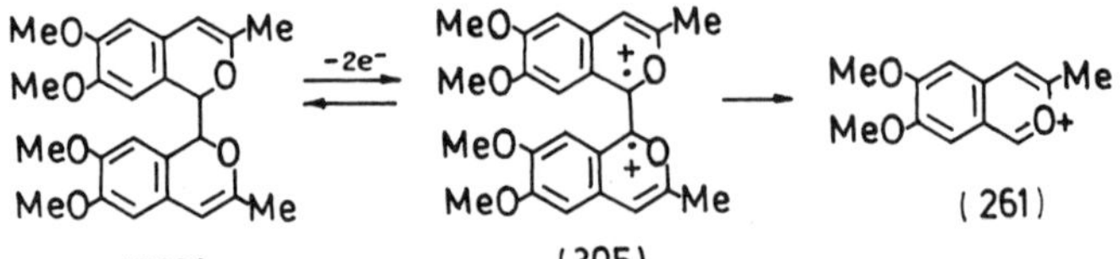

(263)   (305)   (261)

It was found that the oxidation mechanism consists of a reversible two-electron oxidation of **263** to diradical-cations **305**, which are rapidly decomposed to monomeric cations **261**. Interestingly, whereas the chemical oxidation of 1-phenyl-1*H*-isochromene **306** results only in the 1-unsubstituted cation **261**, the electrochemical oxidation gives rise to two types of cations **261** and **154** (81KGS1177).

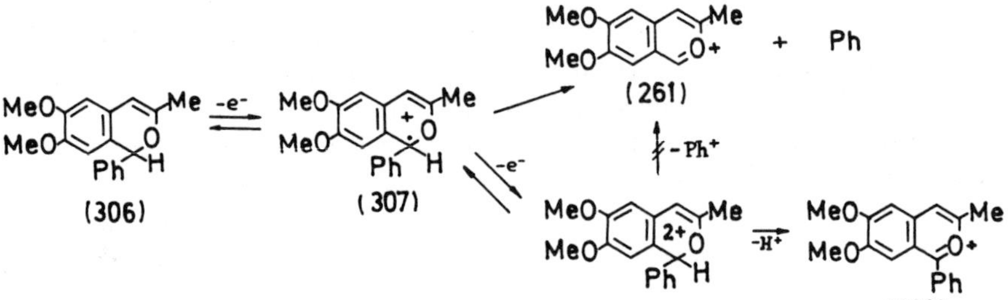

(306)   (307)   (261)   (308)   (154)

In accordance with the data, the formation of dication **308** from **306** includes the transfer of two electrons, but the intermediate cation radical **307** is too short-lived to allow recording of its ESR spectrum. It was assumed that the exclusive formation of **261** on chemical oxidation of **306** is determined by the comparatively low reduction potentials of the electron acceptors used.

C. Theoretical Calculations

As described in Section III, practically all known conversions of 2-benzopyrylium salts are the result of their reactions with nucleophiles, and 2-benzopyrylium, similar to its monocyclic analog [82AHC(Suppl)], behaves chemically more like a carbenium ion, and not like an oxonium salt. This fact can be easily understood if one takes into account resonance formulas **a–f,** where an oxonium form **f** is added for the presence of an alkoxy (hydroxy) substituent in position 6.

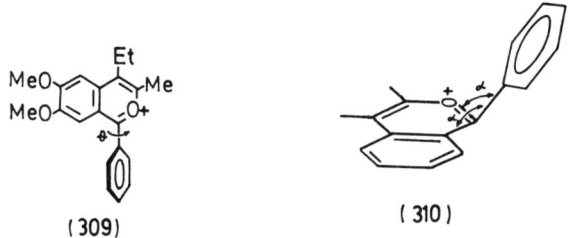

For explanation of experimental results and for correlation of charge densities with NMR data, semiempirical quantum-chemical calculations of benzo[c]pyrylium cation have been employed. Interestingly, the first calculation of 1,3-dimethyl-benzo[c]pyrylium cation by the simple linear combination of atomic orbitals/molecular orbital (LCAO/MO) method (70KGS1308) revealed a preference for the resonance from **a** in which the value of the charge density at C_1 was three times as much as at C_3.

The application of the CNDO/2 method with standard geometry for various substituted 2-benzopyrylium cations results, as one could expect, in the preference for the resonance structures **b–f** with close values of charge densities on C_1 and C_3, and attempts to rationalize on the basis of these calculations yielded statistical agreement for correlations between change densities and ^{1}H-NMR spectral data (90MI3) (cf. Section IV,A,2).

The extended Hückel method has been applied for calculations of the total energy of various rotamers of 1-phenyl-3-methyl-4-ethyl-6,7-dimethoxybenzo[c]pyrylium ion **309** (76MI2). It was found that the rotamer with the 1-aryl ring located at the angle $\theta = 75°$ relative to the plane of 2-benzopyrylium has the minimum energy. The Del Bene-Jaffe CNDO method has been used for correlations of ^{13}C-NMR chemical shifts for some substituted 1-arylbenzo[c]pyrylium with the all-valence charge densities of the various atoms [78ACH(96)273]. In this case, the shifts and charge densities were in a good correlation if an out-of-plane sp^3 valence orientation was assumed on C_1 (cf. Section IV,A,2) (**310**, $\alpha = 109.5°$).

(309)   (310)

References

33JCS555   K. Blount and R. Robinson, *J. Chem. Soc.*, 555 (1933).
34MI1   E. B. Wilson, *Phys. Rev.* **45**, 706 (1934).

49JOC204	R. L. Shriner, H. W. Johnston, and C. E. Kaslow, *J. Org. Chem.* **14**, 204 (1949).
50JA1118	W. E. Doering and J. A. Berson, *J. Am. Chem. Soc.* **72**, 1118 (1950).
50JOC472	A. Müller, M. Meszaros, and K. Körmendy, *J. Org. Chem.* **19**, 472 (1950).
50MI1	H. Henecka, "Chemie der beta-Dicarbonyl-Verbindungen." Akademie-Verlag, Berlin, 1950.
51JOC1064	R. L. Shriner and W. R. Knox, *J. Org. Chem.* **16**, 1064 (1951).
51MI1	S. Wawzonek, *in* "Heterocyclic Compounds" (R. C. Elderfield, ed.), Vol. 2. Wiley, New York, 1951.
52JA3228	E. H. Man, F. S. Frostick, and C. R. Hauser, *J. Am. Chem. Soc.* **74**, 3228 (1952).
54JOC1533	A. Müller, M. Lempert-Sreter, and A. Karczag-Wilhelms, *J. Org. Chem.* **19**, 1533 (1954).
59JA935	R. L. Letsinger and P. T. Lansbury, *J. Am. Chem. Soc.* **81**, 935 (1959).
60CB88	A. Rieche, H. Gross, and E. Hoff, *Chem. Ber.* **93**, 88 (1960).
60CB599	A. T. Balaban and C. D. Nenitzescu, *Chem. Ber.* **93**, 599 (1960).
60JIC379	J. N. Chatterjea and H. Mukherjea, *J. Indian Chem. Soc.* **37**, 379 (1960).
61JA193	R. L. Letsinger and J. D. Jamison, *J. Am. Chem. Soc.* **83**, 193 (1961).
61JOC92	R. L. Letsinger, J. D. Jamison, and A. S. Hussey, *J. Org. Chem.* **26**, 92 (1961).
61JOC2993	R. L. Letsinger and O. Kolewe, *J. Org. Chem.* **26**, 2993 (1961).
61RRC295	A. T. Balaban, G. Mateescu, and C. D. Nenitzescu, *Rev. Roum. Chim.* **6**, 295 (1961).
62JA788	S. Winstein, *J. Am. Chem. Soc.* **84**, 788 (1962).
62JA1315	E. F. Ullman and J. E. Milks, *J. Am. Chem. Soc.* **84**, 1315 (1962).
62T1083	A. T. Balaban, G. D. Mateescu, and M. Elian, *Tetrahedron* **18**, 1083 (1962).
63JA3529	E. F. Ullman, *J. Am. Chem. Soc.* **85**, 3529 (1963).
64ACH(40)217	M. Vajda and F. Ruff, *Acta Chim. Acad. Sci. Hung.* **40**, 217 (1964).
64ACH(40)225	M. Vajda and F. Ruff, *Acta Chim. Acad. Sci. Hung.* **40**, 225 (1964).
64ACH(40)295	M. Vajda, *Acta Chim. Acad. Sci. Hung.* **40**, 295, (1964).
64ACH(41)451	M. Lempert-Sreter and A. Müller, *Acta Chim. Acad. Sci. Hung.* **41**, 451 (1964).
64CRV29	R. P. Barry, *Chem Rev.*, 29 (1964).
64JA2084	E. F. Ullman and J. E. Milks, *J. Am. Chem. Soc.* **86**, 2084 (1964).
64JA3814	E. F. Ullman and J. E. Milks, *J. Am. Chem. Soc.* **86**, 3814 (1964).
64TL1847	H. E. Zimmerman and R. D. Simkin, *Tetrahedron Lett.*, 1847 (1964).
65M369	A. Müller and M. Lempert-Sreter, *Monatsh. Chem.* **96**. 369 (1965).
65URP176592	G. N. Dorofeenko, S. V. Krivun, and V. I. Dulenko, USSR Pat. 176,592 (1965) [*CA* **64**, 9689 (1966)].
66ACH(50)381	M. Lempert-Sreter, *Acta Chim. Acad. Sci. Hung.* **50**, 381 (1966).
66ACH(50)387	A. Müller M. M. El-Sawy, M. Meszaros, and F. Ruff, *Acta Chim. Acad. Sci. Hung.* **50**, 387 (1966).
66DOK(166)359	S. V. Krivun, V. I. Dulenko, L. V. Dulenko, and G. N. Dorofeenko, *Dokl. Akad. Nauk SSSR* **166**, 359 (1966).
66JA1916	M. Byrn and M. Calvin, *J. Am. Chem. Soc.* **88**, 1916 (1966).
66JA4942	E. F. Ullman and W. A. Henderson, *J. Am. Chem. Soc.* **88**, 4942 (1966).
66T(S7)9	C. Toma and A. T. Balaban, *Tetrahedron, Suppl.* **7**, 9 (1966).

66ZOR1492   G. N. Dorofeenko, A. D. Semenov, V. I. Dulenko, and S. V. Krivun, *Zh. Org. Khim.* **2**, 1492 (1966).

66ZOR1499   G. N. Dorofeenko, E. V. Kuznetsov, and S. V. Krivun, *Zh. Org. Khim.* **2**, 1499 (1966).

67ACH(51)107   M. Lempert-Sreter, *Acta Chim. Acad. Sci. Hung.* **51**, 107 (1967).

67ACH(52)261   A. Müller, M. M. El-Sawy, M. Meszaros, and F. Ruff, *Acta Chim. Acad. Sci. Hung.* **52**, 261 (1967).

67MI1   D. Bethel and V. Gold, "Carbonium Ions." Academic Press, London and New York, 1967.

68ACH(57)181   M. Lempert-Sreter and M. Bardy, *Acta Chim. Acad. Sci. Hung.* **57**, 181 (1968).

68CC723   R. Y. Alder, P. S. Boroman, W. R. S. Steele, and D. R. Winterman, *J. C. S., Chem. Commun.*, 723 (1968).

68DOK(181)345   G. N. Dorofeenko, S. V. Krivun, Yu. A. Zhdanov, and E. I. Sadekova, *Dokl. Akad. Nauk SSSR* **181**, 345 (1968).

68TL4379   G. Märkl and H. Baier, *Tetrahedron Lett.*, 4379 (1968).

68UP1   V. I. Dulenko, unpublished results (1968).

69IJC527   N. L. Dutta, M. S. Wadia, and A. A. Bindra, *Indian J. Chem.*, 527 (1969).

69JOC2736   G. A. Reynolds and J. A. Van Allan, *J. Org. Chem.* **34**, 2736 (1969).

69MIP1   J. Korösi and T. Lang, Hung. Pat. 155,572 [*CA* **70**, 115026 (1969)].

69TL711   G. N. Dorofeenko and E. V. Kuznetsov, *Tetrahedron Lett.*, 711 (1969).

69TL2047   J. A. Van Allan and G. A. Reynolds, *Tetrahedron Lett.*, 2047 (1969).

69ZOR191   G. N. Dorofeenko and E. V. Kuznetsov, *Zh. Org. Khim.* **5**, 191 (1969).

70KGS278   G. N. Dorofeenko and G. P. Safaryan, *Khim. Geterotsikl. Soedin.*, 278 (1970).

70KGS1013   G. N. Dorofeenko, G. P. Safaryan, and E. V. Kuznetsov, *Khim Geterotsikl. Soedin.*, 1013 (1970).

70KGS1308   G. N. Dorofeenko, E. I. Sadekova, and V. M. Goncharova, *Khim. Geterosikl. Soedin.*, 1308 (1970).

70KGS(2)196   G. N. Dorofeenko, E. I. Sadekova, and L. N. Pyshkina, *Khim Geterotsikl. Soedin., Sb.* **2**, 196 (1970).

70KGS(2)200   G. N. Dorofeenko, V. G. Korobkova, and S. V. Krivun, *Khim. Geterotsikl. Soedin., Sb.* **2**, 200 (1970).

70KGS(2)207   G. N. Dorofeenko and E. V. Kuznetsov, *Khim. Geterotsikl. Soedin., Sb.* **2**, 207 (1970).

70KGS(2)213   G. N. Dorofeenko, E. I. Sadekova, and L. N. Pyshkina, *Khim. Geterotsikl. Soedin., Sb.* **2**, 213 (1970).

70MI1   C. A. Buehler and D. E. Pearson, "Survey of Organic Syntheses." Wiley (Interscience), New York, 1970.

70MI2   V. M. Vlasov, *Zh. Vses. Khim. Ova* **15**, 708 (1970) [*CA* **74**, 53412 (1971)].

70TH1   E. V. Kuznetsov, Ph.D. Thesis, Rostov University, Rostov on Don, USSR (1970).

70ZOR578   E. V. Kuznetsov and G. N. Dorofeenko, *Zh. Org. Khim.* **6**, 578 (1970).

70ZOR1118   G. N. Dorofeenko, E. I. Sadekova, and V. I. Beletskaya, *Zh. Org. Khim.* **6**, 1118 (1970).

70ZOR1429   S. V. Krivun, G. N. Dorofeenko, and E. I. Sadekova, *Zh. Org. Khim.* **6**, 1429 (1970).

71BSF1362   C. Normant-Chefnay, *Bull. Soc. Chim. Fr.*, 1362 (1971).
71CB2984   K. Dimroth and H. Odenwälder, *Chem. Ber.* **104**, 2984 (1971).
71JCS(C)3559   G. R. Birchall, M. N. Galbraigh, and W. B. Whalley, J. Chem. Soc. C, 3559 (1971).
71JCS(C)3571   R. Chong, R. R. King, and W. B. Whalley, *J. Chem. Soc. C*, 3571 (1971).
71JCS(C)3590   S. Ahmad, W. B. Whalley, and D. F. Jones, *J. Chem. Soc. C*, 3590 (1971).
71JIC707   S. Gulgule and R. N. Usgaonkar, *J. Indian Chem. Soc.* **48**, 707 (1971).
71KGS730   G. N. Dorofeenko, S. V. Krivun, and E. I. Sadekova, *Khim. Geterotsikl. Soedin.*, 730 (1971).
71KGS1437   E. V. Kuznetsov and G. N. Dorofeenko, *Khim. Geterotsikl. Soedin.*, 1437 (1971).
71KGS1601   G. N. Dorofeenko and V. G. Korobkova, *Khim. Geterotsikl. Soedin.*, 1601 (1971).
72BSF3173   H. Khedija, H. Strzelecka, and M. Simalty, *Bull. Soc. Chim. Fr.*, 3173 (1972).
72KGS152   E. V. Kuznetsov, A. I. Pyshchev, and G. N. Dorofeenko, *Khim. Geterotsikl. Soedin.*, 152 (1972).
72KGS454   G. N. Dorofeenko, E. V. Kuznetsov, and E. I. Sadekova, *Khim. Geterotsikl. Soedin.*, 454 (1972).
72MI1   G. N. Dorofeenko, E. I. Sadekova, and E. V. Kuznetsov, "Preparative Chemistry of Pyrylium Salts" (in Russian). Izd. Rostov. Univ., Rostov on Don, USSR (1972) [*CA* **78**, 43238 (1973)].
72MI2   T. L. Gilchrist and R. C. Storr, "Organic Reactions and Orbital Symmetry." Cambridge Univ. Press, London and New York, 1972.
72MI3   M. Vajda and V. Kardi, *Ann. Univ. Sci. Budap. Rolando Eotvos Nominator, Sect. Chim.* **13**, 107 (1972) [*CA* **80**, 107675 (1974)].
72TL4439   G. Märkl and H. Baier, *Tetrahedron Lett.*, 4439 (1972).
73DOK(212)370   L. A. Polyakova, K. A. Bilevich, N. N. Bubnov, G. N. Dorofeenko, and O. Yu. Okhlobystyn, *Dokl. Akad. Nauk SSSR* **212**, 370 (1973).
73KGS881   G. N. Dorofeenko and S. M. Luk'yanov, *Khim. Geterotsikl. Soedin.*, 881 (1973).
73KGS1458   G. N. Dorofeenko, S. V. Krivun, and V. G. Korobkova, *Khim. Geterotsikl. Soedin.*, 1458 (1973).
73URP1   G. N. Dorofeenko, E. V. Kuznetsov, and E. I. Sadekova, USSR Pat. 262,013 (1973) [*CA* **78**, 111123 (1973)].
73URP2   E. V. Kuznetsov, G. N. Doeofeenko, A. V. Bicherov, and D. V. Pruchkin, USSR Pat. 405,886 (1973) [*CA* **79**, 103086 (1973)].
73ZOR1079   G. N. Dorofeenko, E. A. Zvezdina, and V. V. Derbenev, *Zh. Org. Khim.* **9**, 1079 (1973).
74AHC289   C. K. Bradsher, *Adv. Heterocycl. Chem.* **16**, 289 (1974).
74CB3883   J. Körösi and T. Lang, *Chem. Ber.* **107**, 3883 (1974).
74JA480   R. A. Abramovitch and E. P. Kyba, *J. Am. Chem. Soc.* **96**, 480 (1974).
74JA4339   P. A. Burns and C. S. Foote, *J. Am. Chem. Soc.* **96**, 4339 (1974).
74KGS37   G. N. Dorofeenko, V. V. Mezheritskii, and A. L. Vasserman, *Khim. Geterotsikl. Soedin.*, 37 (1974).
74KGS181   E. V. Kuznetsov, D. V. Pruchkin, A. V. Bicherov, and G. N. Dorofeenko, *Khim. Geterotsikl. Soedin.*, 181 (1974).
74KGS342   G. N. Dorofeenko and V. G. Korobkova, *Khim. Geterotsikl. Soedin.*, 342 (1974).

74KGS1575   E. V. Kuznetsov, D. V. Pruchkin, A. V. Bicherov, and G. N. Dorofeenko, *Khim. Geterotsikl. Soedin.*, 1575 (1974).
74KGS1695   Yu. P. Andreitchikov, G. E. Trukhan, V. G. Korobkova, S. N. Lubtchenko, and G. N. Dorofeenko, *Khim. Geterotsikl. Soedin.*, 1695 (1974).
75ACH(85)79 M. Lempert-Sreter, Gy. Oszbach, F. Ruff, and M. Vajda, *Acta Chim. Acad. Sci. Hung.* **85,** 79 (1975).
75KGS25   E. V. Kuznetsov, D. V. Pruchkin, E. A. Muradyan, and G. N. Dorofeenko, *Khim. Geterotsikl. Soedin.*, 25 (1975).
75ZN(B)943   M. Vajda and W. Voelter, *Z. Naturforsch. B: Anorg. Chem., Org. Chem.* **30B,** 943 (1975).
75ZOR1962   S. M. Luk'yanov, L. N. Etmetchenko, A. V. Koblik, O. A. Rakina, and G. N. Dorofeenko, *Zh. Org. Khim.* **11,** 1962 (1975).
76ACS(B)619 B. P. Nilsen and K. Undheim, *Acta Chem. Scand., Ser. B* **30,** 619 (1976).
76JA5581   A. Padwa and A. Au, *J. Am. Chem. Soc.* **98,** 5581 (1976).
76KGS50   E. V. Kuznetsov, I. V. Shcherbakova, and G. N. Dorofeenko, *Khim. Geterosikl. Soedin.*, 50 (1976).
76KGS745   E. V. Kuznetsov, I. V. Shcherbakova, and G. N. Dorofeenko, *Khim. Geterotsikl. Soedin.*, 745 (1976).
76KGS858   E. V. Kuznetsov, D. V. Pruchkin, Yu. D. Smetanin, and G. N. Dorofeenko, *Khim. Geterotsikl. Soedin.*, 858 (1976).
76KGS999   G. N. Dorofeenko, G. F. Safaryan, V. F. Voloshinova, and O. Yu. Okhlobystin, *Khim. Geterotsikl. Soedin.*, 999 (1976).
76MI1   T. A. Tyagunova, *Deposited Doc. VINITI USSR,* 3571 (1976) [*CA* **88,** 114880 (1978)].
76MI2   K. Osapay, M. Farkas, and M. Vajda, *Comput. Chem.* **1,** 125 (1976).
76OMR324   M. Vajda and W. Voelter, *Org. Magn. Reson.* **8,** 324 (1976).
77DOK(236)1364 Yu. P. Andreichikov, N. K. Kholodova, and G. N. Dorofeenko, *Dokl. Akad. Nauk SSSR* **236,** 1364 (1977).
77JOC1111   M. Cushman, J. Gentry, and F. W. Dekow, *J. Org. Chem* **42,** 1111 (1977).
77KGS996   I. V. Shcherbakova, N. N. Potemkina, G. N. Dorofeenko, and E. V. Kuznetsov, *Khim. Geterotsikl. Soedin.*, 996 (1977).
77KGS1138   G. N. Dorofeenko, D. V. Pruchkin, and E. V. Kuznetsov, *Khim. Geterotsikl. Soedin.*, 1138 (1977).
77KGS1176   E. V. Kuznetsov, I. V. Shcherbakova, and G. N. Dorofeenko, *Khim. Geterotsikl. Soedin.*, 1176 (1977).
77KGS1479   E. V. Kuznetsov, D. V. Pruchkin, and G. N. Dorofeenko, *Khim. Geterotsikl. Soedin.*, 1479 (1977).
77KGS1481   E. V. Kuznetsov, I. V. Shcherbakova, and G. N. Dorofeenko, *Khim. Geterotsikl. Soedin.*, 1481 (1977).
77KPS714   G. N. Dorofeenko and E. I. Sadekova, *Khim. Prir. Soedin.*, 714 (1977) [*CA* **88,** 121496 (1978)].
77TH1   D. V. Pruchkin, Ph. D. Thesis, Rostov University, Rostov on Don. USSR (1977).
77ZOR631   E. V. Kuznetsov, I. V. Shcherbakova, V. I. Ushakov, and G. N. Dorofeenko, *Zh. Org. Khim.* **13,** 631 (1977).
78ACH(96)373 M. Farkas, W. Voelter, and M. Vajda, *Acta Chim. Acad. Sci. Hung.* **96,** 373 (1978).
78ACR462   H. C. Van der Plas, *Acc. Chem. Res.* **11,** 462 (1978).

78CPB245	T. Hirota, T. Koyama, T. Nanda, M. Yamato, and T. Matsumura, *Chem. Pharm. Bull.* **26**, 245 (1978).
78JHC731	M. Wozniak, *J. Heterocycl. Chem.* **15**, 731, (1978).
78JOC286	M. Cushman and L. Cheng. *J. Org. Chem.* **43**, 286 (1978).
78JOC822	C. K. Bradsher, G. L. Carlson, N. H. Porter, I. J. Westerman, and T. G. Wallis, *J. Org. Chem.* **43**, 822 (1978).
78JOC2852	M. Shamma and H. H. Tomlinson, *J. Org. Chem.* **43**, 2852 (1978).
78JOC3817	C. K. Bradsher and T. G. Wallis, *J. Org. Chem.* **43**, 3817 (1978).
78KGS1320	E. V. Kuznetsov, D. V. Pruchkin, A. I. Pyshchev, and G. N. Dorofeenko, *Khim. Geterotsikl. Soedin.*, 1320 (1978).
78MI1	M. Shamma and J. L. Moniot, "Isoquinoline Alkaloids Research 1972–1977." Plenum, New York, 1978.
78MI2	K. W. Bentley, *Alkaloids* **7**, 89 (1978).
78RRC1065	A. T. Balaban and M. D. Gheorghiu, *Rev. Roum. Chim.* **23**, 1065 (1978).
78ZOR1643	V. A. Samarskii, V. V. Panov, M. V. Nekhoroshev, O. Yu. Okhlobystin, and V. D. Pokhodenko, *Zh. Org. Khim.* **14**, 1643 (1978).
79AHC2	J. W. Bunting, *Adv. Heterocycl. Chem.* **25**, 2 (1979).
79JOC409	M. Cushman and F. W. Dekow, *J. Org. Chem.* **44**, 409 (1979).
79JOC727	I. J. Westerman and C. K. Bradsher, *J. Org. Chem.* **44**, 727 (1979).
79MI1	A. T. Balaban, *in* "New Trends in Heterocyclic Chemistry" (R. B. Mitra *et al.*, eds.). Elsevier, Amsterdam, 1979.
79MI2	J. Staunton, *in* "Comprehensive Organic Chemistry" (P. G. Sammes, ed.), Vol. 4, p. 607. Pergamon, Oxford, 1979.
79MI3	P. A. Clare, *in* "Comprehensive Organic Chemistry" (P. G. Sammes, ed.), Vol. 4. Pergamon, Oxford, 1979.
79TH1	I. V. Shcherbakova, Ph.D. Thesis, Rostov University, Rostov on Don, USSR (1979).
79ZOR439	E. V. Kuznetsov, A. V. Bicherov, I. V. Shcherbakova, and G. N. Dorofeenko, *Zh. Org. Khim.* **15**, 439 (1979).
79ZOR588	A. V. Bicherov, G. N. Dorofeenko, and E. V. Kuznetsov, *Zh. Org. Khim.* **15**, 588 (1979).
80JHC847	J.-P. Le Roux, P.-L. Desbene, and J.-C. Cherton, *J. Heterocycl. Chem.* **18**, 847 (1980).
80JOC3524	J. M. Hornback and B. Valdman, *J. Org. Chem.* **45**, 3524 (1980).
80JOC5067	M. Cushman, T. C. Chong, and J. T. Valco, *J. Org. Chem.* **45**, 5067 (1980).
80KGS193	A. I. Tolmachev and L. M. Shulezhko, *Khim. Geterotsikl. Soedin.*, 193 (1980).
80KGS1421	I. M. Sosonkin, G. N. Dorofeenko, G. N. Strogov, and G. P. Safaryan, *Khim. Geterotsikl. Soedin.*, 1421 (1980).
80S847	M. Haimova, V. I. Ognyanov, and N. M. Mollow, *Synthesis*, 847 (1980).
80T679	A. R. Katritzky, *Tetrahedron* **36**, 679 (1980).
80T1279	H. Fodor and S. Nagubandi, *Tetrahedron* **36**, 1279 (1980).
81BCJ240	T. Ibata, K. Kitsuhiro, and Y. Tsubokura, *Bull. Chem. Soc. Jpn.* **54**, 240 (1981).
81H221	N. Takao, M. Kamigauchi, M. Sugiura, I. Ninomiya, O. Miyata, and T. Naito, *Heterocycles* **16**, 221 (1981).
81H474	G. D. Pandey and K. P. Tiwari, *Heterocycles* **16**, 474 (1981).

81KGS313	I. V. Shcherbakova, G. N. Dorofeenko, and E. V. Kuznetsov, *Khim. Geterotsikl. Soedin.*, 313 (1981).
81KGS1177	I. M. Sosonkin, G. N. Dorofeenko, G. N. Strogov, G. P. Safaryan, and A. N. Domarev, *Khim. Geterotsikl. Soedin.*, 1177 (1981).
81KGS1608	G. P. Safaryan, I. V. Shcherbakova, G. N. Dorofeenko, and E. V. Kuznetsov, *Khim. Geterotsikl. Soedin.*, 1608 (1981).
81T3425	A. N. Kost, S. P. Gromov, and R. S. Sagitullin, *Tetrahedron* **37**, 3425 (1981).
81ZOR440	E. V. Kuznetsov, A. V. Bicherov, and G. N. Dorofeenko, *Zh. Org. Khim.* **17**, 440 (1981).
82AHC(Suppl)	A. T. Balaban, A. Dinculescu, G. N. Dorofeenko, G. W. Fischer, A. V. Koblik, V. V. Mezheritskii, and W. Schroth, *Adv. Heterocycl. Chem.*, *Suppl.* **2**, 1 (1982).
82JOC492	A. R. Katritzky, A. Chemprarai, and R. C. Patel, *J. Org. Chem.* **47**, 492 (1982).
82JOC498	A. R. Katritzky, R. Awartni, and R. C. Patel, *J. Org. Chem.* **47**, 498 (1982).
82JOC5021	S. Manna and J. R. Falk, *J. Org. Chem.* **47**, 5021 (1982).
82KGS291	T. V. Stupnikova, B. P. Zemskii, R. S. Sagitullin, and A. N. Kost, *Khim. Geterotsikl. Soedin.*, 291 (1982).
82KGS465	E. A. Zvezdina, A. N. Popova, and G. N. Dorofeenko, *Khim. Geterotsikl. Soedin.*, 465 (1982).
82KGS552	I. V. Shcherbakova and E. V. Kuznetsov, *Khim. Geterotsikl. Soedin.*, 552 (1982).
82KGS1184	I. V. Korobka, I. V. Shcherbakova, and E. V. Kuznetsov, *Khim. Geterotsikl. Soedin.*, 1184 (1982).
82KGS1381	I. V. Shcherbakova, O. E. Shelepin, N. A. Kluev, G. N. Dorofeenko, and E. V. Kuznetsov, *Khim. Geterotsikl. Soedin.*, 1381 (1982).
82MI1	W. Zielinski, *Pol. J. Chem.* **56**, 93 (1982).
82TL459	I. K. Stamos, *Tetrahedron Lett.*, 459 (1982).
82ZOR589	I. V. Korobka, A. V. Bicherov, I. V. Shcherbakova, G. N. Dorofeenko, and E. V. Kuznetsov, *Zh. Org. Khim.* **18**, 589 (1982).
83CC7	K. Takeuchi, T. Kitagawa, and K. Okamoto, *J. C. S., Chem. Commun.* 7 (1983).
83KGS274	I. V. Korobka and E. V. Kuznetsov, *Khim. Geterotsikl. Soedin.*, 274 (1983).
83T2869	D. L. Boger, *Tetrahedron* **39**, 2869 (1983).
83TL4911	M. Maleki, A. C. Hopkinson, and E. Lee-Ruff, *Tetrahedron Lett.*, 4911 (1983).
84CC702	P. G. Sammes and R. J. Whitby, *J. C. S., Chem. Commun.*, 702 (1984).
84CC761	R. W. Franckt and R. B. Gupta, *J. C. S., Chem. Commun.*, 761 (1984).
84CJC2435	A. Carty, J. W. Elliott, and G. M. Lenior, *Can. J. Chem.* **62**, 2435 (1984).
84KGS167	V. M. Zolin, N. D. Dmitrieva, Yu. E. Gerasimenko, and A. V. Zubkov, *Khim. Geterotsikl. Soedin.* 167 (1984).
84KGS702	I. V. Shcherbakova, V. G. Brovchenko, and E. V. Kuznetsov, *Khim. Geterotsikl. Soedin.*, 702 (1984).
84KGS1428	N. V. Kholodova, *Khim. Geterotsikl. Soedin.*, 1428 (1984).
84KGS1472	I. V. Korobka, A. I. Voloshina, and E. V. Kuznetsov, *Khim. Geterotsikl. Soedin.*, 1472 (1984).

84MI1   A. R. Katritzky and C. W. Rees, eds., "Comprehensive Heterocyclic Chemistry," Vol. 3, Part 2B, Chapters 2.22, 2.23, and 2.24.
84T4383   G. G. Black and M. Sainsbury, *Tetrahedron* **40,** 4383 (1984).
84TL271   S. Danichefsky and M. Bednarski, *Tetrahedron Lett.,* 271 (1984).
84TL3659   M. E. Jung, R. W. Brown, J. A. Hadenah, and C. E. Strouse, *Tetrahedron Lett.,* 3659 (1984).
84UK1648   V. N. Charushin and O. N. Chupakhin, *Usp. Khim.* **53,** 1648 (1984).
84ZOR224   V. G. Brovchenko, A. V. Bicherov, and E. V. Kuznetsov, *Zh. Org. Khim.* **20,** 224 (1984).
85JCS(P2)1895 Z. Dega-Szafran, M. Szafran, and A. R. Katritzky, *J. C. S., Perkin Trans. 2,* 1895 (1985).
85KGS910   I. V. Korobka, Yu, V. Revinskii, and E. V. Kuznetsov, *Khim. Geterotsikl. Soedin.,* 910 (1985).
85KPS815   I. V. Shcherbakova, V. G. Brovchenko, and E. V. Kuznetsov, *Khim. Prir. Soedin.,* 815 (1985).
85LA2116   G. Bringmann and J. R. Jansen, *Liebigs Ann. Chem.,* 2116 (1985).
85MI1   A. R. Katritzky, "Handbook of Heterocyclic Chemistry." Pergamon, New York, 1985.
85MI2   Yu. A. Nikolukin, S. L. Bogsa, and V. I. Dulenko, *Deposited Doc. VINITI USSR,* 7387-B (1985) [*CA* **105,** 39864 (1986)].
86CRV781   D. L. Boger, *Chem. Rev.* **86,** 781 (1986).
86KGS125   V. G. Brovchenko and E. V. Kuznetsov, *Khim. Geterotsikl. Soedin.,* 125 (1986).
86KGS276   I. V. Korobka and E. V. Kuznetsov, *Khim. Geterotsikl. Soedin.,* 276 (1986).
86KGS460   Yu. B. Vysotskii, B. P. Zemakii, E. A. Zemskaya, V. I. Dulenko, L. V. Dulenko, E. V. Kuznetsov, and G. N. Dorofeenko, *Khim. Geterotsikl. Soedin.,* 460 (1986).
86KGS999   Yu. A. Nikolukin, L. V. Dulenko, and V. I. Dulenko, *Khim. Geterotsikl. Soedin.,* 999 (1986).
86MIP1   T. Lang, J. Korösi, P. Botka, T. Hamori, G. Zolyomi, T. Balogh, G. Symmogi, and T. Lang (Mrs.), Hung. Pat. 37,739 (1986) [*CA* **105,** 152712 (1986)].
86NKK1622   E. Iguchi, S. Itoh, and H. Takahashi, *Nippon Kagaku Kaishi,* 1622 (1986).
86UP1   I. V. Shcherbakova and E. V. Kuznetsov, unpublished results (1986).
86UP2   I. V. Korobka and E. V. Kuznetsov, unpublished results (1986).
86UP3   S. Kut'ko, I. V. Korobka, and E. V. Kuznetsov, unpublished results (1986).
87KGS600   V. I. Dulenko and Yu. A. Nikolukin, *Khim. Geterotsikl. Soedin.,* 600 (1987).
87KGS1032   I. V. Shcherbakova, L. Yu. Ukhin, V. V. Komissarov, E. V. Kuznetsov, A. V. Polyakov, A. I. Yanovskii, and Yu. T. Struchkov, *Khim. Geterotsikl. Soedin.,* 1032 (1987).
87MI1   N. V. Volbushko, A. V. Metelitza, G. P. Safaryan, and E. V. Kuznetsov, *in* "Organic Luminophores," Abstr. All Nat. Conf. Kharkov, USSR, 1987.
87MI2   A. T. Balaban, *in* "Organic Synthesis: Modern Trends" (O. Chizhov, ed.), p. 263. Blackwell, Oxford, 1987.

87RRC417   I. V. Shcherbakova, E. V. Kuznetsov, L. Yu, Ukhin, M. D. Gheorghiu, A. Meghea, J. Herdan, and A. T. Balaban, *Rev. Roum. Chim.* **32**, 417 (1987).

87T409   A. T. Balaban, M. D. Gheorghiu, I. V. Shcherbakova, E. V. Kuznetsov, and I. A. Yudilevich, *Tetrahedron* **43**, 409 (1987).

87TH1   A. V. Biherov, Ph.D. Thesis, Rostov University, ROSTOV On Don (1987).

87TL3143   C. Uncuta, M. D. Gheorghiu, and A. T. Balaban, *Tetrahedron Lett.*, 3143 (1987).

87UP1   I. V. Shcherbakova, I. A. Yudilevich, and E. V. Kuznetsov, unpublished results (1987).

87UP2   M. D. Gheorghiu, A. T. Balaban, I. V. Shcherbakova, and E. V. Kuznetsov, unpublished results (1987).

88KGS1185   Yu. A. Zhdanov, S. V. Verin, I. V. Korobka, and E. V. Kuznetsov, *Khim. Geterotsikl. Soedin.*, 1185 (1988).

88MI1   I. V. Shcherbakova and E. V. Kuznetsov, in "Abstracts of the 3rd National Congress on Chemistry," Vol. 1, p. 163. Bucharest, Roumania, 1988.

88T6217   I. V. Shcherbakova, E. V. Kuznetsov, I. A. Yudilevich, O. E. Compan, A. T. Balaban, A. H. Abolin, A. V. Polyakov, and Yu. T. Struchkov, *Tetrahedron* **44**, 6217 (1988).

88UP1   I. V. Shcherbakova and E. V. Kuznetsov, unpublished results (1988).

88UP2   I. V. Shcherbakova, S. V. Verin, A. V. Bicherov, and E. V. Kuznetsov, unpublished results (1988).

89KGS454   Yu. A. Zhdanov, V. G. Brovchenko, N. A. Kluev, and E. V. Kuznetsov, *Khim. Geterotsikl. Soedin.*, 454 (1989).

89KGS750   S. V. Verin, E. V. Kuznetsov, and Yu. A. Zhdanov, *Khim. Geterotsikl. Soedin.*, 750 (1989).

89KGS1039   S. V. Verin and E. V. Kuznetsov, *Khim. Geterotsikl. Soedin.*, 1039 (1989).

89KGS1564   N. V. Kholodova, V. G. Brovchenko, A. I. Pyshchev, and E. V. Kuznetsov, *Khim. Geterotsikl. Soedin.*, 1564 (1989).

89KPS75   I. V. Shcherbakova, S. V. Verin, and E. V. Kuznetsov, *Khim. Prir. Soedin.*, 75 (1989).

89MI1   I. V. Shcherbakova and E. V. Kuznetsov, *in* "Abstracts of the 6th European Simposium on Organic Chemistry," p. 22. Belgrade, Yugoslavia, 1989.

89MI2   S. V. Verin and E. V. Kuznetsov, "Aromatic Nucleophilic Substitution." Novosibirsk, USSR, 1989.

89RRC1425   C. Uncuta, T.-S. Balaban, A. Petride, F. Chiraleu, and A. T. Balaban, *Rev. Roum. Chim.* **34**, 1425 (1989).

89TH1   V. G. Brovchenko, Ph.D. Thesis, Rostov University, Rostov on Don (1989).

89TH2   S. V. Verin, Ph.D. Thesis, Rostov University, Rostov on Don (1989).

89UP1   S. V. Verin and E. V. Kuznetsov, unpublished results (1989).

89ZOR164   I. V. Shcherbakova and E. V. Kuznetsov, *Zh. Org. Khim.* **25**, 164 (1989).

90KGS315   S. V. Verin, D. E. Tosunyan, E. V. Kuznetsov, and Yu. A. Zhdanov, *Khim. Geterotsikl. Soedin.*, 315 (1990).

90KGS(ip1)   Yu. A. Zhdanov, S. V. Verin, and E. V. Kuznetsov, *Khim. Geterotsikl. Soedin.* (1990) (in press).

90KGS(ip2)   S. V. Verin, D. E. Tosunyan, P. I. Zakharov, V. K. Shevtzov, and
              E. V. Kuznetsov, *Khim. Geterotsikl. Soedin.* (1990) (in press).
90KGS(ip3)   S. V. Verin, D. E. Tosunyan, and E. V. Kuznetsov, *Khim. Geterotsikl.
              Soedin.* (1990) (in press).
90MI1   I. V. Shcherbakova, I. A. Yudilevich, and E. V. Kuznetsov, *J. Phys.
              Org. Chem.* **3,** (1990) (in press).

Thiadiazines with Adjacent Sulfur and Nitrogen Ring Atoms

REGINALD E. BUSBY

Chemistry Department, Brunel University,
Uxbridge, Middlesex UB8 3PH, *England*

I. Introduction

Apart from the various tautomeric forms, there are four kinds of parent thiadiazine structures, having molecular formula $C_3H_4N_2S$, in which the sulfur atom is adjacent to at least one ring nitrogen atom (structures I–IV). In order to name a particular tautomer, the "extra" hydrogen requires designation. Thus, using Chemical Abstracts nomenclature, structure I is named as 6H-1,2,3-thiadiazine, and structures V, VI and VII are named as

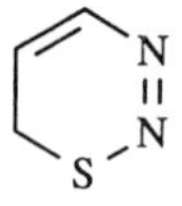

1,2,3-   1,2,4-   1,2,5-   1,2,6-
  I   II   III   IV

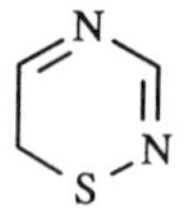

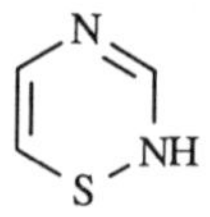

  V   VI   VII   VIII

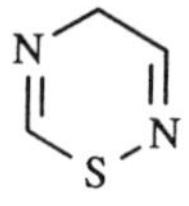

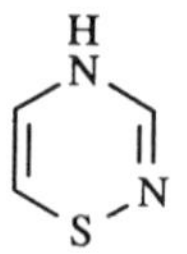

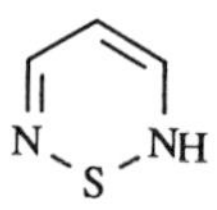

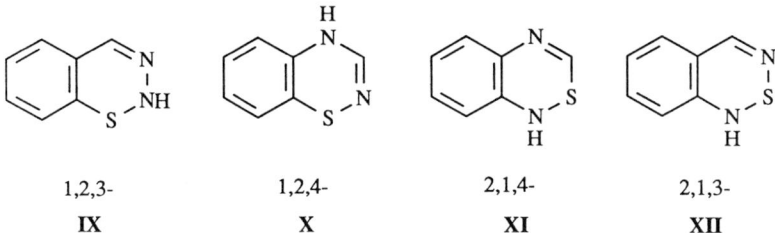

1,2,3-	1,2,4-	2,1,4-	2,1,3-
IX	X	XI	XII

$2H$-1,2,4-thiadiazine, $4H$-1,2,4-thiadiazine, and $2H$-1,2,6-thiadiazine, respectively. If the sulfur atom exhibits a valency other than 2, then the valency is indicated by including the small Greek letter λ after the numeral that specifies the ring sulfur atom, followed by a superscript numeral to indicate the valency of the sulfur atom. For instance, structure VIII is named $3H$-1λ^4,2,6-thiadiazine. The four thiadiazine systems can give rise to benzo-derivatives, i.e. structures IX–XII; examples of all four types are known.

Thiadiazines have been reviewed, in part, several times previously. A review of thiadiazine 1,1-dioxides covered the literature up until December, 1968 (70CRV593). The review did not include the very large number of 1,2,4-benzothiadiazines which have figured into the quest for pharmacologically active products, since these were thought to deserve a separate review. Apart from those benzo-derivatives reported, other condensed systems were not included. Reviews have also appeared in Volume 44 of this publication (88AHC81), in large compendia (61MI1; 79MI1; 84MI1), and elsewhere (70SST473). Since 1,2,6-thiadiazine 1,1-dioxides, which include the sulfamide moiety, have received comprehensive attention in Volume 44 of this publication (88AHC81), only those 1,2,6-thiadiazines that are not cyclic sulfamides are included here (Section V). This review includes advances covered by Chemical Abstracts up to the issue of July 17, 1989. Except in the cases of the lesser known 1,2,3- and 1,2,5-thiadiazines, reports on publications are included that have appeared since the coverage of the previous reviews mentioned.

While 1,2,4- and 1,2,6-thiadiazines have been extensively investigated, there is relatively little known about 1,2,3-thiadiazines, and 1,2,5-thiadiazines are the least known. Apart from 1,2,5-thiadiazines, the thiadiazines most commonly exist as S,S-dioxides. This is in part due to their ease of formation, their stability, and the potential of the sulfonamide group to impart biological activity.

II. 1,2,3-Thiadiazines

In their more common 1,1-dioxide forms, these thiadiazines are sometimes called cyclic sulfonylhydrazides.

A. SYNTHESES

1. Monocyclic 1,2,3-Thiadiazines

There are only a few reported members of this series. The first mentioned is 3,4,5,6-tetrahydro-3-phenyl-2H-1,2,3-thiadiazine 1,1-dioxide, which is prepared in low yield by cyclization of N-3-chloropropylsulfonyl-N'-phenylhydrazine with base (62JPR56). The well-known property of azoalkenes to undergo [4 + 2] cycloaddition reactions with dienophiles can be exploited to give 1,2,3-thiadiazines. Thus, N-phenylazostilbene, PhCH=C(Ph)N=NPh, reacts slowly with sulfine 1 at room temperature to yield the trans (44%) and cis (13%) isomers 2 and 3

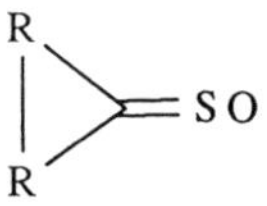

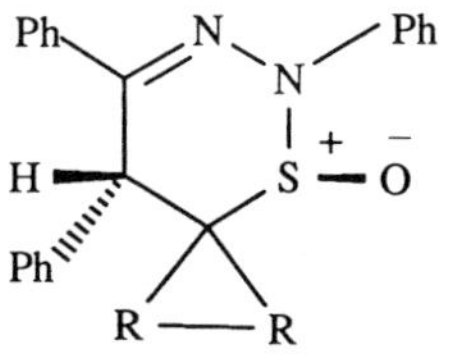

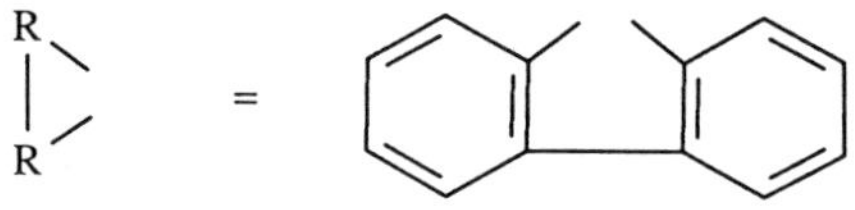

together with the 1,3,4-thiadiazine isomer 4 (5%). The structure of 2 was confirmed by X-ray crystallography. The minor product 4 was found to isomerize, presumably by a cycloreversion–recyclization process, to 2 and 3 [81JCS(P1)2322].

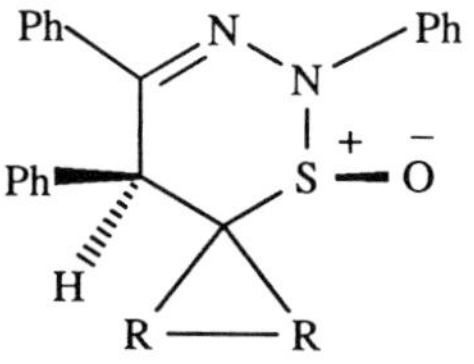

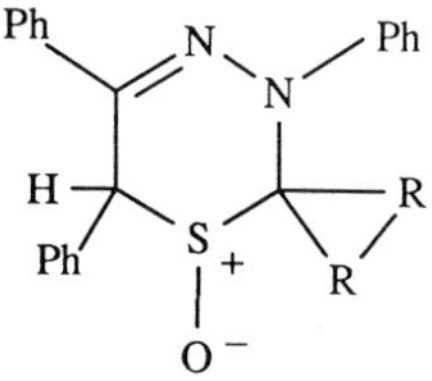

2. Condensed 1,2,3-Thiadiazines

a. *1,2,3-Benzothiadiazines.* Cyclization using hydrazine or hydrazones by formation of the N—S bond is frequently used for making these thiadiazines. The first recorded members of this class were made by the method used by Schrader, e.g, thiadiazine 7 was prepared from cyanosulfonyl chloride 5 (R = H) (17JPR180). Hydrazine has been replaced by its hydrate in this preparation (62BEP615374). Also, 5 (R = H) can be made from saccharin by the action of phosphorus pentachloride (63FRP2166). Intermediate 6 (R = OEt) has been isolated and ring-closed in dilute aqueous-ethanolic hydrochloric acid to give 7. Interestingly, isopropyl-

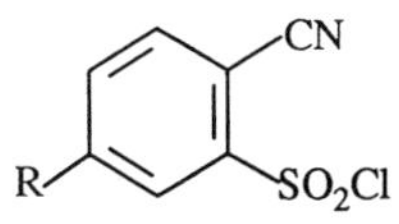

(5) 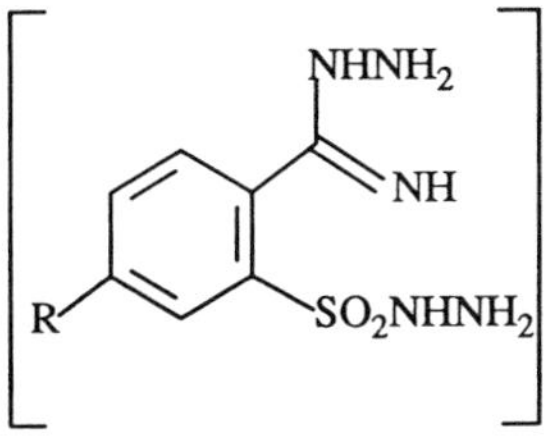   (6) 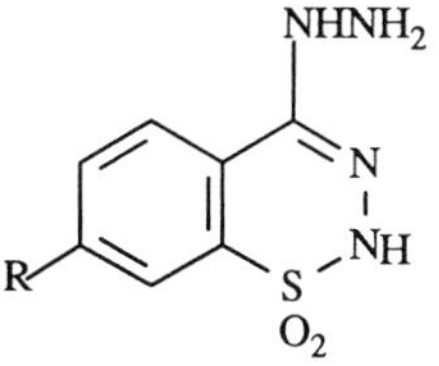   (7)

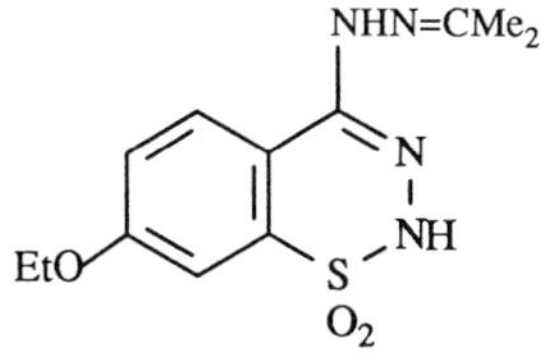

(8) 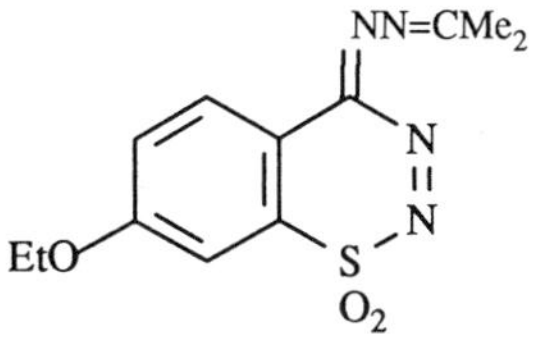   (9)

idene derivative **8** may be oxidized with hydrogen peroxide to azine **9** (62HCA996).

Rather than yielding sulfonylhydrazide, the action of hydrazine on *o*-esters of benzenesulfonyl chloride **10** effects cyclization to benzothiadiazine **11** (62JOC1703). The action of the phosphorus pentachloride on the sodium salt of hydrazone **12** gives up to an 80% yield of benzothiadiazine **13** (R = H) (69CC33), which can also be made from the reaction of hydrazine with *o*-formylbenzenesulfonyl chloride (79MI1).

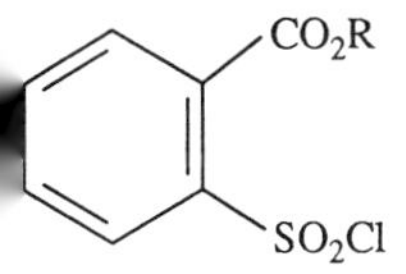

(10)   (11) 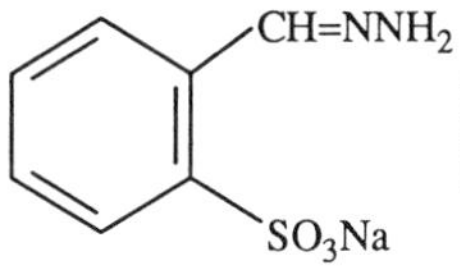   (12) 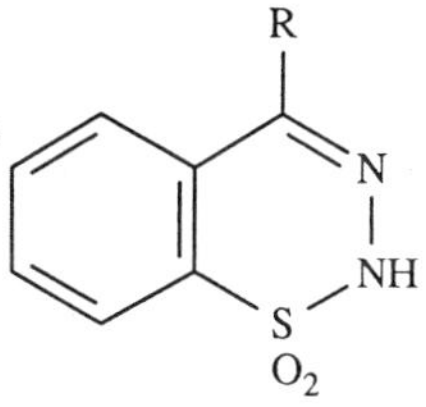   (13)

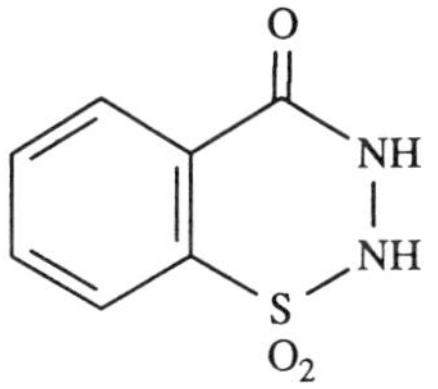

The availability of *o*-benzoylbenzenesulfonyl chlorides, prepared by diazotization of *o*-aminobenzophenones followed by reaction with sulfur dioxide in the presence of cuprous ions, can be exploited in the reaction with hydrazine or methylhydrazine to give 1,2,3-benzothiadiazine **14**. Product **14** (R^1 = Me) can also be obtained by methylation of **14** (R^1 = H) with methyl iodide and sodium hydroxide. The reduced form **15** results from catalytic hydrogenation of **14** (R^1 = H, R^2 = H) (68JHC453). Hydrazobenzene reacts with 2-chlorosulfonylbenzoyl chloride in the presence of triethylamine to give diphenyl derivative **16** (79MI1). 3,1,2-Benzothia-

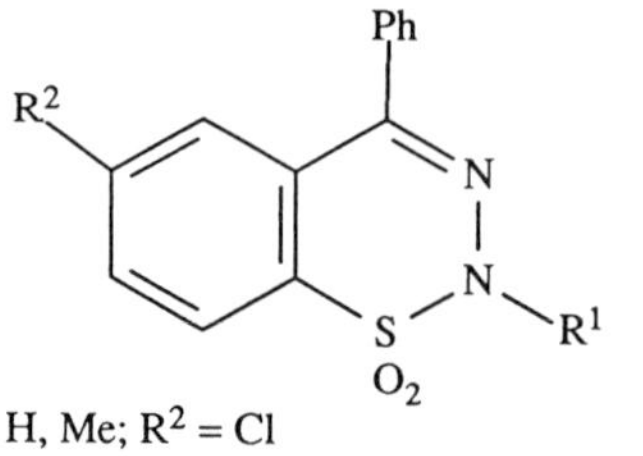

R¹ = H, Me; R² = Cl

(14)

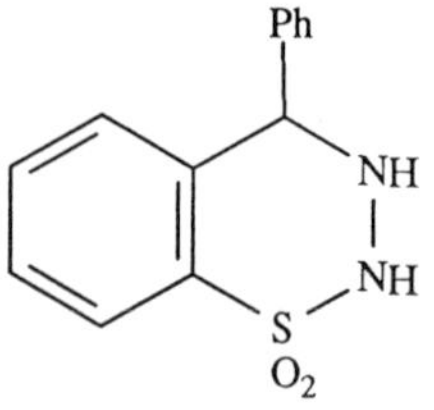

(15)

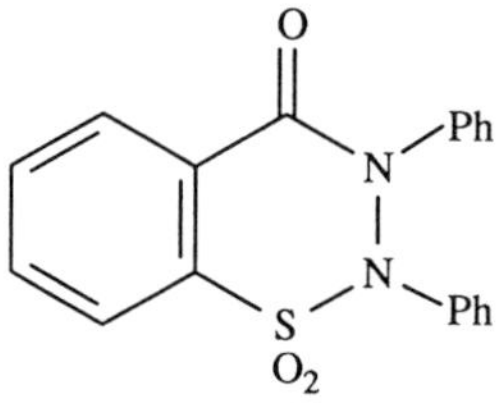

(16)

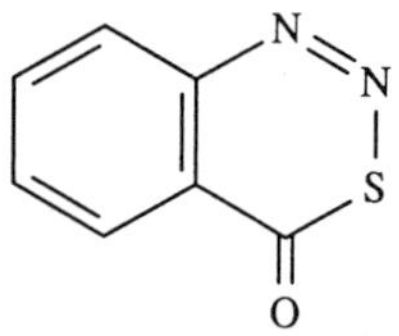

(17)

diazine **17** is obtained by diazotization of thioanthranilic acid under aprotic conditions (72JA8505).

Benzothiete **18** when heated, undergoes reversible ring-opening to give **19**. It is this latter *o*-quinonoidal form that is thought to account for the reactivity of **18** when it is heated with heterodienophiles such as R¹—N=N—R¹ (R¹ = COR, where R = OEt, OCH₂Ph, OCMe₃, Ph) to give the reduced 1,2,3-benzothiadiazines (**20**). This represents an interesting regiospecific thermally allowed [8 + 2] cycloaddition (86TL5703).

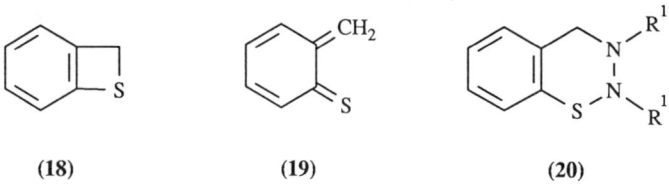

(18)   (19)   (20)

b. *Miscellaneous Condensed 1,2,3-Thiadiazines.* Diazotization of *peri*-aminosulfinate **21**, in an attempt to prepare perylene by reaction with the intermediate dehydronaphthalene, yields naphthothiadiazine **22**. Reduction of **22** with zinc and acetic acid gives the dihydro compound **23**, which can be oxidized by lead tetraacetate back to **22** (67LA96). A number of 1,2,3-thiadiazines (**24**) (*n* = 4 − 5) and their thiadiazinium perchlorates result from the treatment of aryl hydrazones of 2-thiocyanatocycloalkene-1-aldehydes with 70% perchloric acid. The products are stable, colorless or yellow salts, while the free bases are more deeply colored (73S225).

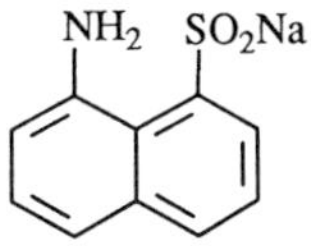

NH₂ SO₂Na

(21)

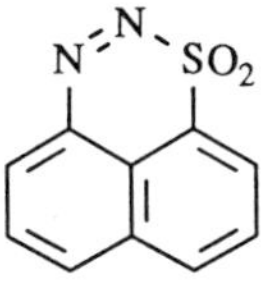

(22)

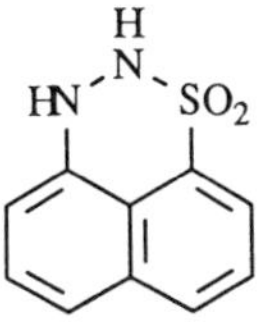

(23)

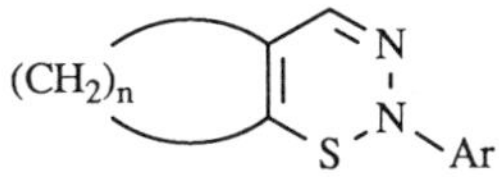

(CH₂)ₙ

(24)

B. CHEMICAL AND PHYSICAL PROPERTIES

1. General Survey

X-ray diffraction studies on crystals of fluorenethiadazine **2** indicate a multiplanar structure for the 1,2,3-thiadiazine ring with the sulfur atom situated out of the general plane of the other ring atoms [81JCS(P1)2322].

The IR spectrum of **8** has been reported (63FRP2166), and the UV spectra of some 1,2,3-benzothiadiazines have been reviewed (70CRV593). Mass spectrometry has been used to establish the structure of 1,2,3-thiadiazines **24**, and the fragmentation pattern enabled the alternative isothiazol-N-imines to be ruled out (75OMS579).

Not a great deal is known about the reactivity of 1,2,3-thiadiazines. Reactivity similar to that of sulfonamides would be expected. Indeed the ring sulfonamide NH-group is acidic, and the solubilities in alkali of 3,4,5,6-tetrahydro-3-phenyl-2H-1,2,3-thiadiazine 1,1-dioxide (62JPR56) and **11** (62JOC1703) have been quoted as the main evidence for the presence of a 1,2,3-thiadiazine ring. 4-Hydrazino-2H-1,2,3-benzothiadiazine 1,1-dioxide (**7**) is readily soluble in acids or bases and, when heated, undergoes ring fission to give bis(o-carboxyphenyl)disulfide. It is possible that the intermediate, o-sulfobenzoic acid, is reduced to the disulfide by hydrazine formed during the hydrolysis (17JPR180).

2H-1,2,3-Benzothiadiazines can be methylated at N-2, e.g., **14** (R^1 = H) can give **14** (R^1 = Me) (see Section II,A,2,a). Acetylation also takes place at the N-2 position (68JHC453). Hydrogenation of 1,2,3-benzothiadiazine 1,1-dioxides gives 3,4-dihydro-derivatives (79MI1).

2. *Reactions Involving Nitrogen Loss*

Some thiadiazines readily lose nitrogen, e.g., when pyrolyzed **13** (R = H) gives sultine **25** in poor yield (79MI1). Loss of nitrogen can also be induced by irradiation, as in the case of naphthothiadiazine **22** which gave thietan **26** in 25% yield (67LA96). Thiadiazine **17**, although more stable than the diazonium salt from which it is made (see Section II,A,2,a), loses nitrogen when either photolyzed or heated (at higher temperatures it also loses carbonyl sulfide) to give 3*H*-1,2-benzodithiole-3-one and other products, indicating the probable intermediate formation of benzyne (72JA8505). The action of an equimolar amount of chlorine on **13** (R = H) is accompanied by evolution of nitrogen, giving a compound thought to be **27**. Hydrolysis of **27** gives the pseudo-acid **28**, which can be converted with sodium hydroxide to the sodium salt of *o*-sulfobenzaldehyde (69CC33).

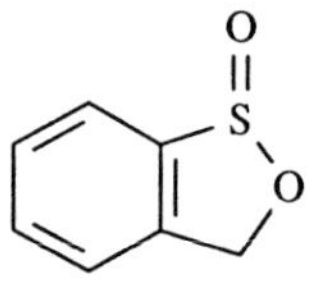

(25) 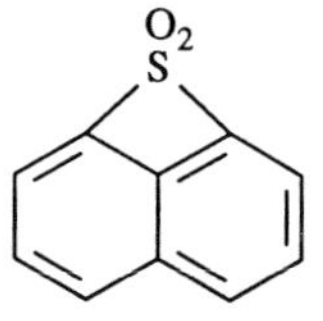   (26)

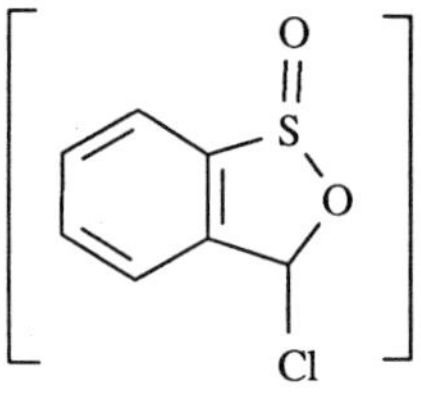

(27) 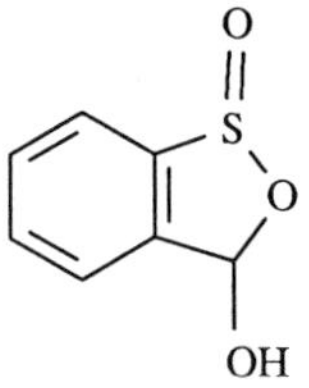   (28)

C. Applications

The 1,2,3-benzothiadiazine 1,1-dioxide (**7**, R = OEt) has diuretic and antihypertensive activity (62HCA996)—pharmacological features more common to a number of 1,2,4-benzothiadiazine 1,1-dioxides (see Section III,D).

III. 1,2,4-Thiadiazines

1,2,4-Benzothiadiazines have been selectively reviewed since 1982, although occasional refences have sometimes been made to publications in earlier years. Monocyclic 1,2,4-thiadiazines have been reviewed more comprehensively. Monocyclic 1,2,4-thiadiazines, however, were reviewed previously (70CRV593; 79MI1; 84MI1) as were 1,2,4-benzothiadiazines (79MI1; 84MI1), but the latter reviews, as with this present one, have necessarily been selective because of the vast increase in publications since the discovery of chlorothiazide (6-chloro-7-sulfamoyl-2H-1,2,4-benzothiadiazine) in 1957 [62AG(E)235].

A. Syntheses

1. *Monocyclic 1,2,4-Thiadiazines*

Monocyclic 1,2,4-thiadiazines **29** have been claimed to result from the treatment of thiamine hydrochloride and related thiazolium compounds

with hydroxylamine in strong alkali. The products are said to be thiamine antagonists (70MI1). Compound **29** is the only reported monocyclic thiadiazine with a dicovalent sulfur ring atom.

a. *By formation of One Bond.* The exploitation of sulfoacetic acid derivatives has enabled analogs of barbituric acid (1,2,4-thiadiazine-3,5-diones) to be made. Thereby, a useful synthesis of **33** can be effected from diester **30** by heating it with liquid ammonia in a sealed tube at 75°C to give diamide **31** (59JOC1983). This, when heated with potassium cyanate, gives the urea derivative **32,** which in turn, on refluxing in anhydrous pyridine, cyclizes to **33** (70CRV593). Attempts to use this method to prepare 6-monoalkyl- and 6,6-dialkylthiadiazines via alkylation of the malonate analog **30** met with mixed success (63JMC603; 61JOC3461). Derivatives with 4-substituents can be made from carbethoxymethylsulfonamide by reaction with, e.g. methylamine, to give **34,** which can be cyclized to **35** (61JOC3461). Another route to the 1,3-dione is by cyclization of **36** to monoimine **37** followed by hydrolysis to **33** (62DIS65). Lawson and Tink-

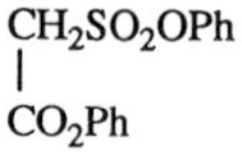

CH$_2$SO$_2$OPh

CO$_2$Ph

(30)

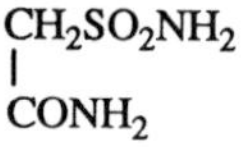

CH$_2$SO$_2$NH$_2$

CONH$_2$

(31)

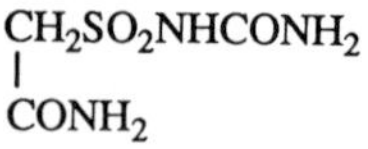

CH$_2$SO$_2$NHCONH$_2$

CONH$_2$

(32)

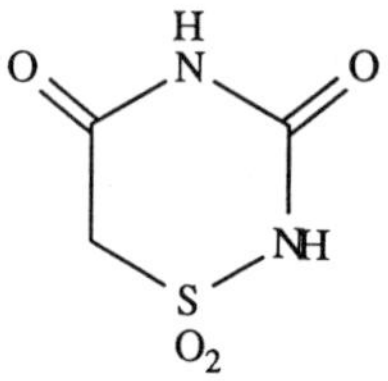

(33)

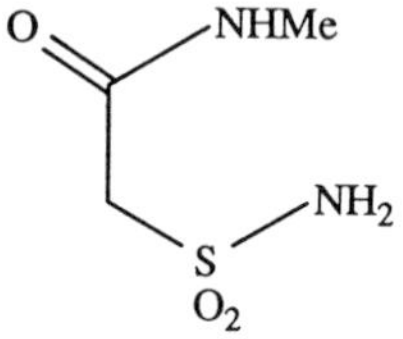

(34)

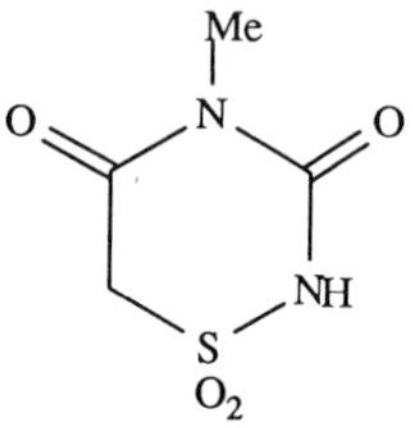

(35)

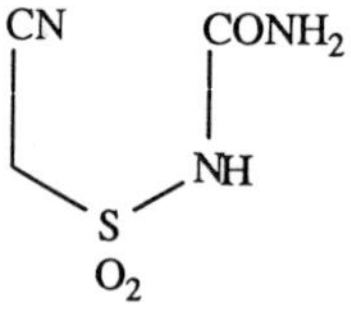

(36)

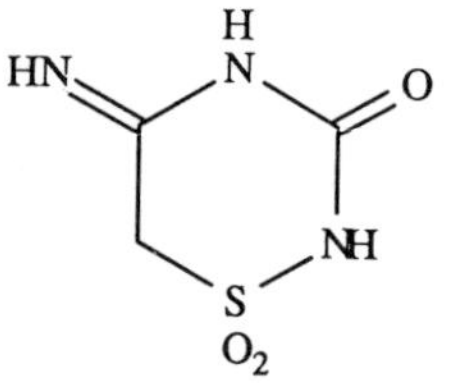

(37)

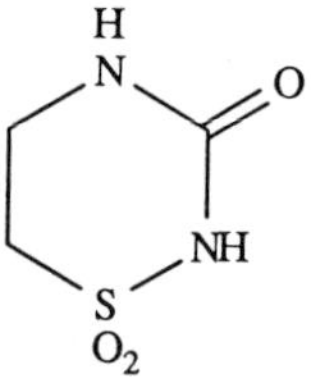

(38)

ler's review (70CRV593) provides a useful insight into some of the pitfalls associated with the synthesis of **33** from sulfoacetic acid derivatives.

2-Ureidoethanesulfonamide, made from taurinamide hydrochloride and potassium cyanate, ring-closes in refluxing anhydrous pyridine (possibly via the isocyanatosulfonamide) to give 3-oxo-thiadiazine **38** (59JOC1983).

Intramolecular conjugate addition of β-sulfonylstyrenes **39** (74JMC549) and **40** (R^2 = Ar) (72BCJ1893) affords the 4H- and 5,6-dihydro-1,2,4-thiadiazines **41** and **42**, respectively. Compound **39** is produced by sulfonyl-

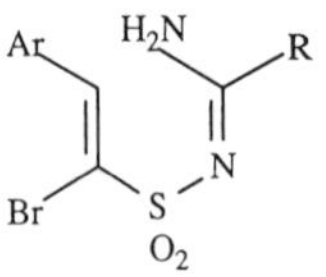

(39)

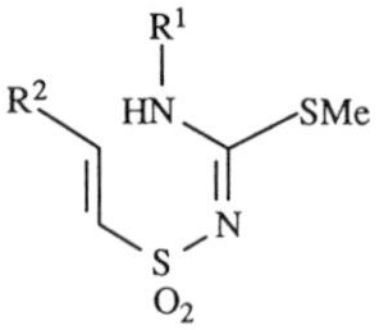

(40)

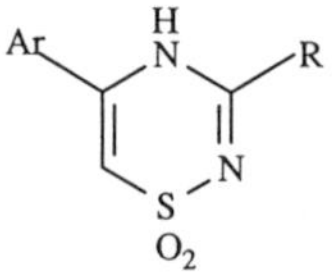

(41)

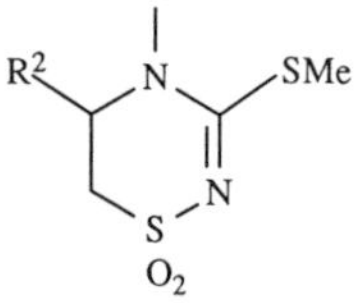

(42)

ation of acetamidines or benzamidines with α-bromostyrene-α-sulfonyl chloride and **40** by the action of *trans*-styrenesulfonyl chloride on Smethylisothiourea. Many 5,6-dihydro thiadiazines (**42**, R^1 = H) can be made by the latter means followed by facile replaceability of the SMe group in position 3 by nucleophiles (73BSF985; 73BSF2361; 74BSF1395) (see IIIC2a). This method was also used to prepare a cimetidine derivative (**43**) in which the cyanoguanidine moiety is replaced by the [3-(4*H*-1,2,4-thiadiazinyl)]amino group (86JMC44). Further examples are afforded by the incorporation of a variety of substituents, such as the [(piperi-

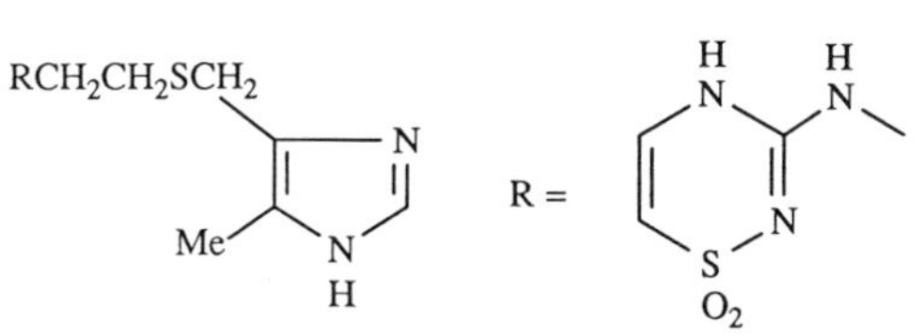

(43) 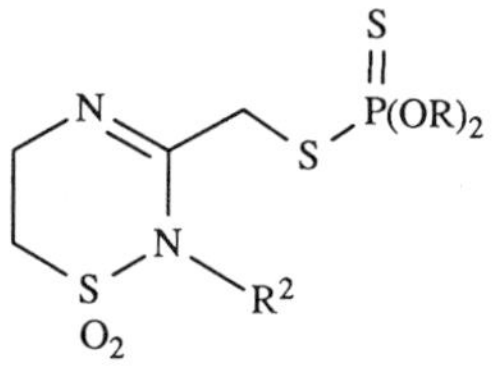   (44)

dinomethyl)phenoxy]propylamino group, into the 3-position of 4*H*-1,2,4-thiadiazine 1,1-dioxide [85JAP(K)60112781]. Use can be made of the nucleophilicity of *O,O*-dialkyldithiophosphate ions by reacting them with 3-chloromethyl-1,2,4-thiadiazines to yield products **44,** which are claimed to be fungicides, acaricides, nematocides, and insecticides (83GEP3208187).

Heterocyclic systems that contain the sulfoximide group have been previously reviewed (75CSR189; 76PS309; 77MI1; 80CSR477; 86TH1). A thiadiazine that includes this group can be made from ylide **45**, which ring-closes to cyclic sulfoximide **46** when treated with methoxide ion (77JOC952).

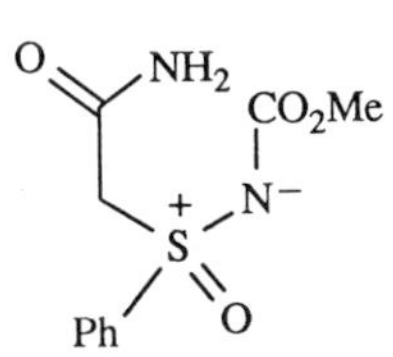

(45) 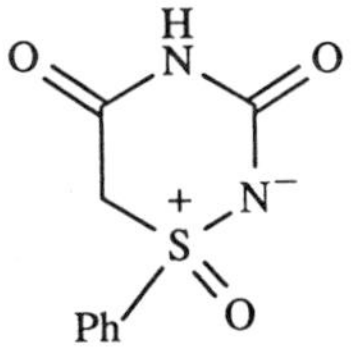   (46) 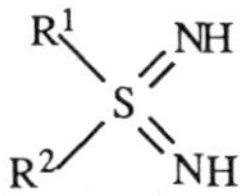   (47) 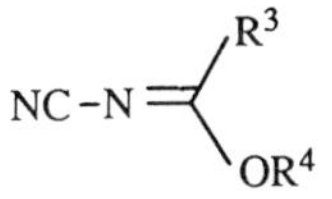   (48)

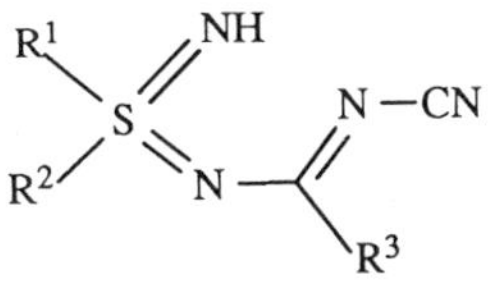

(49) 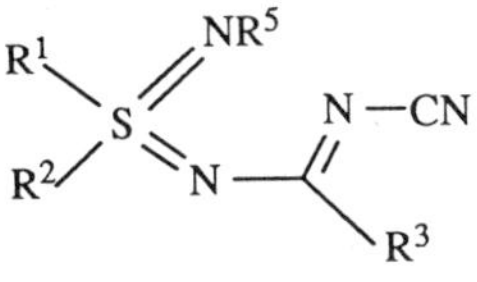   (50) 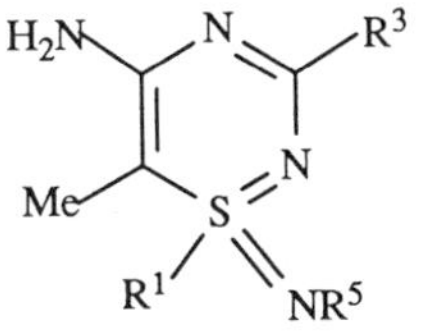   (51)

S,S-Dialkylsulfur diimides **47** (R^1 = Me, Et; R^2 = Et) condense with *N*-cyanoimidates **48** (R^3 = Ph, 2-thienyl) to give **49**. Substitution of the imido group of **49** by electrophiles (R^5X) leads to the *N,N'*-disubstituted *S,S*-dialkylsulfur diimides **50** (e.g., R^5 = SO$_2$Me). Cyclization to **51** is achieved by treatment of **50** with sodium hydride in dimethylsulfoxide (DMSO) (88CB1689).

b. *By Formation of Two Bonds.*
i. *From [5 + 1] atom fragments.* The 2*H*-1,2,4-thiadiazine **53** is formed by the action of ammonia on chlorosulfonylethylbenzimidoyl chloride **52**. The latter can be made from benzoyltaurine and thionyl chloride (47JA1393).

Some cyclic sulfoximines (1λ^6,2,4-thiadiazine 1-oxides) are featured in a patent that claims they have anxiolytic properties. Thus, **56** is said to be made by reacting sulfoxime **54** with benzonitrile and butyllithium to give **55**, followed by treatment with the methyl acetal of dimethylformamide (DMF) (85EUP161143). Ring formation can also be accomplished by treating **55** with (a) triethyl orthochloro- and orthobromoacetates to give **57** (R = CH$_2$Cl, CH$_2$Br) (45 and 65% yields, respectively), (b) *N,N'*-carbonyldiimidazole to give **58** (70% yield), (c) cyanogen bromide to **57** (R = NH$_2$) (21% yield), (d) acid chlorides or anhydrides via acylation of the imino group followed by treatment with either sodium hydroxide/ethanol or 1,5-diaza-bicyclo[4.3.0]non-5-ene (DBN) to give **57** (R = CH$_2$Cl, CH$_2$Ph, COOMe, CH$_2$CH$_2$COOMe, CF$_3$, Ph, 2-tolyl, 4-methoxyphenyl, and 2-furyl) (76, 50, 50, 51, 77, 66, 63, 64, and 63% yields, respectively), and (e) 2,2-dimethoxypropane to give **59** (71% yield) (86TH1).

The introduction of a ring carbonyl group into the 3-position with the aid of 1,1'-carbonyldiimidazole is again illustrated by the reaction of the latter with sulfonamide ester **60,** followed by ring closure to give thiadiazine **61,** which has modest dihydroorotose inhibiting properties (84JMC228).

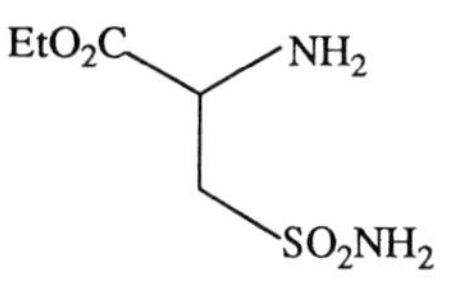

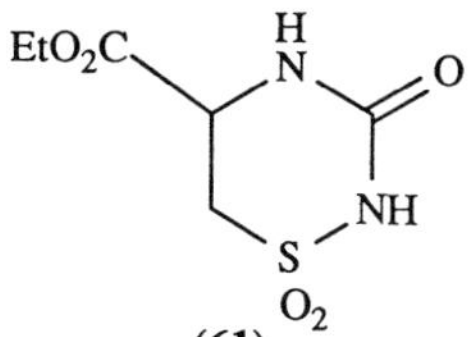

The antibacterial compound Taurolidine (or Tauroline or Drainsept) **62** is made by reacting 38% formaldehyde with taurinamide in the presence of sodium bicarbonate. Also, refluxing *N*-[2-(butylamino)ethylsulfo-nyl]aniline with 40% formaldehyde in aqueous ethanol yields **63** (R = Ph, R^1 = Bu). The parent compound **63** (R = R^1 = H) is known as Taurultam, an antibacterial and antifungal agent (66FRP1458701).

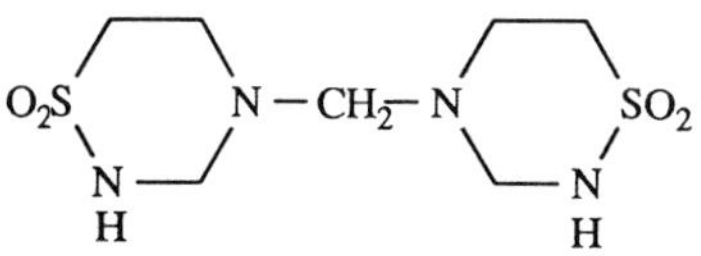

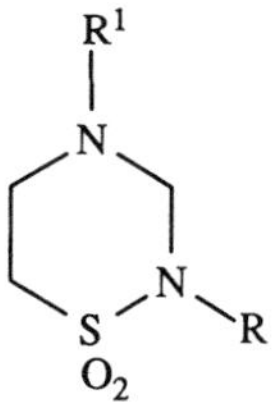

ii. *From [4 + 2] atom fragments.* An intriguing Diels–Alder [4 + 2] cycloaddition of 1,3-diazabutadienes **64** (Scheme 1) with sulfene (gener-ated *in situ* from methanesulfonyl chloride and triethylamine) yields **65** (87TL2641).

iii. *From [3 + 3] atom fragments.* Reaction of vinylsulfonyl chloride with amidines (Scheme 2) gives the 4*H*-thiadiazines **66** (73BSF985).

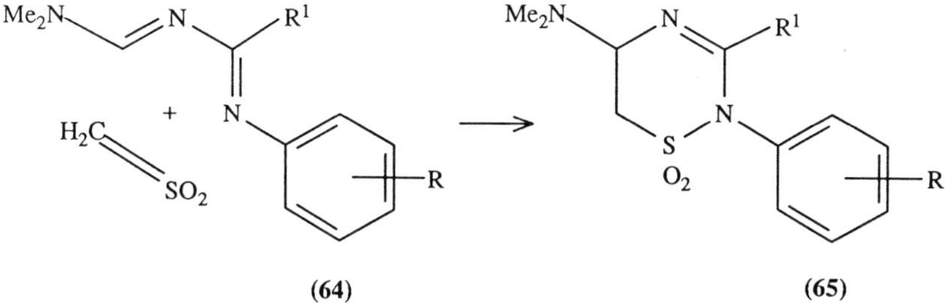

SCHEME 1

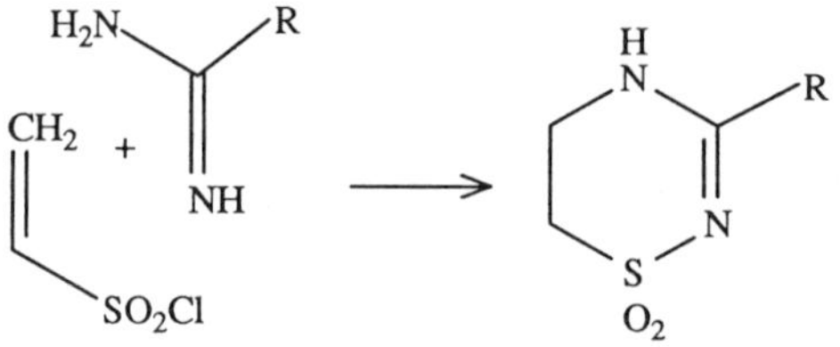

(66)

Scheme 2

c. *By Formation of Four Bonds.* The knowledge that 2-oxosulfon-amides can react bifunctionally with aldehydes and ketones to give hetero-cycles can be exploited to yield 1,2,4-thiadiazines, by including ammo-nium acetate among the reactants. For instance, **67** and benzaldehyde, when warmed with ammonium acetate, give **68** (64%). Although the reac-

(67)   **(68)**   **(69)**   **(70)**

(71)   **(72)**   **(73)**

tion is quite general for aldehydes, the use of ketones can lead to self-condensation of the ketosulfonamide, rather than the formation of the thiadiazine. More reactive ketones such as acetone and cyclohexanone do nevertheless react with **69** to yield **70** (39%) and spiran **71** (70%), respec-tively. Warming **69** with phenyl isothiocyanate in alkali leads to **72** (30%), whereas with methyl isocyanate in cold alkali, **73** (53%) is the product (85LA579).

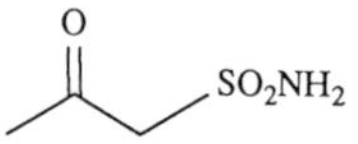

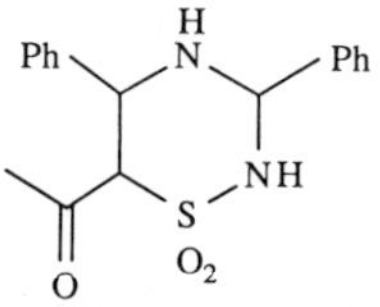

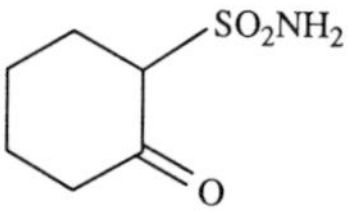

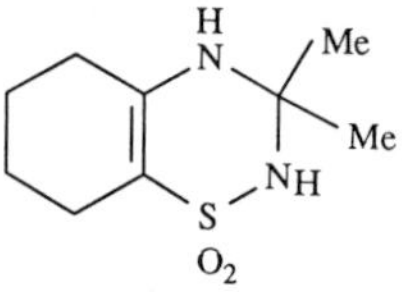

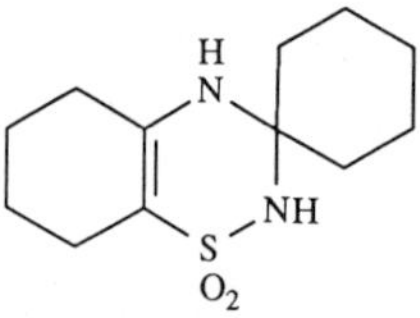

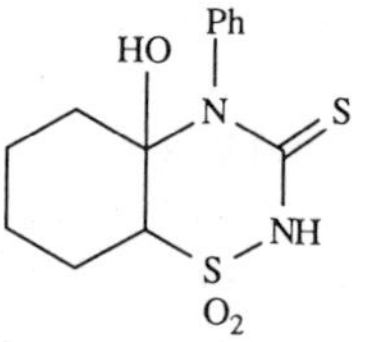

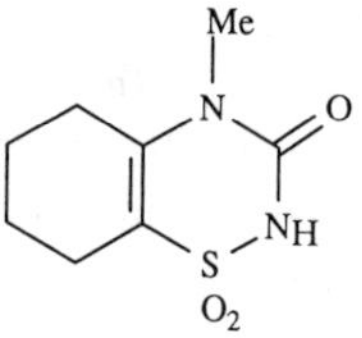

d. *By Ring Expansion.* The oldest method of preparing 1,2,4-thiadiazines is by oxidation of 2-iminothiazolidines, which can be made from 1,2-dibromoethane and a thiourea (1883M131; 04M682). For exam-ple, thiadiazinone **75** can be obtained from **74** on treatment with potassium chlorate and hydrochloric acid. Structure determination of both **74** and **75**

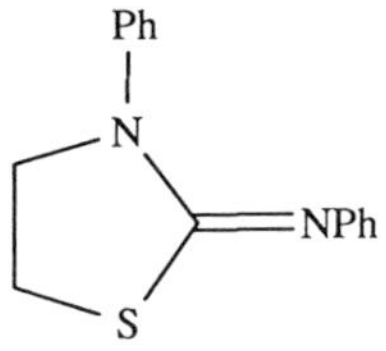

(74) 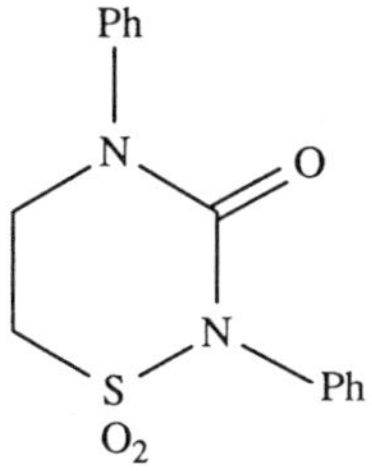   (75)

follows from the hydrolysis of **75** with barium hydroxide to give aniline and
N-phenyltaurine.

e. *From Other Heterocycles.* Oxathiazines **76** (R^1 = *N*-morpholinyl,
R^2 = H, R^3 = PhCH$_2$; R^1 = *N*-morpholinyl, R^2 = H, R^3 = Ph;
R^1 = *N*- piperidinyl, R^2 = H, R^3 = PhCH$_2$) can be transformed in high
yield to 1,2,4-thiadiazine 1,1-dioxides **77** by the action of ammonia in
ethanol at room temperature (88S521).

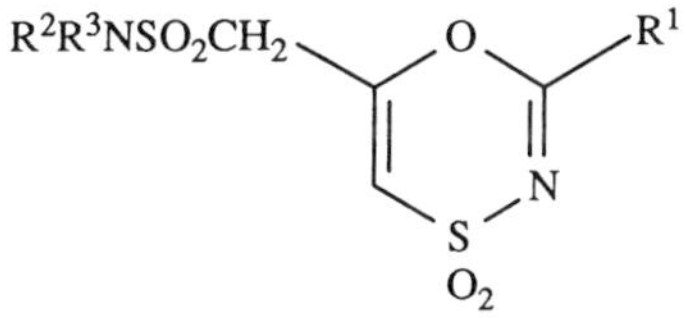

(76) 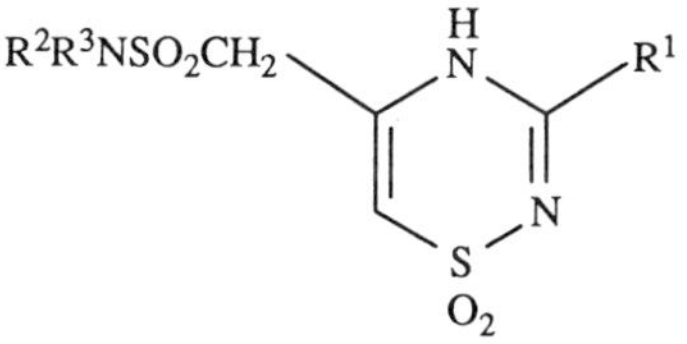   (77)

2. *1,2,4-Benzothiadiazines*

a. *By Formation of One Bond.* This continues to be a popular method
for synthesizing 1,2,4-benzothiadiazines. Very often intermediates are
designed to include an acyl moiety at the *o*-amino group of *o*-
aminobenzenesulfonamides. This allows subsequent ring-closure to be
accomplished by attack at the acyl carbonyl carbon atom by the sulfonamide nitrogen, followed by loss of water to give the product. For instance, *o*-aminobenzenesulfonamide can be reacted with diethyl malonate
to give **78**, which is then cyclized and hydrolyzed to give 4*H*-1,2,4-
benzothiadiazines **79,** some of which have anti-inflammatory activity
(81FES905). Acylation of *o*-aminobenzenesulfonamides can be accomplished with anhydrides and the derivatives then ring-closed; e.g., reaction
of 2-amino-3-methyl-1,5-benzenedisulfonamide with acetic anhydride,
then heating the product yields 7-(*N*-acetylsulfamoyl)-5-methyl-2*H*-1,2,4-

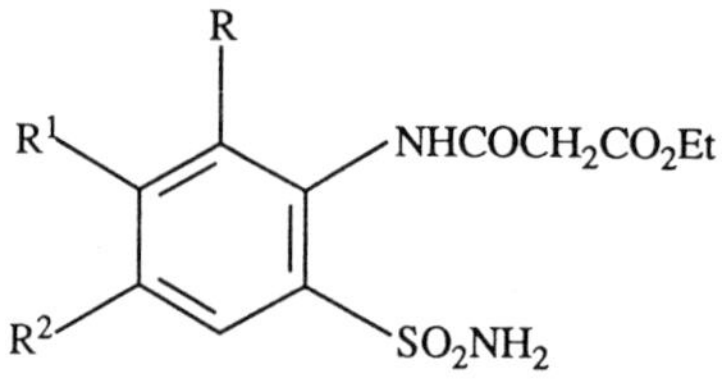

(78)

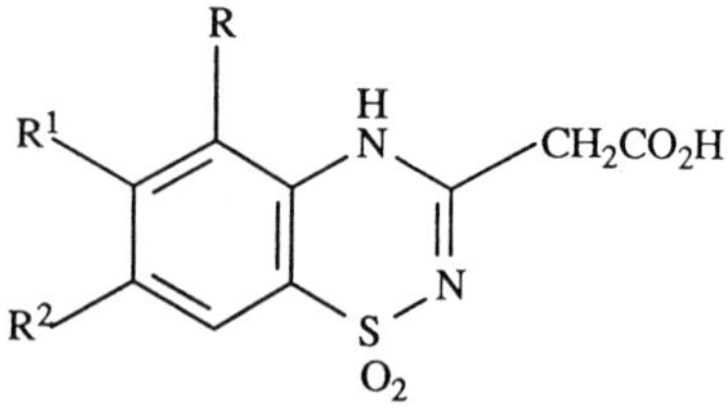

(79)

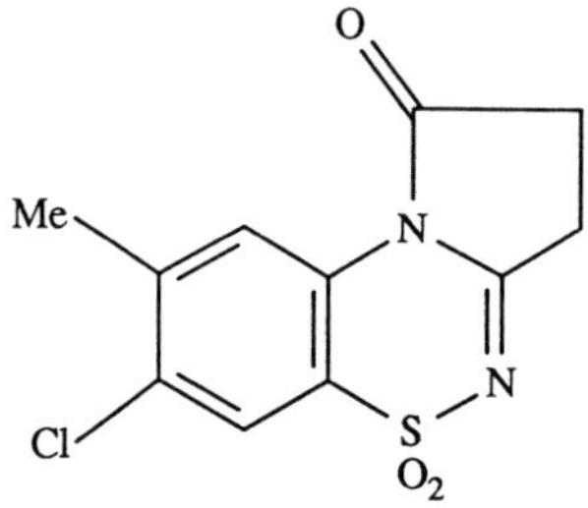

(80)

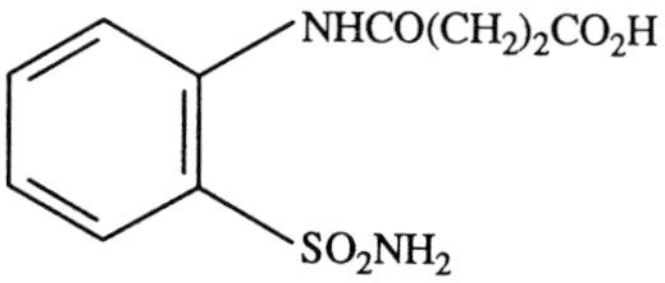

(81)

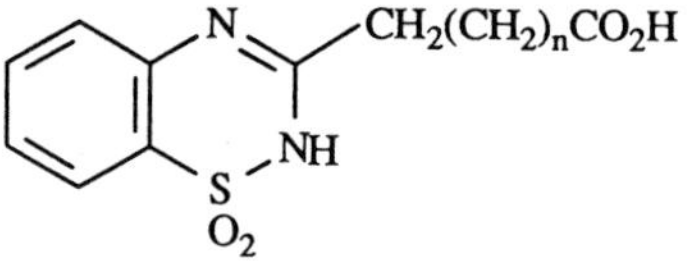

(82)

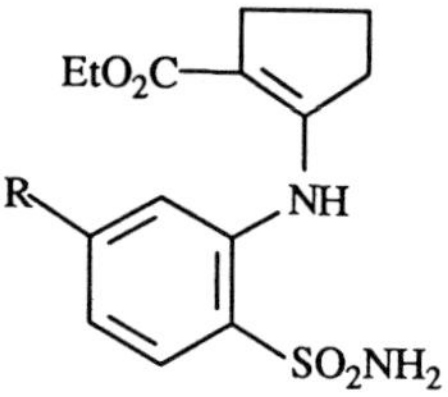

(83)

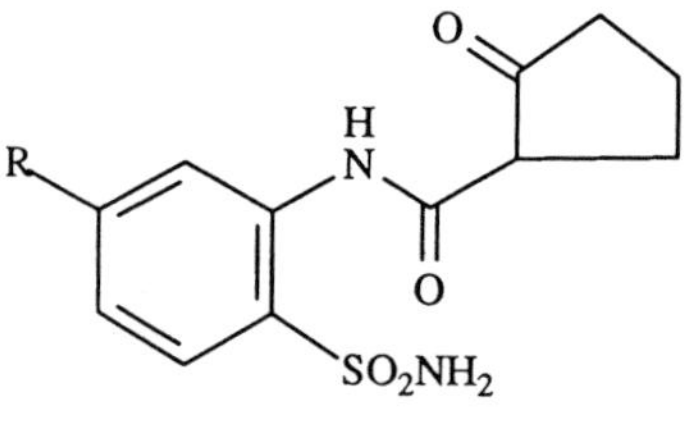

(84)

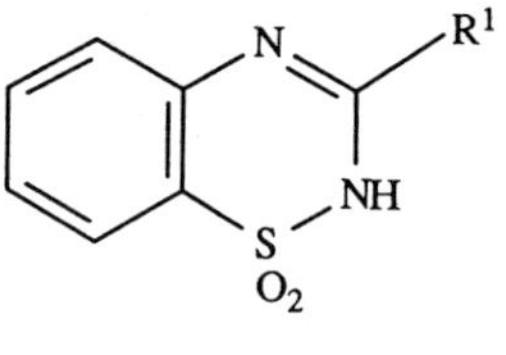

(85)

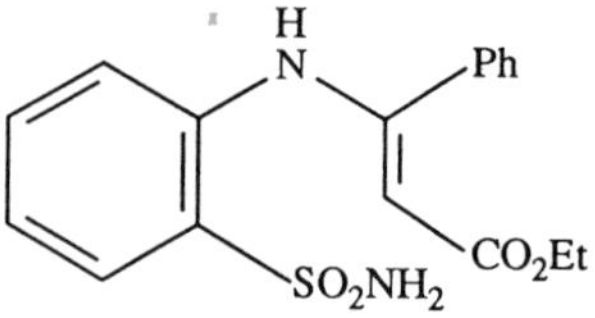

(86)

benzothiadiazine 1,1-dioxide (84MI2). Reaction with anhydrides of diprotic acids can give tricyclo compounds; e.g., **80** results from the reaction of the appropriate *o*-aminobenzenesulfonamide with succinic anhydride (79MI1). However, heating *o*-aminobenzenesulfonamide with succinic anhydride at 120°C yields (80%) acyl derivative **81,** which can be cyclized by sodium methoxide in refluxing methanol to **82** ($n = 1$) (86MI1). The same compound **82** ($n = 1$) and its butanoic homolog **82** ($n = 2$) as well as a number of other derivatives that show pronounced anti-inflammatory activity can be made from the corresponding ω-esters by reaction with sodium hydroxide (85AP903). Reaction of *o*-aminobenzenesulfonamides with β-ketoesters, such as ethyl 2-cyclopentanone carboxylate, yield compound **83** and/or anilide **84** together with the corresponding ketimine. Both **83** (R = H) and **84** (R = H) are ring-closed with sodium ethoxide to **85** [$R^1 = (CH_2)_4COOH$]. In reaction with ethyl 2-benzoylacetate, *o*-aminobenzenesulfonamide yields **86,** which cyclizes on treatment with sodium ethoxide to give **85** (R^1 = Ph) (66%) (84ZOR589; 84MI3).

Ethyl chloroformylformate in the presence of triethylamine reacts with *o*-aminobenzenesulfonamides to yield *N*-carbalkoxycarbonylacyl derivatives **87,** and these can be cyclized in high yield with sodium methoxide to give the 2*H*-1,2,4-benzothiadiazines **88** (R^1 = OEt) (83MI1). Alternatively, **88** (R^1 = OEt) can be made directly (see Section III,A,2,b). By reacting **88** (R^1 = OEt) with ammonia or primary amines, R^2NH_2

(87)   (88)

(89)   (90)

(R^2 = H, Me, Bu, cyclohexyl, $PhCH_2$), the corresponding amide derivatives of **88** result. Diuretic activity has been shown in **88** (R^1 = OH, NHR^2) (83MI1). The crotonyl or cinnamoyl group can be introduced at the *o*-amino group of *o*-aminobenzenesulfonamides by acylating the latter with the appropriate acid chloride to give **89** (R^2 = Me, Ph), followed by ring closure with base to benzothiadiazines **90** (84KGS907).

The introduction of longer, fluorinated chains with terminal trifluoromethyl groups in attempts to improve pharmacological activity and lipophilicity and reduce biodegradability, is illustrated by the cyclization of **91** to thiazide **92** (R^1 = $CF_3(CF_2)_n$; n = 0, 2, 4; R = Cl, Me, OMe). In fact, only compounds with relatively short chains (**92**, R^1 = $CF_3(CF_2)_n$; n = 0, 2; R = Cl) exhibit notable diuretic activity, although some of the long-chain compounds show significant antihypertensive activity (86BSF871).

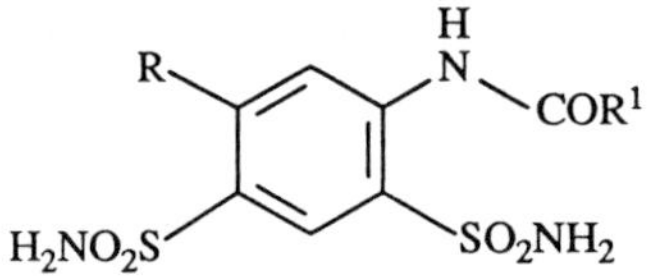

(91) 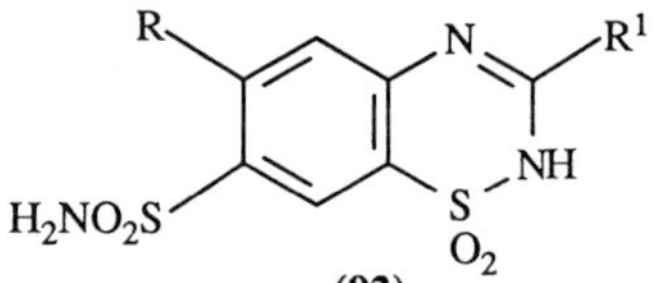   (92)

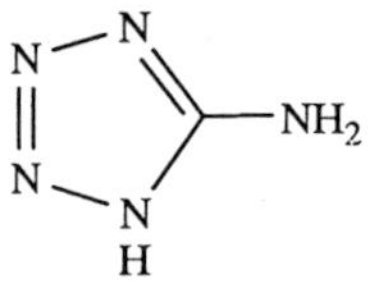

(93) 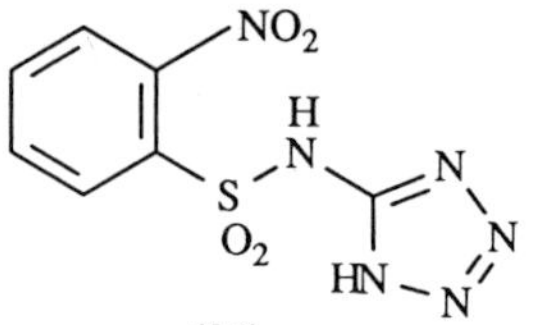   (94)

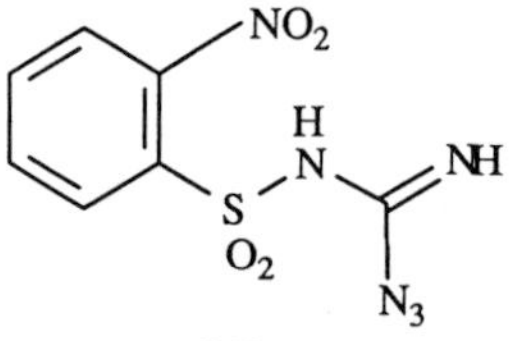

(95) 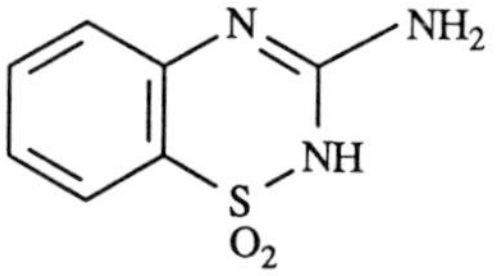   (96)

The reaction of *o*-nitrobenzenesulfonyl chloride with aminotetrazole (**93**) in an attempt to make **94** gave instead the ring-opened isomeric sulfonylcarbamimidic azide **95** (76%). Treatment of **95** with sodium dithionite in aqueous potassium hydroxide solution yielded benzothiadiazine **96** in quantitative yield—possibly by ring closure of the first-formed *o*-aminobenzenesulfonylcarbamimidic azide (87JHC1531).

Other well-established methods include the condensation of amine hydrochlorides (NRR^{1}H.HCl) with 3-oxo thiadiazines **97** (R^2 = R^3 = Me, R^4 = H) in the presence of phosphorus pentoxide and *N,N*-dimethylcyclohexylamine to give the 3-amino thiadiazines **98** (82CS248). The 3-oxo compounds **97** are prepared by a well-known method using chlorosulfonyl isocyanate and the appropriate aniline derivative to give chlorosulfonylureas **99**, which need not be isolated, but rather heated with aluminum chloride to give **97** [79JCS(P1)1043]. These 3(4*H*)-one 1,1-dioxides (**97**) are required for building fused quinazolinobenzo-thiadiazines, which may possibly possess, as do some of the quinazolin-ones themselves, a broad spectrum of biological activity. To this end, 6-alkyl- or 6-arylquinazolino[3,2-*b*][1,2,4]benzothiadiazine-13(6*H*)-one 11,

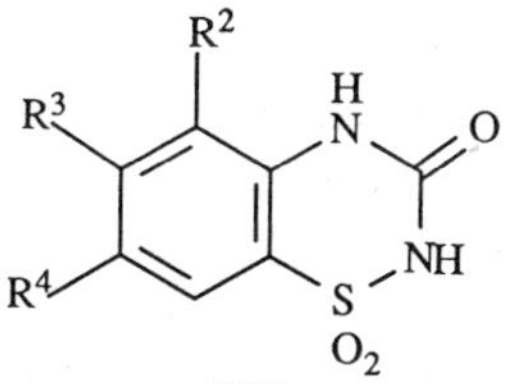

(97) 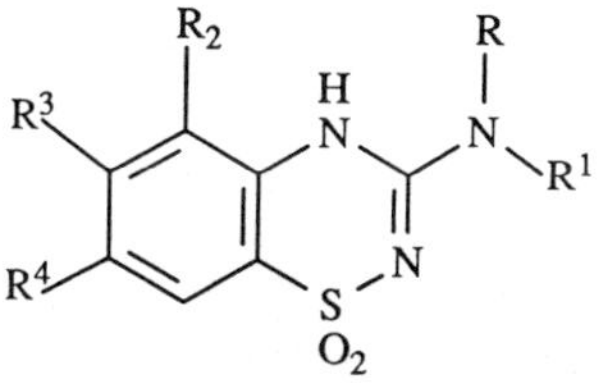   (98)

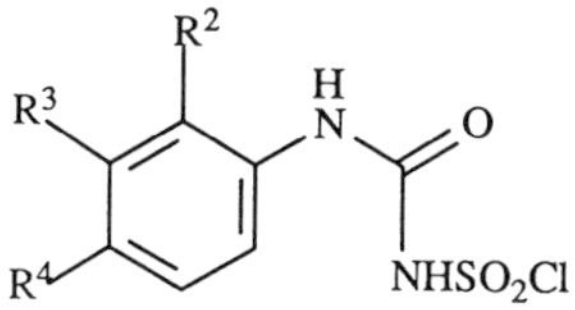

(99) 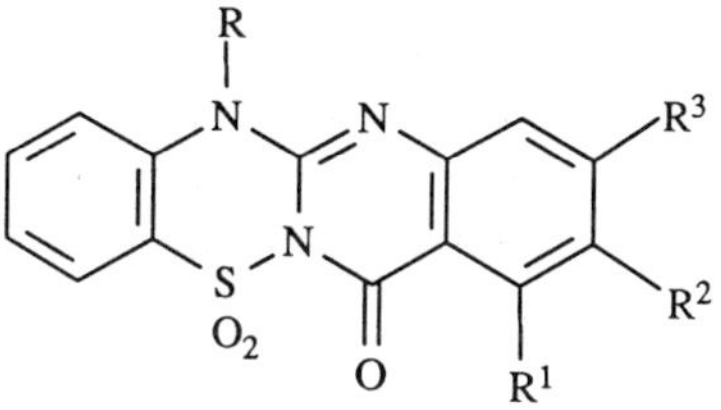   (100)

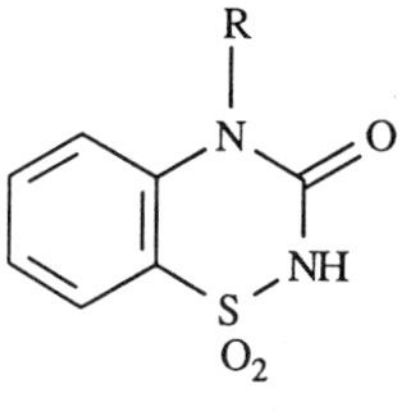

(101) 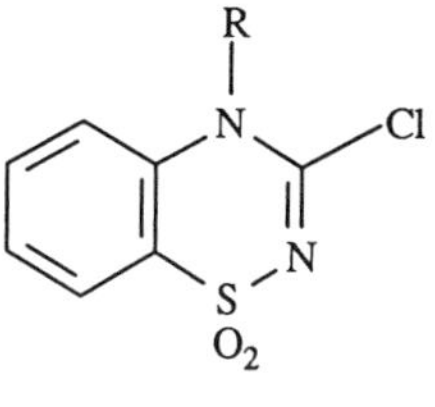   (102) 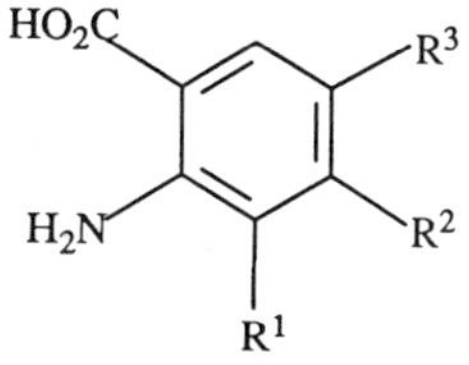   (103)

11-dioxides (**100**) are prepared by first treating **101** (R = Me) with phosphorus pentachloride to give **102** (R = Me), followed by heating with anthranilic acid derivatives **103** (86MI2). The technique of ring closing *o*-ureidoben zenesulfonamides, as illustrated by the action of phosphoryl chloride on 4-ureidotoluene-3-sulfonamide to give 3-chloro-6-methyl-2*H*-1,2,4-benzothiadiazine 1,1-dioxide, which includes a relatively inert chlorine atom (79MI1), has found no use since 1979. Cyclization of *o*-alkyl(or aryl) thiophenylureas to 1-alkyl(or aryl)-3,4-dihydro-3-oxo-1*H*-1,2,4-benzothiadiazines by bromine and sodium methoxide can be followed by hydrolysis, to give acyclic sulfoxides, or by oxidation with permanganate in acetonitrile to yield stable 1-alkyl- (or aryl) 3,4-dihydro-3-oxo-1*H*-1,2,4-benzothiadiazine 1-oxides (68LA223).

b. *By Formation of Two Bonds.*
i. *From [5 + 1] atom fragments.* This represents the most popular means of synthesizing 1,2,4-benzothiadiazines. Methods include either the incorporation of carbon at position 3 or of sulfur at position 1 to give the thiadiazine. The inclusion of carbon has been accomplished by the use of esters, formimidates, carboxylic acids, aldehydes or acetals, enolic ethers, *N*,*N*-disubstituted glycine esters, 1,1'-carbonyldiimidazole, phosgene, thiophosgene, urea, thiourea, and alkyl or aryl isothiocyanates.

3-Carboethoxy-2*H*-1,2,4-benzothiadiazine 1,1-dioxides (**88**, R = Cl, Br; R^1 = OEt) can be prepared from cyclocondensation of 4-halogeno 2-aminobenzenesulfonamides with diethyl oxalate in the presence of sodium methoxide (83MI1). A novel method using ethyl carboethoxyformimidate **104** in reaction with *o*-aminobenzenesulfonamide also gives **88** (R = H; R^1 = OEt) by an easier route than previously (76KGS479) despite the seemingly modest yield (44%) (85T2625).

(104)   (105)

The use of polyphosphoric acid trimethylsilyl ester as a means of condensing aryl, arylalkyl, and arylalkenyl carboxylic acids with *o*-aminobenzenesulfonamide in a one-pot synthesis gives high yields of 3-substituted 4*H*-1,2,4-benzothiadiazine 1,1-dioxides **105** (R = Ph, 84%; 4-MeC$_6$H$_4$, 88%; 4-MeOC$_6$H$_4$, 89%; 4-ClC$_6$H$_4$, 92%; 4-O$_2$NC$_6$H$_4$, 93%; 3-O$_2$NC$_6$H$_4$, 96%; PhCH$_2$CH$_2$, 89%; cyclohexyl, 87%; PhCH=CH, 91%) [83S851; cf. 85JAP(K)6025984]. When toluidinedisulfonanilides are heated with formic or acetic acids, ring-closure occurs to give the corresponding 2-aryl-1,2,4-benzothiadiazine 1,1-dioxides (83MI2).

Aldehydes are still used to prepare diuretics (82USP4338435) and antihypertensives (87USP4634689), e.g., when **106** is heated with 6-bromohexanal, thiadiazine **107** (X = Br) results. Substituted sulfonamide **106** can apparently be made from the corresponding disulfonyl chloride by sequential amidation with methylamine and then benzylamine. The presence of the bromine moiety in **107** allows **107** [X = PO-(OH)CH$_2$COOH] to be made by reaction with diethyl carboethoxymethylphosphonite [(EtO)$_2$PCH$_2$COOEt] and subsequent saponification of the phosphinate and acetate groups. Hence, condensation in trifluo-

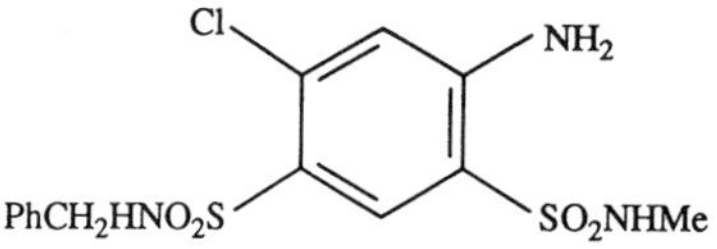

(106)

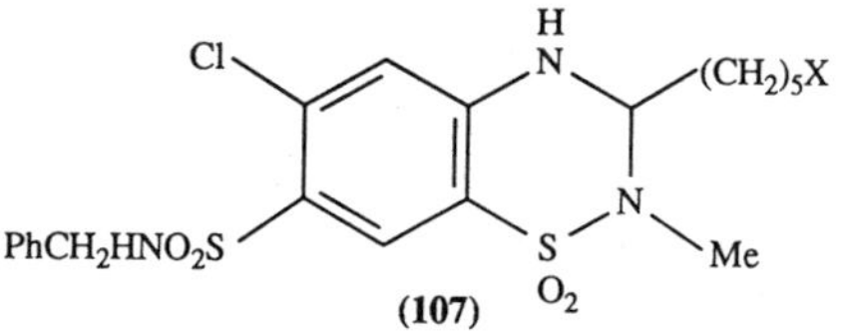

(107)

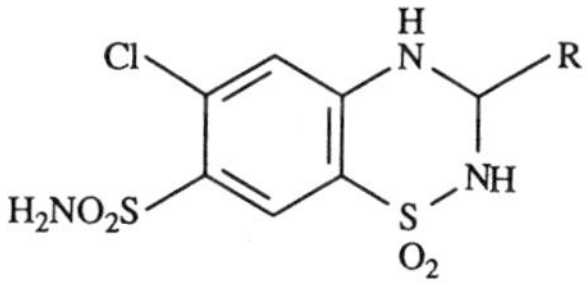

(108)

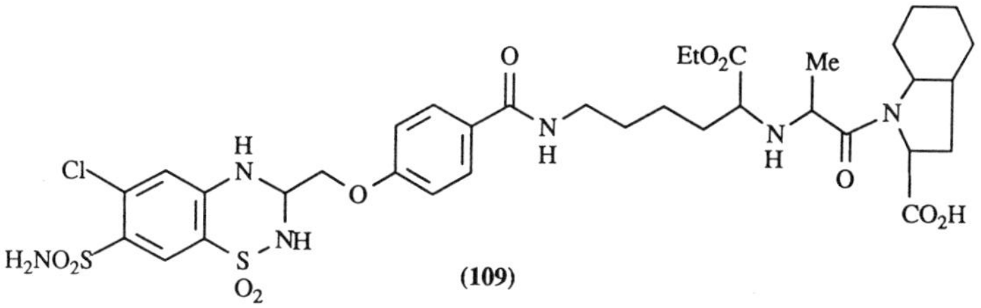

(109)

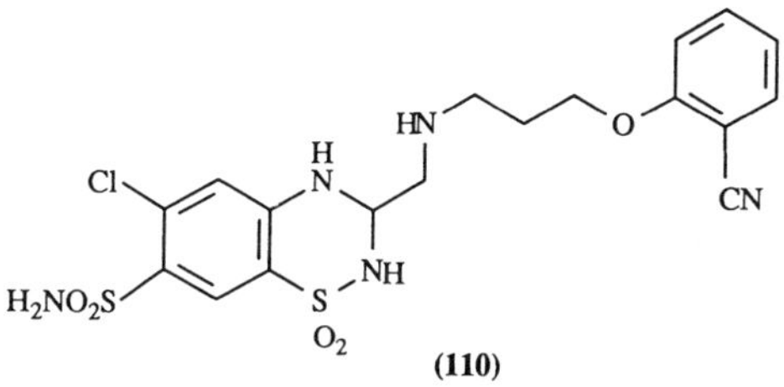

(110)

roacetic acid with N-cyclopentylglycine 1,1-dimethyl ethyl ester yields **107**
[X = PO(OH)CH$_2$CON(cyclopentyl)CH$_2$COOH] (87USP4634689). Such
chain-lengthening and introduction of phosphorus oxy-acid ester groups,
as well as amino acid or dipeptide moieties, are typical of a number of
patented processes that claim to produce molecules with more efficient
targeting prospects. Acetals have been featured in the attainment of similar goals. Thus, **108** (R = 4-CH$_2$OC$_6$H$_4$COCl) is said to result from the
reaction of 5-chloroaniline-2,4-disulfonamide with 4-(EtO)$_2$CHCH$_2$-
OC$_6$H$_4$COOH, followed by condensation with 1-{N-[5-aminopentyl-
1(S)-ethoxycarbonyl]-(S)-alanyl}octahydro-1H-indole-(S)-carboxylic acid
[made from benzyl 1-pyruvoyloctahydro-1H-indole-2(S)-carboxylate and
H-Lys(COOCH$_2$Ph)-OEt] to give **109,** which is claimed to have antihypertensive and diuretic activity (84USP4468396; cf. 85USP4559340;
86USP4584285). Also, N-[3-(2-cyanophenoxy)-1-propyl]aminoacetalde-
hyde dimethyl acetal hydrobromide can be refluxed with 3-chloroaniline-
4,6-disulfonamide in ethanol in the presence of p-toluenesulfonic acid
to yield **110** (88GEP3632544).

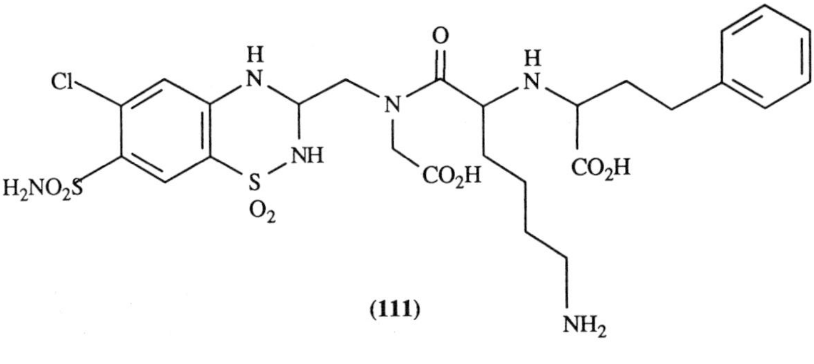

(111)

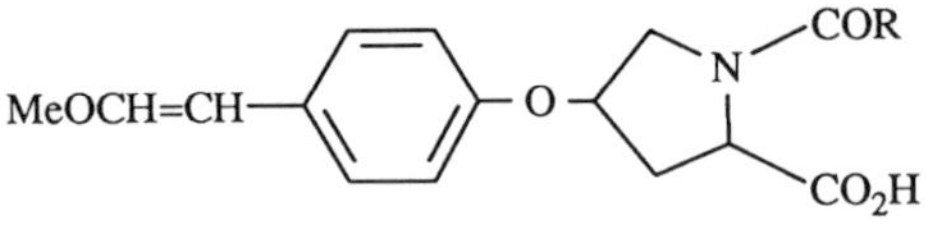

(112)

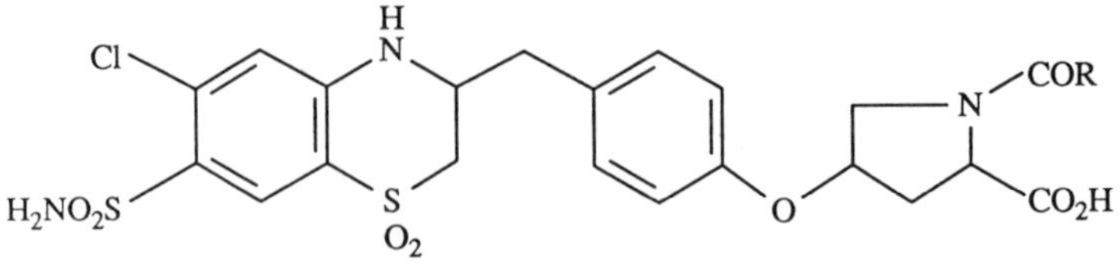

(113)

Further examples can be drawn from the preparation of dipeptides, such as **111,** that include the 1,2,4-benzothiadiazine moiety and are used in attempts to impart antihypertensive activity to the compound. These are claimed to be prepared, for instance, by the cyclocondensation of N-(2,2-diethoxyethyl)-N-{N^{α}-(2,2,2-trichloroethoxycarbonyl)-N^{α}-[(1S)-1-(ethoxycarbonyl)-3-phenylpropyl]-N^{ε}-carbobenzoxy-L-lysyl}glycine ethyl ester with 4-amino-6-chloro-1,3-benzenedisulfonamide followed by sequential deprotection with zinc dust, saponification, and acidification with hydrochloric acid (87USP4666906). Enolic ethers may be used to supply the ring carbon atom at position 3 as claimed in the use of 4-(2-methoxyvinyl)-phenoxyproline derivative **112** [R = CH(Me)CH$_2$SH], which cyclocondenses with 4-amino-6-chloro-1,3-benzenedisulfonamide to give **113** [R = CH(Me)CH$_2$SH] (85USP4536501). The well-tried method of using 1,1'-carbonyldiimidazole can be demonstrated by a patent claiming that 1,2,4-benzothiadiazines, such as **115,** can be made from the substituted o-aminobenzenesulfonamides **114** (88EUP268990). The

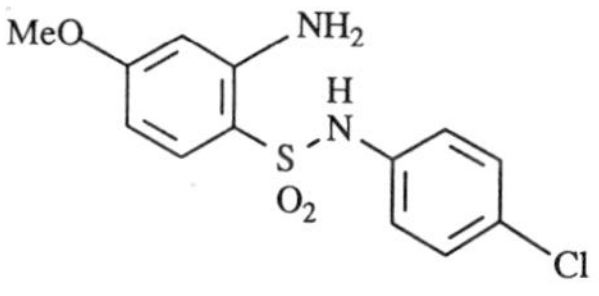

(114) 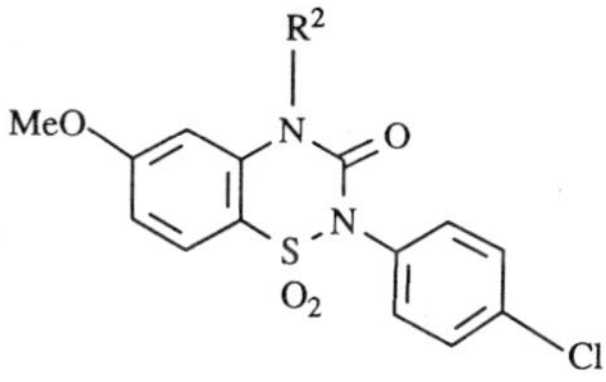   (115) 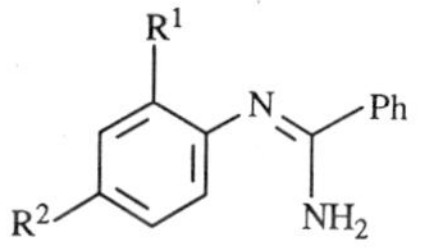   (116)

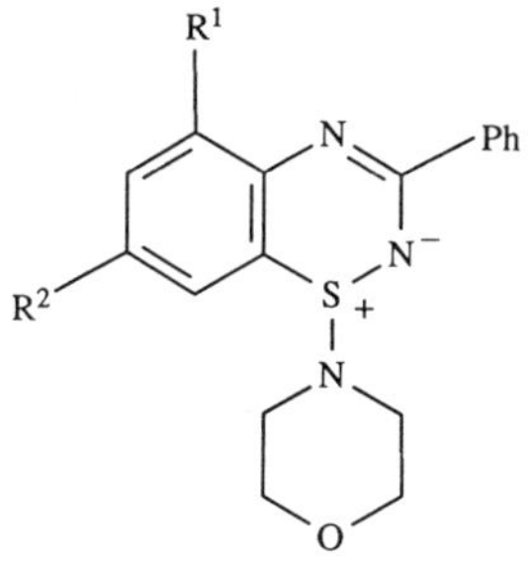

(117) 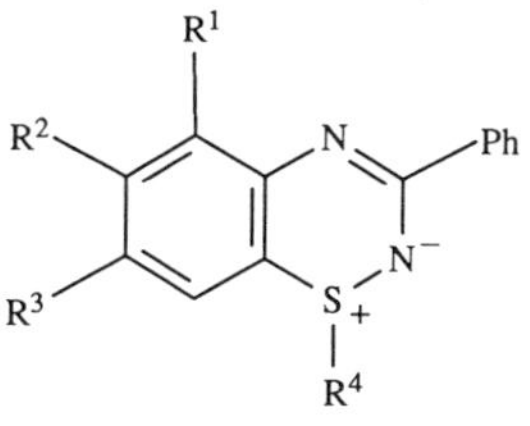   (118)

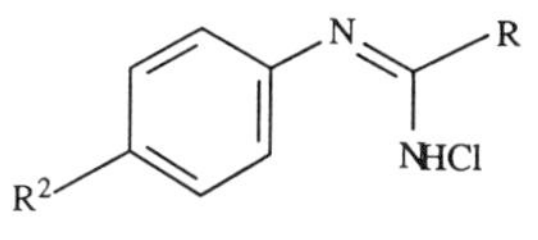

(119) 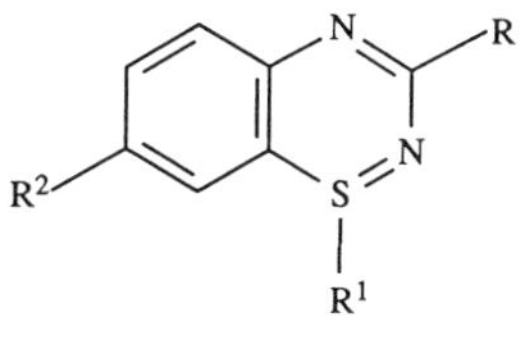   (120) 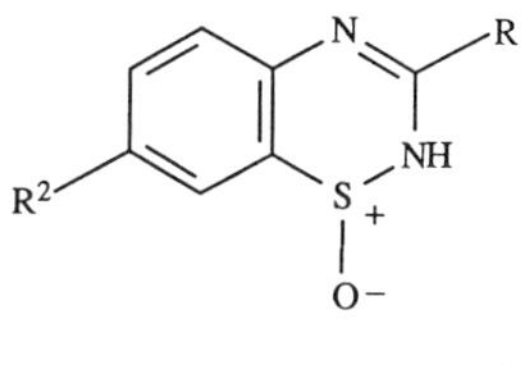   (121)

method of reacting alkyl or aryl isothiocyanates with o-aminobenzene-sulfonamides to yield 3-alkylimino or 3-arylimino-3,4-dihydro-1,2,4-benzothiadiazine 1,1-dioxides (79MI1) has no application since 1979. Also, the reactions of o-aminobenzenesulfonamide with thiophosgene (79MI1) or thiourea (62FES320) that yield 3,4-dihydro-1,2,4-benzothiadiazine-3-one 1,1-dioxide or its 3-mercapto tautomer have not been widely applied in nonpatent work.

There are a number of ways that sulfur can be inserted to form the 1,2,4-benzothiadiazine ring system. Among some successful syntheses of open-chain sulfimides involving the reaction of N-arylbenzamidines 116 and 4-phenylthiomorpholine (with N-chlorosuccinimide at $-20°C$), the latter was replaced by 4,4′-thiobismorpholine, thereby yielding ylidic thia-diazines, i.e., $1\lambda^4,2,4$-benzothiadiazine 117 (R^1 = H, R^2 = Me; 60%). It was thought that since a better leaving group on the sulfur atom might favor the formation of these cyclic ylides, sulfenyl chlorides (R^4SCl) might prove more effective than sulfenamides. This was born out by the larger yields of $1\lambda^4,2,4$-benzothiadiazines 118 (e.g., R^1 = R^2 = R^3 = H, R^4 = 4-MeC$_6$H$_4$; 86%) [83JCS(P1)49]. This new route to these cyclic ylides adds to the few already known (68LA223; 73ZOR2038). Substituted chloroamidines 119 are cyclized with sulfur dichloride to S-chloro-1,2,4-benzothiadiazines 120 (R^1 = Cl), which then can be either aminated with, say morpholine to give 120 (R^1 = 1-morpholino), or hydrolyzed to 1-oxides 121 (84ZOR196).

Guanidines 122 (R = Ph, X = H; R = Aryl, X = Me or OMe or Cl) react with thionyl chloride in the presence of triethylamine and with or

(122)   (123)

(124)

(125)   (126)

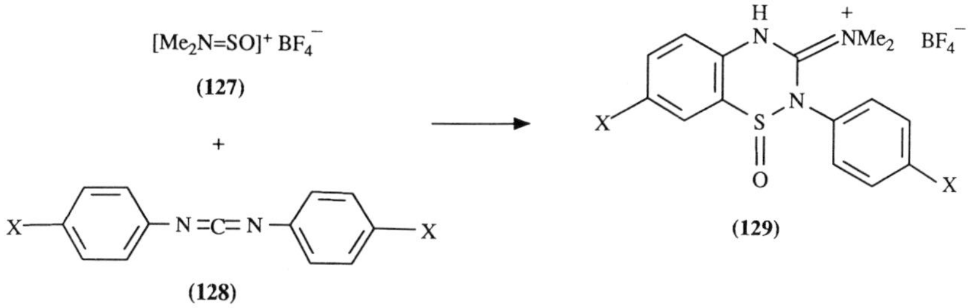

SCHEME 3

without Lewis acids to give 1-oxides **123.** This reaction is thought to proceed via the possible intermediate **124** rather than the alternative **125,** particularly since a Friedel–Crafts-type reaction does not appear to take place because the product can form slowly when no Lewis acid is present. The starting materials, **122,** can be formed from the corresponding carbodiimides by reaction with dimethylamine. 1-Oxides **123** may be oxidized on treatment with hydrogen peroxide to give high yields of their corresponding 1,1-dioxides. The reduction of **123** with zinc and hot acetic acid does not lead to the deoxygenated thiadiazine, but rather to benzthiazoles **126** (87PS41). *N*-methyl-*N*-sulfinylmethanaminium tetrafluoroborate **127** reacts with diarylcarbodiimides **128** to give the 1-oxide tetrafluoroborates **129** (X = H, MeO, Br) (Scheme 3). It is suggested that intermediate **130**

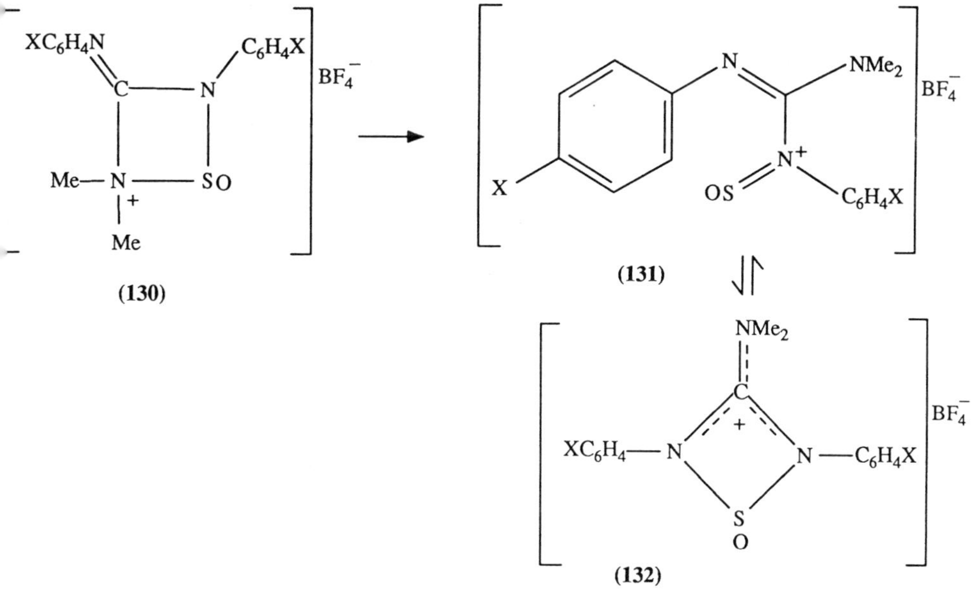

SCHEME 4

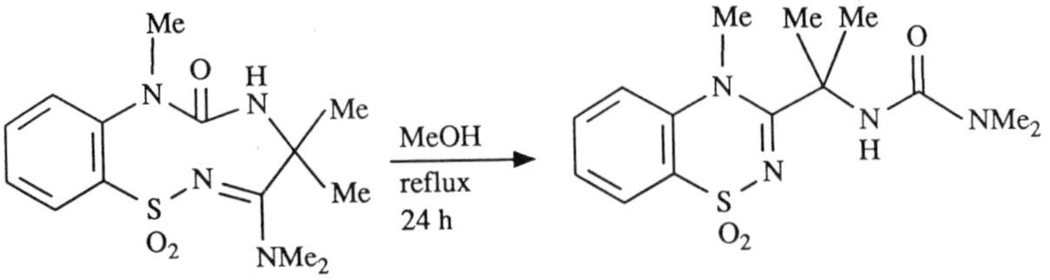

(133)   **(134)**

SCHEME 5

(Scheme 4) is first formed by cycloaddition and then ring opened to give rise to the reversible system (Scheme 4) involving **131** and **132**, but that product **129** is most likely formed from **131** (84LA904). The reaction of *N*-arylamidines [ArN=C(Ar')NH$_2$] with *N*-sulfinylsulfonamides (R^{2}SO$_2$N=SO; R^2 = Me, Ph, *p*-tolyl) gives 1-sulfonylimido-2*H*-1,2,4-benzothiadiazines, which can be hydrolyzed to the corresponding 1-oxide (79MI1) (see Section III,C,1,a,iii).

   c. *By Ring Contraction.* An unusual isomeric rearrangement of 1,2,5,7-benzothiatriazonines, such as **133**, can be accomplished in refluxing methanol to give 1,2,4-benzothiadiazine **134** (78%) (Scheme 5). A mechanism is tentatively suggested for this intriguing rearrangement. However, the rearrangement may have little value in the preparation of 1,2,4-benzothiadiazines, since the benzothiatriazonine (**133**) is itself made from the former type of system by reacting 2,2-dimethyl-3-dimethylamino-2*H*-azirine with 3,4-dihydro-4-methyl-2*H*-1,2,4-benzothiadiazine 1,1-dioxide. The structure of reduction product **135**, obtained from **134** by reaction with sodium borohydride in 2-propanol, was confirmed by X-ray analysis (84C316).

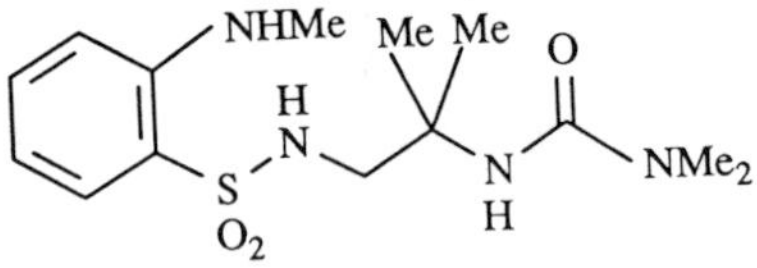

(135)

   d. *By Ring Expansion.* Photolysis of azidobenzothiazolium salts can lead to a variety of products that sometimes include 1,2,4-benzothiadiazines; e.g., when **136** is irradiated in hydrofluoroboric acid, three products (**137, 138,** and **139**) are obtained. 1,2,4-Benzothiadiazine **137** is the major product (67% of on overall 83% yield) (87HCA2045).

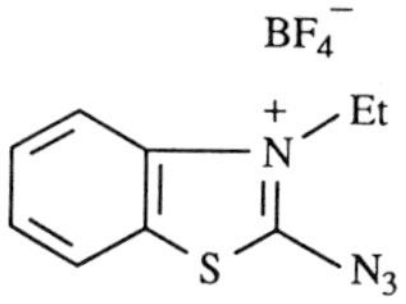

(136) 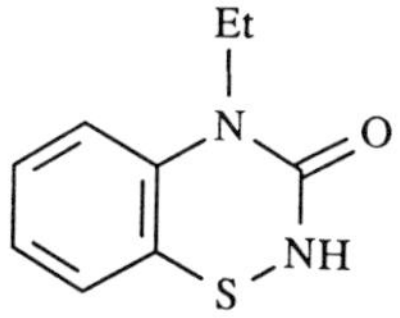   (137)

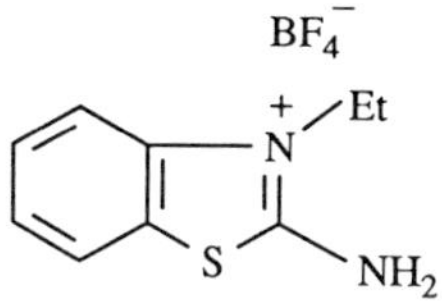

(138) 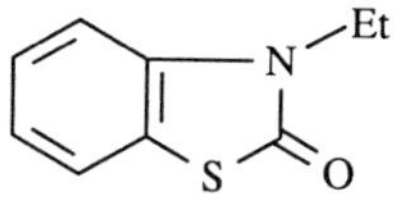   (139)

3. Other Condensed 1,2,4-Thiadiazines

The *N,N'*-disubstituted *S,S*-(1,4-tetramethylene)sulfur diimides (**140**; R^3 = 2-thienyl, Ph, 4-ClC$_6$H$_4$, 4-MeC$_6$H$_4$, R^5 = SO$_2$Me, COPh, COMe, SO$_2$Ph) can be cyclized with sodium hydride to thieno[2,1-f][1λ^6,2,4]thia-diazines **141** (cf. **50** and **51** in Section III,A,1,a) (88CB1689).

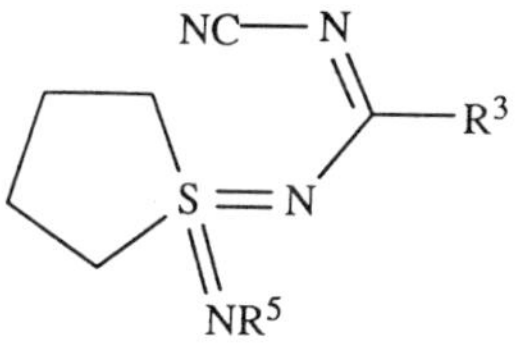

(140) 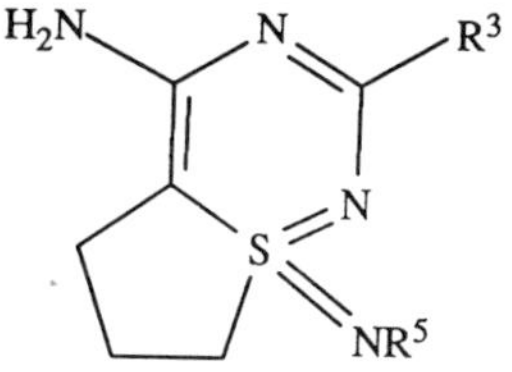   (141) 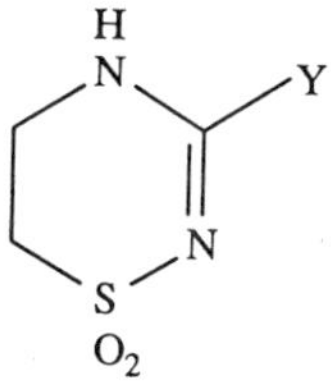   (142)

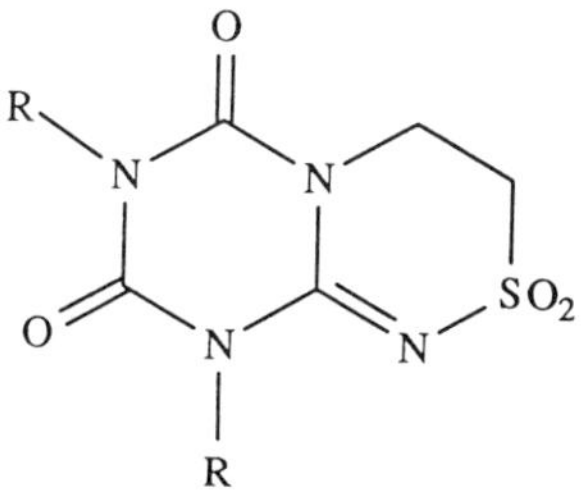

(143) 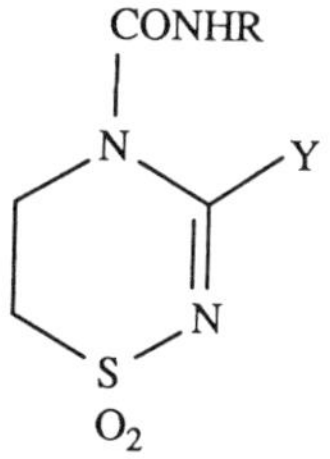   (144) 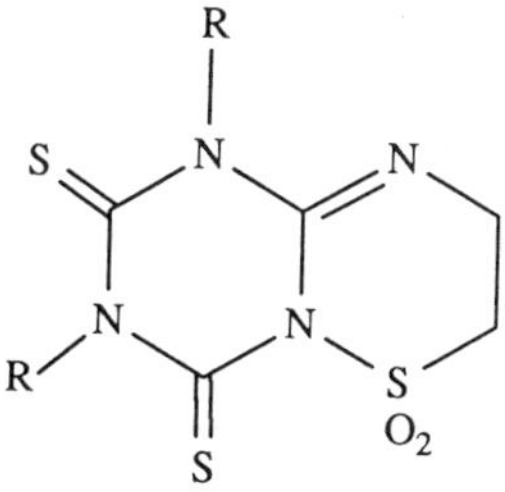   (145)

The reaction of 4,5-dihydro-*4H*-1,2,4-thiadiazine 1,1-dioxides **142**, which have a leavable group in the 3-position (Y = OR^1, SR^1, NH$_2$), with two molecules of an alkyl isocyanate (RNCO) produces the condensed system **143**, while the alkyl isothiocyanate yields the different condensed system **145**. The former reaction apparently proceeds through intermediate **144**, i.e., by reaction at the 4-position. On the other hand, the first molecule of the isothiocyanate may, by analogy, react at the 2-position via a corresponding intermediate (74BSF1917).

B. STRUCTURE

1. *Theoretical Methods*

The electronic structures of hydrochlorothiazide **146** and related substances have been determined by the complete neglect of differential overlap (CNDO/2) method (83AF688). The calculation confirms that the diuretic thiazides are electron-accepting, whereas diazoxide **147**, a non-

(146)   (147)   (148)

diuretic substance, is not. Interestingly, the presence of a sulfamoyl group at position 7 of 1,2,4-benzothiadiazine 1,1-dioxide causes a negative formal charge to arise at that position. Such a finding is compatible with the bond lengths and angles determined by an X-ray crystallographic study that suggests canonical structure **148** makes the most significant contribution to the hydrochlorothiazide hybrid (72AX(B)2340). In contrast, the presence of a chlorine atom at position 7, as in diazoxide **147**, gives rise to a positive formal charge at that position. The energy level of the lowest unoccupied molecular orbital (LUMO) appears to be lower when the sulfamoyl group is present, as in hydrochlorothiazide **146**.

Because of the lipophilic nature of the biological membrane and the importance of size and charge distribution, the formal charge at position 7 of the 1,2,4-benzothiadiazine 1,1-dioxide systems and two other parameters, Hansch's hydrophobic π parameter and the van der Waals volume (as calculated by Bondi's method), are included in a structure–activity relationship study. The van der Waals volume alone bears a close relationship to diuretic activity. All three parameters are correlated in a regression equation. A model is proposed for the active site of hydrochlorothiazide and the tubular membrane which consists of a lipophilic hole and an electrostatic reaction site at position 7 (83AF688).

The difference in hydrogen-acceptor histamine antagonist activity of substituted cimetidines such as **43** has been related to the orientation of the dipole moment of various cyanoguanidine replacement groups and the lipophilcity of the compound (87MI1,87MI2).

2. *Molecular Dimensions*

As mentioned previously, an X-ray crystallographic study of hydrochlorothiazide suggests a structure very much like **148**. Thus, the N4–C4a bond length, 134.4 pm, implies a degree of double-bond character, while the N2–C3 and C3–N4 bond lengths, 147.3 and 146.9 pm, respectively, are normal for C–N single bonds. All the ring atoms except N2 lie close to the general plane of the molecule so that the thiadiazine ring may be regarded as adopting an envelope conformation with the N2 atom at the flap. The four oxygen atoms attached to the two sulfur atoms also lie outside the general plane of the molecule [72AX(B)2340].

An X-ray crystallographic study of 1-oxide fluoroborate **129** (X = OMe) shows each of the two benzene rings is planar, while the thiadiazine ring is twisted. Structure **129** depicts a double bond between the C3 ring atom and the dimethylamino nitrogen atom. However, although structure **129** more closely resembles the actual structure than any other canonical form, there appears to be some double bond character between the C3 ring atom and the other two nitrogen atoms to which it is attached. This is indicated by the following bond lengths: C3–NMe$_2$, 132.0 pm; C3–N4, 133.9 pm; N2–C3, 136.3 pm. There is not as much double-bond character between N2 and the methoxyphenyl ring (145.1 pm), and this difference is substantiated by the larger angle of 39° between the *N*-(methoxy)phenyl ring plane and the plane that includes N2, S, C3, and the C1′ atom of the methoxyphenyl group. Compare this to the smaller angle of 18° between the plane that includes N2, C3, N4, and the nitrogen atom (N3′) of the dimethylamino group and the plane which includes C3, N3′, and the two carbon atoms of the dimethylamino group (84LA904).

3. *Molecular Spectra*

The electron ionization (EI) (70 eV) mass spectra of seven dihydrothiazides have been investigated. Only hydrochlorothiazide **146** (**149**, X = Cl,

Y = H, R = H) (M$^+$. 297, 55%) and hydroflumethazide 149 (X = CF$_3$, Y = H, X = H) (M$^+$. 331, 90%) gave prominent molecular ion peaks. The remainder, bendrofluazide 149 (X = CF$_3$, Y = H, R = CH$_2$Ph), cyclothiazide 149 (X = Cl, Y = H, R = norborn-5-en-2-yl), cyclopenthiazide 149 (X = Cl, Y = H, R = cyclopentylmethyl), polythiazide 149 (X =Cl, Y = Me, R = CH$_2$SCH$_2$CF$_3$), and methylchlorthiazide 149 (X = Cl, Y = Me, R = CH$_2$Cl) gave much smaller M$^+$. intensities (5–8%). Fragmentation of the molecular ions of hydrochlorothiazide and hydroflumethazide occurs probably via the M-1 ion followed by loss of HCN (base peak, M-28), SO or SO$_2$, and then HCN. Loss of SO$_2$ via the possible intermediate M-1 ion may involve a benzimidazole. In the case of bendrofluazide, loss of the benzyl radical gives rise to the base peak, and the tropylium ion occurs in 40% abundance; the presence of its 3-substituent probably induces formation of the ion m/z 118 (25%), [HN≡CCH$_2$Ph]$^+$. As with bendrofluazide, cyclopenthiazide and polythiazide lose their 3-substituents to give the base peaks m/z 296 and 310, respectively. However, simple loss of the 3-substituent from cyclothiazide (M-93, 7%) is less important since here the norborn-5-en-2-yl substituent appears to undergo a retro Diels–Alder reaction and loss of a hydrogen atom to give its base peak m/z 66 (C$_5$H$_6^+$. The formation of the [HN≡CR]$^+$ fragment is also apparent in the spectra of cyclopenthiazide (m/z 110, 46%) and cyclothiazide (m/z 120, 78%), but not in the spectrum of polythiazide. The spectrum of methylchlorthiazide exhibits low-intensity molecular ions (m/z 359, 4%; m/z 361, 29%) which fragment mainly by loss of the 3-substituent, CH$_2$Cl. In contradistinction to a Chemical Rubber Company list of mass spectra, chlorthiazide manifests a molecular ion peak (87M13).

In EI mass spectrometers, 1-phenyl-substituted 1,2,4-thiadiazine 1-oxides can undergo migration of the phenyl group from the sulfur atom to nitrogen, a common feature in the mass spectra of aryl alkyl sulfoximes (77JOC952).

1,2,4-Thiadiazines exhibit an intense end-absorption in their UV spectra, but otherwise, neither they nor their monoanions give rise to any distinctive UV absorption. However, their dianions show an absorption maximum near 240 nm, which is similar to that of the first ionization of 5,5-diethylbarbituric acid (70CRV593). 1,2,4-Benzothiadiazines, however, often show three distinct absorptions, as in for instance 97, whose maxima lie in the ranges 214–222, 242–251, and 290–314 nm [79JCS(P1)1043], and also 150, which absorbs in the ranges 250–260, 262–275, and 271–300 nm (86JOC1967). The UV spectra of 3,4-dihydro-7-sulfamoyl-1,2,4-benzothiadiazine dioxides show three distinct absorption maxima of decreasing intensity at 224–226, 264–273, and 302–325 nm, which correspond to the spectra of aniline-2,4-disulfonamides (60JOC970).

(149)   (150)   (151)

If carbonyl groups are present, IR spectroscopy can be used to assign the CO group to a cyclic amide, urea, or carbamate. The ring SO_2 group gives rise to absorptions in the ranges 1120–1180 and 1310–1381 cm^{-1} in the spectra of 1,2,4-thiadiazine 1,1-dioxides, while the corresponding ranges are 1150–1180 and 1230–1330 cm^{-1} in the spectra of 1,2,4-benzothiadiazine 1,1-dioxides.

^{1}H and ^{13}C NMR of 1,2,4-thiadiazines are useful in confirming structure. A study of ^{1}H and ^{13}C-NMR (and IR) spectra of Taurolidine **62**, Taurultam **63**, and Tauroflex [2% aqueous solution of Taurolidine containing 5% poly(vinyl)pyrrolidone] in aqueous solution suggests there are two successive equilibria leading to Taurultam methylol **151**. It is this methylol derivative that is supposed to be the active component against bacteria and their endotoxins (82MI1).

In cyclic sulfoximides such as 3-ethoxy-4,5-dihydro-3-oxo-1-phenyl- and 3,4,5,6-tetrahydro-3,5-dioxo-1-phenyl-1H-1λ^6,2,4-thiadiazine 1-oxides **(46)**, the C6 proton signals appear at much higher fields [4.35 and 4.9 ppm (2H)] than for aromatic systems [benzene, 7.27 ppm; thiophene, 7.2 (C2-H) and 6.95 ppm (C3-H)]. Similarly, the ^{13}C-NMR signals of the C6 atom appear at higher fields. These facts suggest a degree of ylidic character (77JOC952), which can be demonstrated by the reactivity of such compounds towards electrophiles (see Section III, C,3,f).

4. Tautomerism

Studies of the tautomeric state of 1,2,4-thiadiazines have been more or less confined to the biologically active 1,2,4-benzothiadiazines. Ultraviolet spectroscopic studies suggest that the 4H-tautomer of 1,2,4-benzothiadiazines, e.g. **152**, is preferred in ethanol (60JOC970); in alkali, the anion dominates. Extended Hückel molecular orbital calculations (70MI2) and ^{13}C-NMR studies (79T2151) confirm this view (Scheme 6).

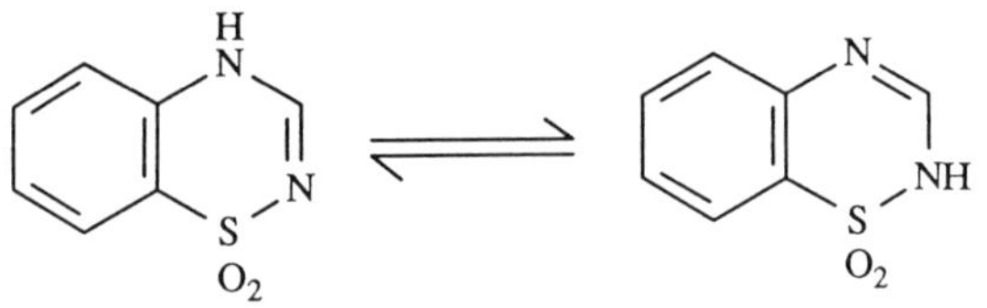

(152)   (153)

SCHEME 6

In general, when hydroxyl groups might seem possible, as in barbituric acid analog **33**, the carbonyl tautomers are preferred (84MI1), possibly because of the more favorable antiparallel orientation of the CO and NH dipoles.

C. REACTIVITY

1. *Reactivity of the Ring*

a. *General Survey.* The 1,2,4-thiadiazine ring is not aromatic. The presence of sulfonamide and sulfinamide groups within the ring impart properties of those functions to the compound. Thus, most 1,2,4-thiadiazines are acidic and relatively stable compounds.

i. *Acidity.* The measurement of the pK_a values of barbituric acid analog **33** and its 4- and 6-substituted derivatives, together with an NMR study, indicates the hydrogen at the 2-position is the most acidic (70CRV593). Compound **33** does resemble barbituric acid, as it is strongly acidic pK_a 2.7), sparingly soluble, and has a high melting point (79MI1).

The pK_a values of five thiazide-type compounds have been measured more accurately, than was previously possible, by a solvent-extraction method using a porous Teflon separator. In all cases but polythiazide **149** (X = Cl, Y = Me, R = $CH_2SCH_2CF_3$), which is monoprotic, and flumethazide **154** (R = CF_3) (see Scheme 7), the most acidic proton (pK_{a1} range 8.52–9.82) resides on the nitrogen atom of the 7-sulfamoyl group. The additional acidic proton pK_{a2} range (10.00–10.88) is attached to the N2 nitrogen atom (86MI3). The pK_a values (7.70–10.61) of another series of seven 2H-1,2,4-benzothiadiazine 1,1-dioxides, which are either unsubstituted in the 7-position or carry a methyl group at that position, have also been determined (85MI1). 3-Imino-1,2,4-benzothiadiazines are weakly acidic (pK_a 9–11) (79MI1).

ii. *Alkylation and acylation.* Methylation of 3-mercapto-4*H*-1,2,4-benzothiadiazine 1,1-dioxide with methyl iodide and sodium bicarbonate gives the 3-methylthio derivative, further methylation of which with diazomethane yields the 4-methyl-3-methylthio derivative. Otherwise, direct treatment of the starting material with diazomethane gives the 2-methyl-3-methylthio derivative together with a smaller amount of the 4-methyl-3-methylthio derivative, a finding that leads to speculation as to the particular tautomeric form or forms of the original mercapto compound.

Alkylation of chlorothiazide **154** (R = Cl) with dimethyl sulfate or with allyl bromide in aqueous or alcoholic alkali yields the 4-methyl and 4-allyl derivatives, respectively (60JOC970). Chlorothiazide-type compounds and their 3-oxo compounds, but not their 3,4-dihydro compounds, can be

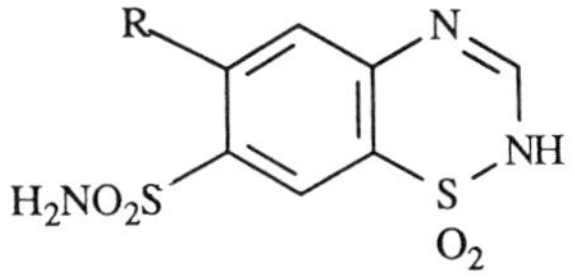

(154)

alkylated to give both the 2- and 4-alkyl derivatives upon heating with trimethyl- or triethylorthoformates. Methylation of the 3-oxo compounds in alkali with either methyl iodide or dimethyl sulfate leads to the isomeric 3-methoxy and 2-methyl-3-oxo derivatives, whereas diazomethane yields the 3-methoxy-4-methyl and 2,4-dimethyl derivatives. The 3-amino or 3-iminodihydro (which appear, on spectral evidence, to be in the amino form) derivatives are methylated with methyl iodide and potassium carbonate to give the 4-methyl derivative, and further methylation takes place at the exocyclic nitrogen atom. The reaction of diazomethane on the 3-amino compound yields both the 2- and 4-methyl derivatives (79MI1). Generally, 2-alkylated hydrothiazide compounds show increased diuretic activity, whereas the 4-alkylated counterparts do not.

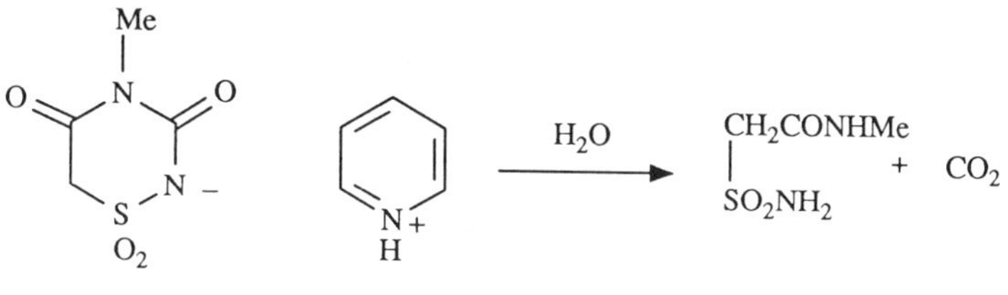

(155)

SCHEME 7

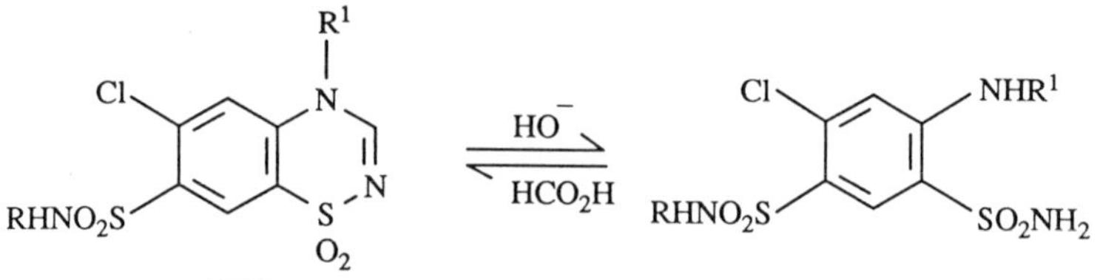

(156)

SCHEME 8

Alkylation of cyclosulfoximide **58** with alkyl iodides in DMF/potassium carbonate gives the N- and O-alkyl derivatives in decreasing ratios as the alkyl group is changed from methyl to ethyl and then to isopropyl (Me, 9.6 : 0.4; Et, 3.8 : 6.2; i-Pr, 0.0 : 10.0). A steric factor appears to be responsible for these dramatic results. The O-allyl derivative, which comprises 36% of the product from the reaction of allyl bromide on **58**, may be converted to the N-isomer by heating in a type of Claisen Rearrangement with palladium acetylacetonate at 220°C (86TH1). Acylation of chlorothiazide **154** (R = Cl) with acetic or butyric anhydride yields the 7-acylsulfamoyl derivative (60JOC970).

iii. *Hydrolysis.* The 1,2,4-thiadiazines are susceptible to hydrolysis. Thus, 3,4,5,6-tetrahydro-3-oxo-2,4-diphenyl-2H-1,2,4-thiadiazine 1,1-dioxide (**75**, R = R^1 = Ph) is hydrolyzed by barium hydroxide to give aniline and N-phenyltaurine (1883M131; 04M682; 14M137; 14M151), and 5,6-dihydro-3-phenyl-4H-1,2,4-thiadiazine 1,1-dioxide (**53**) is also hydrolytically cleaved by 20% HCl to taurine and benzoic acid (47JA1393). Attempts to free 3,4,5,6-tetrahydro-4-methyl-3,5-dioxo-2H-1,2,4-thiadiazine 1,1-dioxide in aqueous media from its pyridinium salt (**155**) result in its hydrolysis (Scheme 7) (70CRV593).

The 4H-chlorothiazide-type compounds (**156**, (R = H, R^{2}CO; R^1 = methyl, allyl, H) undergo controlled hydrolysis in alkali with thiadiazine ring fission (Scheme 8). The products recyclize to **156** on treatment with formic acid (60JOC970). The 2H-benzothiadiazines (**157**, R = H, Me, nC$_3$H$_7$, nC$_5$H$_{11}$, CH$_2$Cl, Ph, 2-ClC$_6$H$_4$, 4-Cl C$_6$H$_4$; R^1 = H, Cl, I; R^2 = H, F, Cl, Br, CF$_3$, Me, OMe, NO$_2$, NH$_2$) are also cleaved with hot alkali (Scheme 9) to give the corresponding disulfamoylaniline. Acid hydrolysis

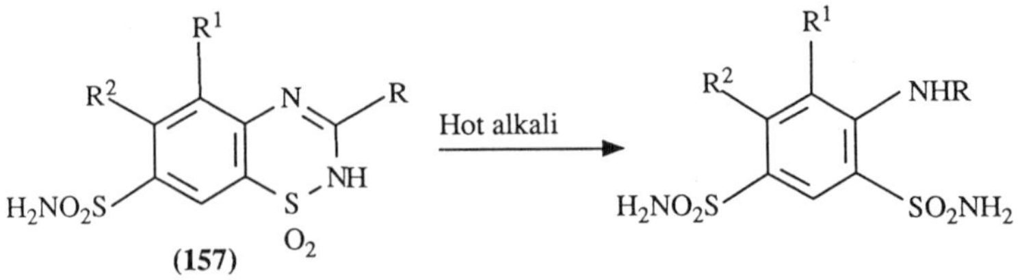

(157)

SCHEME 9

of these compounds is a slower process (60JOC970). Generally, the corresponding dihydro compounds are more difficult to hydrolyze (79MI1), while the presence of an alkyl substituent in positions 2 or 4 greatly increases the vulnerability to hydrolysis (60JOC970). Hydrolysis of 1-alkyl (or aryl) -3,4-dihydro-3-oxo-1H-1,2,4-benzothiadiazines leads to the corresponding acyclic sulfoxides (64LA189). The 1-chloro 1-oxides (**158**) are hydrolyzed to the corresponding 3-aryl-2H-1,2,4-benzothiadiazine 1,1-dioxides.

(**158**)   (**159**)

iv. *Other ring-opening reactions.* Chlorothiazide (**154**, R = Cl) undergoes ring cleavage when reacted with anhydrous dimethylamine to give **159**, which, when treated with mild alkali, is converted back to chlorothiazide. Primary amines give rise to a mixture of products in this reaction, but the ring remains intact on treatment with anhydrous ammonia. When the 2-position of chlorothiazide is substituted by a methyl or *p*-chlorophenyl group, the thiadiazine ring becomes demonstrably susceptible to cleavage when treated with chlorosulfonic acid and yields the corresponding 2-substituted sulfamoyl-5-chloroaniline-4-sulfonyl chloride (60JOC970).

v. *Reduction and oxidation reactions.* Catalytic hydrogenation of chlorothiazide **154** with ruthenium on charcoal gives the 3,4-dihydro compound, hydrochlorothiazide **146** (60JOC970). This reduction is reported to be possible with sodium borohydride (58E463), although the use of this reducing agent in methanol can apparently lead to cleavage of the 2,3 bond of **146**. Hydrochlorothiazide is oxidized back to chlorothiazide on treatment with alkaline permanganate. Reductive cleavage of the 2,3 bond is also possible with trimethylamine/borane in acetic acid. Hydrochlorothiazide **146** is stable to catalytic hydrogenation or to reduction with lithium aluminum hydride (79MI1). The 1-alkyl(or 1-aryl) -3,4-dihydro-3-oxo-1H-1,2,4-benzothiadiazines (see Section III,A,2,a) are oxidized to the corresponding 1-oxides with permanganate in acetonitrile (64LA189). 1-Arenesulfonylimido-2H-1,2,4-benzothiadiazines can be oxidized to the corresponding sulfoximes, i.e., 1-arenesulfonylimido-2H-1,2,4-benzothiadiazine 1-oxides, with alkaline permanganate (68LA223). Oxida-

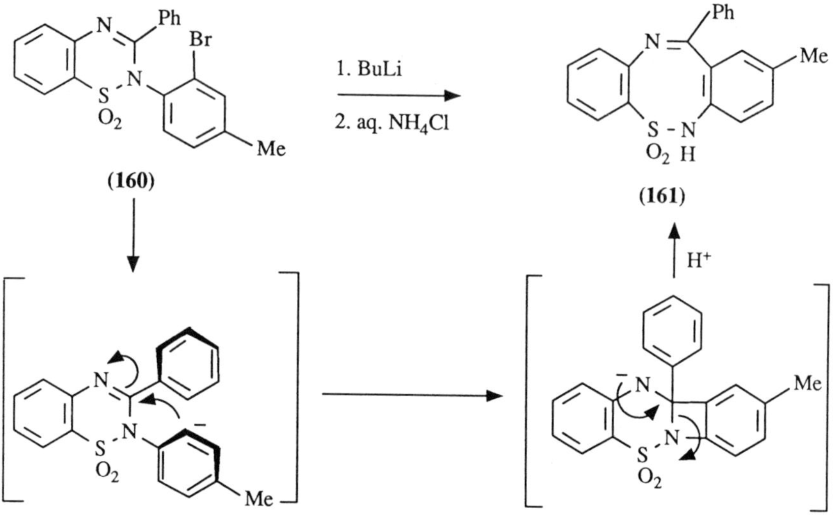

SCHEME 10

tion of 2H-1,2,4-benzothiadiazine 1-oxides with hydrogen peroxide leads to the corresponding 1,1-dioxide (68LA223).

b. *Ring Expansion.*   Ring expansion is not an uncommon property of thiadiazines. When the concept of [1,3]-migrations of sulfonyl and carbonyl functions to neighboring aryl anions was applied within the field of 1,2,4-benzothiadiazine chemistry, it was found that appropriately designed precursor heterocycles such as **160** can give rise to the 1,2,6-dibenzothiadiazocine dioxide **161** (Scheme 10). In the possible rearrangement intermediate **162,** the carbanionic function lies conveniently behind the C=N dipole. The relatively flexible thiadiazine ring and the propeller conformations of the two aryl substituents allow off-nodal plane overlap of the nonbonding aryl-anion orbital and the π^* orbital of the imine function. This feasible speculation might then be consummated to give the product

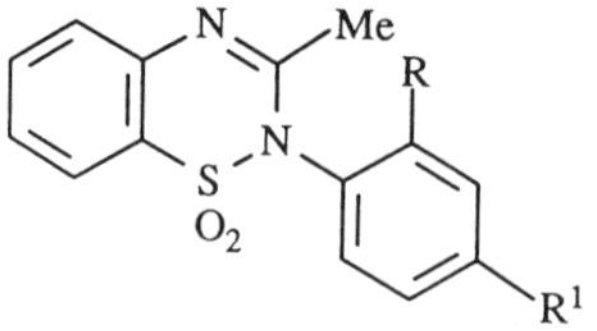

R = H, Me

R^1 = H, Me, OMe, C$_6$H$_5$CO$_2$Me-4

(164)   (165)

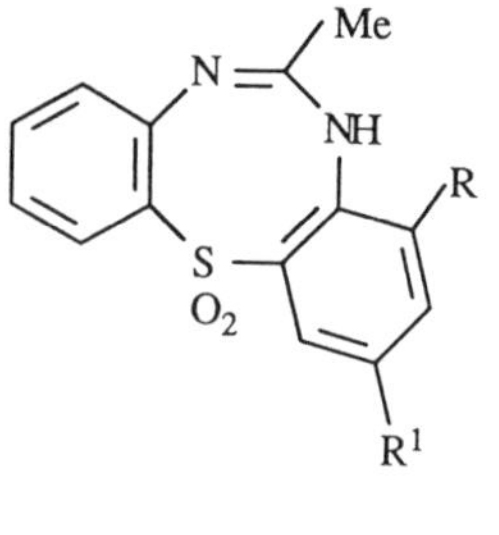

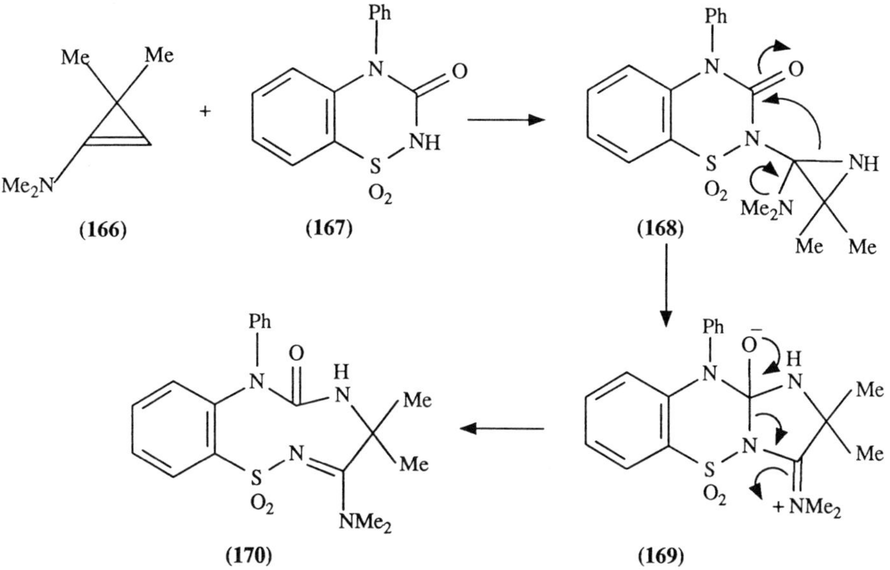

SCHEME 11

via intermediate **163** (83T2073). In contrast, irradiation of **164** gives 1,4,6-dibenzothiadiazocine **165** (86JOC1967).

While methods were being developed for using azirines in ring-expansion reactions, the reaction of aminoazirines **166** with 2*H*-1,2,4-

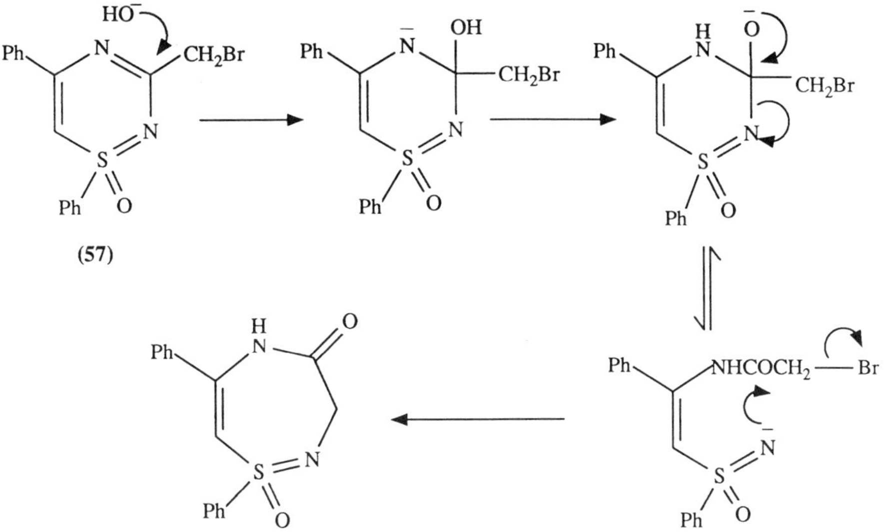

SCHEME 12

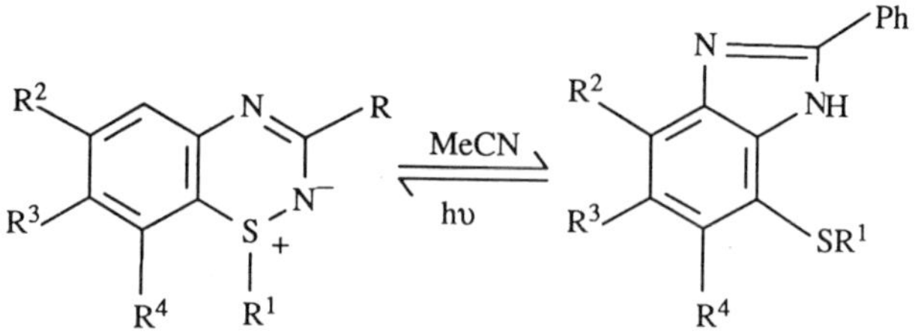

(171)   **(172)**

Scheme 13

benzothiadiazine 1,1-dioxide **167** in chloroform was found to yield 1,2,5,7-benzotriazonine-6-one 1,1-dioxide **170**. A possible mechanism (Scheme 11) involving aziridine intermediate **168** and zwitterion **169** was presented. The structure of **170** was confirmed by X-ray crystallographic analysis (84H1667).

In an extension of the ring-expansion reaction of quinazoline-3-oxides to give benzodiazepines, 3-chloromethylbenzothiadiazines have been shown to undergo a ring expansion, on treatment with sodium hydroxide, to give benzothiadiazepinones (72CB757). Monocyclic analog **57** (R = CH$_2$Br) similarly undergoes ring expansion to the thiadiazepine, but with lower yields. A feasible mechanism suggested for the reaction of the benzo analog (72CB757) may also apply to the monocyclic reaction (Scheme 12) (86TH1).

c. *Ring Contraction.* Photochemical rearrangement of 1-aryl- and 1-alkyl-1λ^4,2,4-benzothiadiazines **171** (R^1 = Ph) gives benzimidazoles **172** (Scheme 13). This reaction may proceed by an intermediate nitrene, particularly since irradiation of the related benzothiadiazine (**173**) in the presence of DMSO yields the intercepted product **175** very probably via nitrene **174** (Scheme 14) [83JCS(P1)55]. Reduction of 1-oxides **176** with zinc and acetic acid gives high yields of benzothiazoles **177** (87PS41).

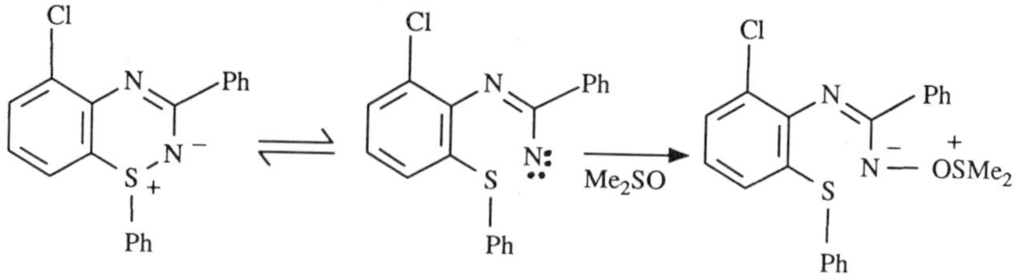

(173)   **(174)**   **(175)**

Scheme 14

(176)   (177)

d. *Rearrangement.*   Some   1-aryl-1λ^4,2,4-benzothiadiazines   (**171,** R^1 = aryl, $R^2 = R^3 = R^4 = H$) undergo an isomeric, thermal [1,4]-sigmatropic shift of the aryl group in *o*-dichlorobenzene at 180°C to give the corresponding 4-aryl-1-phenyl-4*H*-1,2,4-benzothiadiazines. This shift appears to be mechanistically similar to that observed in the thiabenzene series [83JCS(P1)55].

e. *Radical Reactions.*   Formation of radical **179** has been claimed upon treatment of **178** with sodium in benzene or tetrahydrofuran (THF). The same radical is apparently formed from **180** by oxidation with lead dioxide in the presence of potassium carbonate (81ZOR2619).

(178)   (179)   (180)

f. *Miscellaneous Reactions.*   The analog of barbituric acid **33** (Section III,A,1,a) includes a reactive methylene group in the 6-position, which is said to be nitrosated or coupled with aryldiazonium salts (62DIS65; 79MI1). Chlorination of 2*H*-1,2,4-benzothiadiazine 1-oxides (**121,** R = Ar, R^2 = H) yields the 1-chloro derivatives **158,** which may be reacted with amines such as morpholine to give 1-oxide derivatives, i.e., 3-aryl-1-(*N*-morpholino)-1*H*-1λ^6,2,4-benzothiadiazine 1-oxides. The alkyl groups of 3-alkoxy-1,2,4-benzothiadiazine 1,1-dioxides migrate on heating to give 2-alkyl-3,4-dihydro-3-oxo-1,2,4-benzothiadiazine 1,1-dioxides. As in the Claisen rearrangement, an α,β-unsaturated group is inverted in the product (cf. Section III,C,1,a,ii for a corresponding migration in cyclosulfoximides) (79MI1).

The ylidic character of some cyclosulfoximides (see Section III,B,3 for NMR evidence of ylidic character) can be demonstrated by the ease of electrophilic attack at the 6-position. Thus, 3-ethoxy-4,5-dihydro-5-oxo-1-phenyl-1H-1,2,4-thiadiazine 1-oxide is easily brominated in 87% yield at the 6-position at room temperature. The same compound can be nitrated (43% yield) with acetyl nitrate at the same position (77JOC952). Judging from the smaller ^{1}H- and ^{13}C-NMR shifts to higher field of the C6 proton and [13]C6 signals, cyclosulfoximides **57** and **58** should be less susceptible to electrophilic attack at the 6-position. Indeed, nitration of the thiadazine ring of **57** does not occur, but bromination at C6 of **57** is nevertheless quite facile. One explanation for this latter observation is that bromination may occur by an addition–elimination mechanism rather than by direct electrophilic attack (86TH1).

2. Reactivity of Substituents

a. *Nucleophilic Replacement Reactions.*   Good leaving groups in position 3 of 1,2,4-benzothiadiazine 1,1-dioxides are still used for modifying moieties in that position by reacting them with appropriate nucleophiles. For instance, 3-chloro derivatives continue to be reacted with ammonia (86S864), amines (87MI2), ethanethiol [85JAP(K)6097970], and pyrazoles [85IJC(B)1295]. Frequently, 3-chloro-4H-1,2,4-benzothiadiazine 1,1-dioxides are made from 2H-1,2,4-benzothiadiazine-3(4H)-one 1,1-dioxides and phosphorus pentachloride. Methylthio groups are also used as leaving groups in the 3-position for incorporating side chains [85JAP(K)60112781].

b. *Formation of Tricyclic Derivatives.*   3-Hydrazino-substituted 1,2,4-benzothiadiazine 1,1-dioxides can be diazotized and cyclized to give tetrazolo derivatives, thus **181** yields **182** [85JAP(K)6078990].

(181)   (182)

D. APPLICATIONS

The review by Schittler *et al.* [62AG(E)235] is still quite useful as an introduction to the 7-sulfamoyl-1,2,4-benzothiadiazines 1,1-dioxides (the thiazides and hydrothiazides) and their diuretic activity. This early work is surveyed before and after 1957, when it was realized that the formyl derivative of 6-amino-4-chlorobenzene-1,3-disulfonamide cyclizes rapidly to chlorothiazide **154**, a very effective diuretic of low toxicity. Such high potency is found only in compounds that include a 7-sulfamoyl group, although some 1,2,4-benzothiadiazines without this group show antihypertensive activity. Thus, 7-chloro-3-methyl-1,2,4-benzothiadiazine 1,1-dioxide (diazoxide, **147**) is used in the treatment of hypertension. Landquist's review (79MI1) is helpful in providing an interim picture of the developments in biological applications along with a description of the chemistry of 1,2,4-benzothiadiazines. Some 1,2,4-benzothiadiazine 1-oxides have wide-ranging pharmacological properties and have been included in previous patents (e.g., 75GEP2451566; 74GEP2321786). During the period covered by this review, a greater variety of applications has appeared (see Table I). The main area of interest is in the search for antihypertensives. Although compounds with potential diuretic activity still figure significantly in publications, there are signs that interest in this field is beginning to tail off. However, with the advent of computerized molecular graphics, the study of structural-activity relationships may well receive a boost. Since so many diuretics have been prepared and biologically tested, they might represent a rich field for receptor–donor studies. The availability of data from X-ray crystallographic analysis, semiempirical programs, lipophilicity, and dipole-moment measurements, etc. has already led to the formulation of some structural-activity relationships (see Sections III,B,1 and III,B,2). Variation of 3-substituents represents the main focus of workers in the search for improved biological activity in the 1980s. Although substitution at the 2-position can lead to enhanced potency in hydrochlorothiazides, comparatively little attention has been given to it. The inclusion of peptide and dipeptide moieties in 3-substituent side-chains is a recent development, as is the preparation of quinazolino-1,2,4-benzothiadiazines and systems in which the thiadiazine moiety has been fused with other heterocycles. Some idea of the extent to which application claims range is given in Table I.

The main successful applications of monocyclic 1,2,4-thiadiazines in earlier years have been the use of bis[4-(3,4,5,6-tetrahydro-1,1-dioxo-2H-1,2,4-thiadiazinyl)]methane (**62**, Taurolidine, Taurolin, Drainsept) as an antibacterial agent and 3,4,5,6-tetrahydro-2H-1,2,4-thiadiazine 1,1-dioxide (**63**, Taurultam) as both an antibacterial and antifungal agent (66FRP1458701). Claimed uses of 1,2,4-thiadiazines are listed in Table II.

TABLE I

Claimed Applications of 1,2,4-Benzothiadiazines

Activity	References
Agricultural bactericidal	85JAP(K)6078990[a]
Agricultural fungicidal	85JAP(K)6097970[b]
Analgesic and antiinflammatory	86MI2[c]
Antidepressant	84GEP3321969
Antiglaucoma	86USP4616012[d,e,f] 87USP4634698[e]
Antiglaucoma and anti-hypertensive	85USP4556655[f]
Antihypertensive	83EUP88350[f], 83EUP95584[f], 84EUP121830, 84USP4468396[f], 85EUP153755[f], 85USP4559340[f], 86GEP3427309[f], 86SAP00083[g], 86USP4584285, 87USP4634689[d], 87USP4666906[e], 88GEP3632544[g]
Antihypertensive and diuretic	86BSF871[f], 87JAP(K)62195371[f]
Antiinflammatory	85AP903[h], 87MI4[i], 87MI5[h]
Antiulcer and central nervous system agent	85JAP(K)6092278[j]
Anxiolytic	86GEP3433037[k]
Bactericidal	83FES466[h], 84MI2[g], 85JAP(K)6025984[g], 83MI4,
Bone resorption inhibitor	88EUP268990[g]
Cardiovascular effects	83FES738[h], 88PHA37,
Color photography	83JAP(K)58111943, 83JAP(K)58117542[g]
Diuretic	82USP4338435[f], 83MI1[g]
Histamine hydrogen-receptor antagonist	84EUP105732[f]
Inhibition of angiotensin-converting enzyme	85USP4536501[f]
Inhibition of peptic ulcer	85JAP(K)6072868[b]
Plant bacterical, fungicidal, and herbicidal	82CS248[g]
Psychotropic	85JAP(K)60156680[g]
Reduction of blood pressure	85EUP161498
Taurine antagonist	82JMC113, 85JAP(K)6072868

[a] Tetrazolo tricyclic system.

[b] 2,3-Disubstituted 2H-1,2,4-benzothiadiazines.

[c] Quinazolino.

[d] 2,3-Disubstituted 3,4-dihydro-1,2,4-benzothiadiazines.

[e] 3-Substituted 3,4-dihydro-1,2,4-benzothiadiazines with peptide(s) in 3-side-chain.

[f] 3-Substituted 3,4-dihydro-1,2,4-benzothiadiazines.

[g] 3-Substituted 2H-1,2,4-benzothiadiazines.

[h] 3-Substituted 4H-1,2,4-benzothiadiazines.

[i] Imidazo tricyclic system.

[j] Unsubstituted thiadiazine ring.

[k] Triazolo tricyclic system.

TABLE II
CLAIMED APPLICATIONS OF 1,2,4-THIADIAZINES

Activity	Reference
Antihistamine	85JAP(K)60112781
Anxiolytic	85EUP161143
Fungicide, acaricide, nematocide, and insecticide	83GEP3208187
Hydrogen receptor histamine antagonist	86JMC44
Inhibition of dihydroorotase	84JMC228

IV. 1,2,5-Thiadiazines

These compounds are relatively rare.

A. SYNTHESES

1. Monocyclic 1,2,5-Thiadiazines

Reaction of carbon disulfide with N,N'-dialkylethylenediamines, $RNHCH_2CH_2NHR$, gives γ-alkylaminodithiocarbamic acids **183**, the zwitterions of which can be oxidized with iodine/potassium iodide in alkali to yield 1,2,5-thiadiazines **184** (R = Et, i-Pr, cyclohexyl) (49JOC946; 50USP2514200; 51USP2537633).

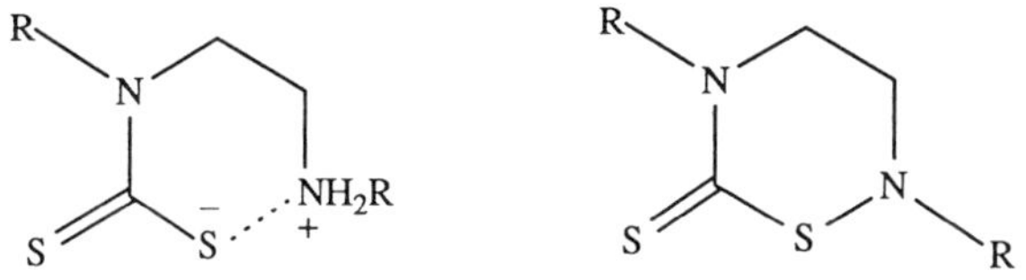

(183)   (184)

Ring expansion of 1,2,5-thiadiazole hydrochloride **185**, using no more than a molar amount of cyanide ion, gives 1,2,5-thiadiazine **187**, possibly via ring-opened intermediate **186**. Although product **187** is stable in the

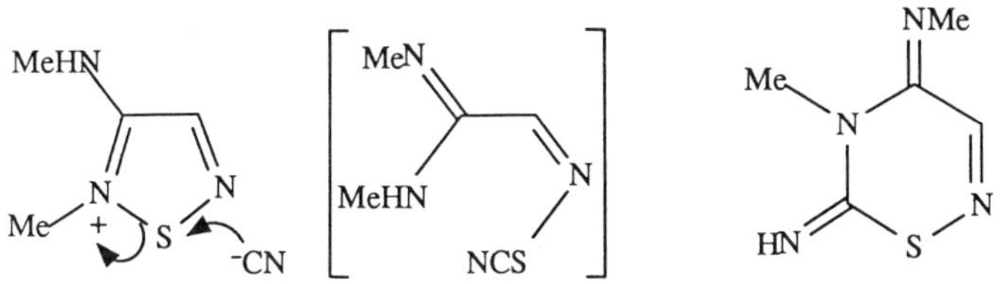

(185)   (186)   (187)

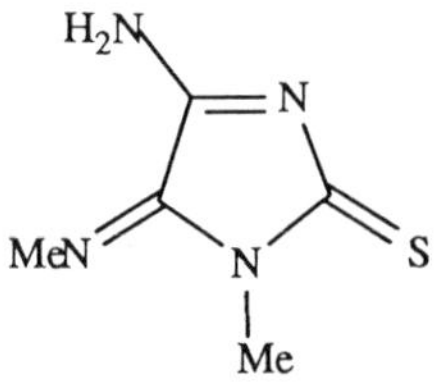

(188)

cold, the reaction is reversed, on standing at room temperature for several days, to give a 10% yield of thiadiazole **185**. The use of an excess of cyanide ion gives an 80% yield of imidazole **188** via an exothermic reaction (79TL1281).

1,2,6-Thiadiazine-6-thione **190** is the main product of the benzoylation of 5-hydroxyaminothiazolidine-2-thione **189**. The other products are benzamide **191** and benzoic anhydride (79TL1281).

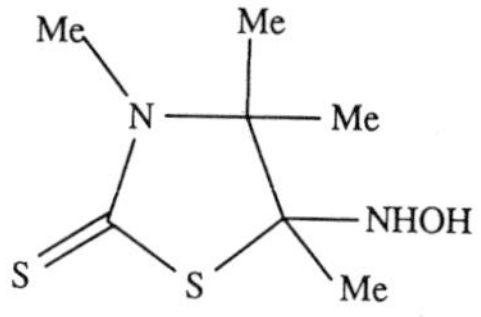

(189) 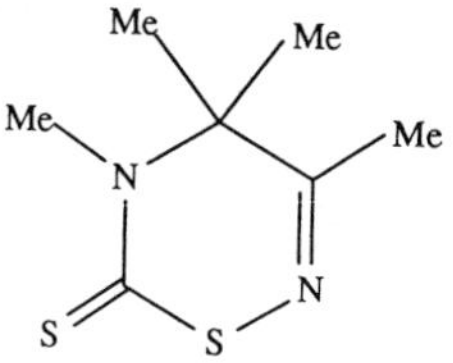   (190) 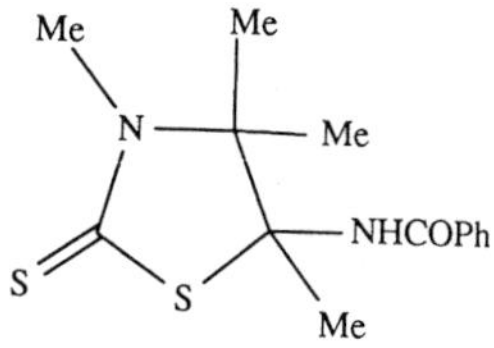   (191)

2. 2,1,4-Benzothiadiazines

3-Carbomethoxyamino-2,1,4-benzothiadiazines, such as **192**, can be made by the action of methyl isothiocyanatomethanoate on 2-nitroaniline, followed by reductive ring-closure with sodium dithionite. Oxidation with 3-chloroperbenzoic acid gives the corresponding 2-oxide **193** (71-

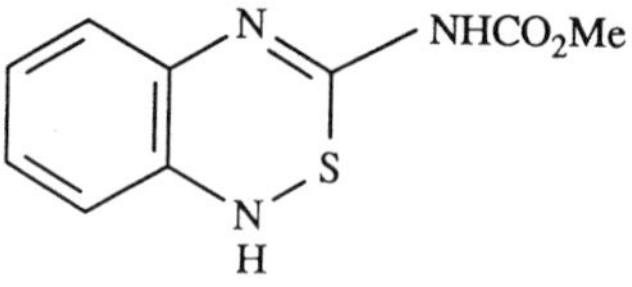

(192) 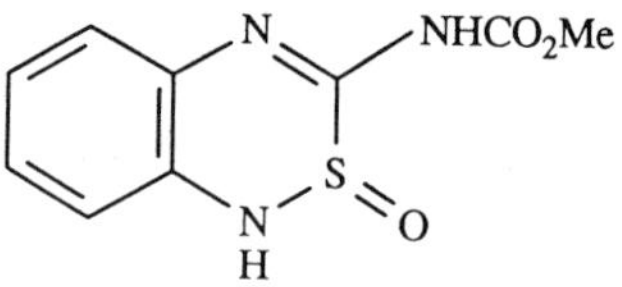   (193)

GEP2108461). Similar methods have been used to make 7-benzoyl and 7-phenylsulfonyl derivatives of **193**, which are said to be anthelmintics (72GEP2215733), and also 6-phenylthio (76GEP2438099) and 7-phenylsulfonato derivatives of **192** (77GEP2541742).

B. Structure

The structure of **190** has been confirmed by X-ray crystallography. The ring system is in a boat conformation with the S1 and C4 atoms at the flaps and deviating from the plane of the other four ring-atoms by 22.9 and 49 pm, respectively (87KGS1280).

C. Reactivity

Not much is known about the properties of 1,2,5-thiadiazines. However, there is one outstanding property: the contraction of the 1,2,5-thiadiazine ring, which is shown both by monocyclic 1,2,5-thiadiazines and 2,1,4-benzothiadiazines. For instance, 1,2,5-thiadiazine **187,** which is stable at low temperatures, tends to revert to 1,2,5-thiadiazolium cyanide **185** on standing at room temperature for several days (79TL1281). In this contraction, the ring sulfur atom is retained. Usually the sulfur atom is extruded to give imidazole derivatives, e.g., 1,2,5-thiadiazine **184** (R = *i*-Pr), when heated, yields thiazolidinethione **194** (49JOC946). In the 2,1,4-

Me₂HC–N⏜N–CHMe₂

(194)

(195)

(196)

benzothiadiazine series, extrusion of S or SO is used as a means of making benzimidazoles, some of which are said to be anthelmintics. Thus, **195** (R = H, R¹ = PhOSO₂), when refluxed with triphenylphosphine, gives benzimidazole **196** (R = H, R¹ = PhOSO₂) (77GEP2541751); **195** (R = H, R¹ = PhCO) similarly yields **196** (R = H, R¹ = PhCO) (74BRP1350277).

V. 1,2,6-Thiadiazines

1,2,6-Thiadiazines 1,1-dioxides, which include the sulfamide moiety, are not included here since they have already been reviewed (88AHC81).

This article, therefore, covers only 1,2,6-thiadiazines that do not include the 1,1-dioxide group.

A. SYNTHESES

1. *Monocyclic 1,2,6-Thiadiazines*

Established methods for making 1,2,6-thiadiazines that have no 1,1-dioxide group include reactions of (i) sulfur diimides with malonyl chlorides and 1,3-diaminopropanes, (ii) sulfur dichloride with 1,3-diaminopropanes, and (iii) thionyl chloride with N,N-1,5-dinucleophiles such as 1,3-diaminopropanes or β-amino α,β-unsaturated aldimines and ketimines to give S-oxides (79MI1; 84MI1).

By Formation of Two Bonds.
i. *From [5 + 1] atom fragments.* Conjugated imides (azabutadienes) **197** can react with thionyl chloride in pyridine or benzene-triethylamine at room temperature to give good yields of 1,2,6-thiadiazine 1-oxides **198** (R^1 = aryl, R^2 = H or Me, R^3 = aryl) (79CC891), (R^1 = aryl, R^2 = Br or Cl, R^3 = aryl) [83JCS(P1)2273].

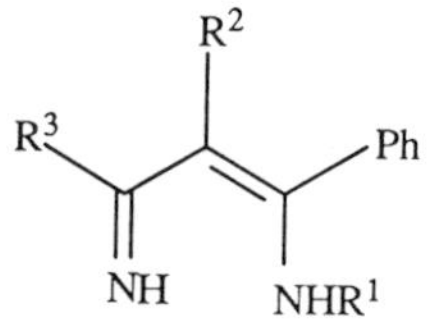

            (197) 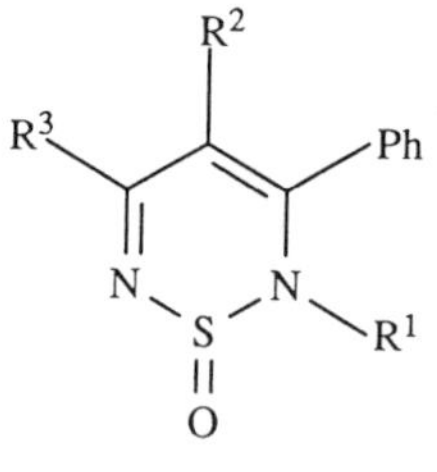   (198)

Thiadiazine **199** is formed from the reaction between either sulfur dichloride and dichloromalononitrile or $N,2,2$-trichlorocyanoacetimidoyl chloride and sulfur (Scheme 15). Product **199** is very unstable, but treatment with formic acid transforms it into **200** (Scheme 15). The chlorine atoms of **200** can be successively replaced by reaction with ammonia, primary or secondary amines, alkoxide, and other nucleophiles (73TL4489).

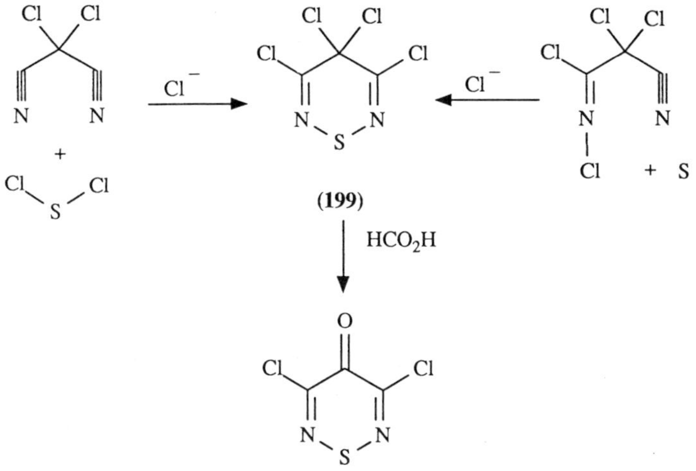

(199)

HCO$_2$H

(200)

Scheme 15

Sulfenyl chlorides, such as piperidine-1-sulfenyl chloride, are more convenient to use than sulfur dichloride for the inclusion of sulfur in heterocyclic rings. They are easily purified, do not disproportionate so readily, and are not accompanied by unwanted chlorinated byproducts. Thus, 1,3-diaminopropane reacts with three equivalents of piperidine-1-sulfenyl

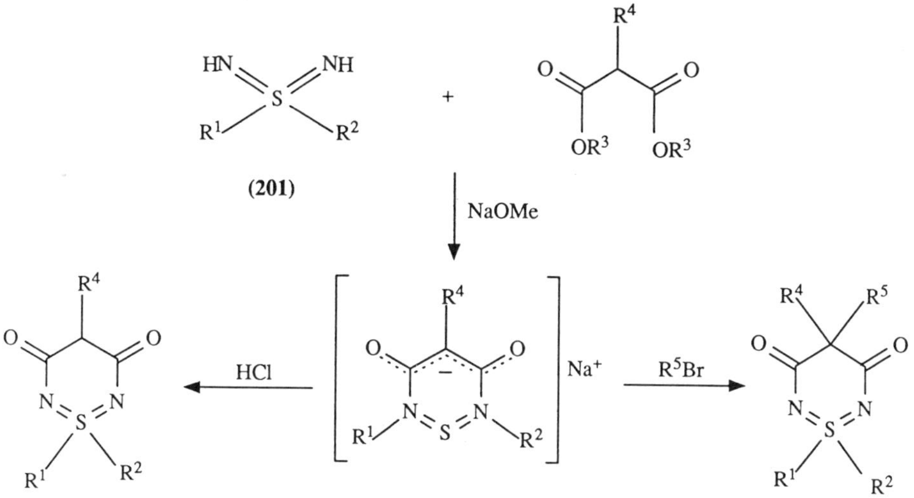

(201)

NaOMe

HCl

Na$^+$ R^{5}Br

(203)   **(202)**   **(204)**

Scheme 16

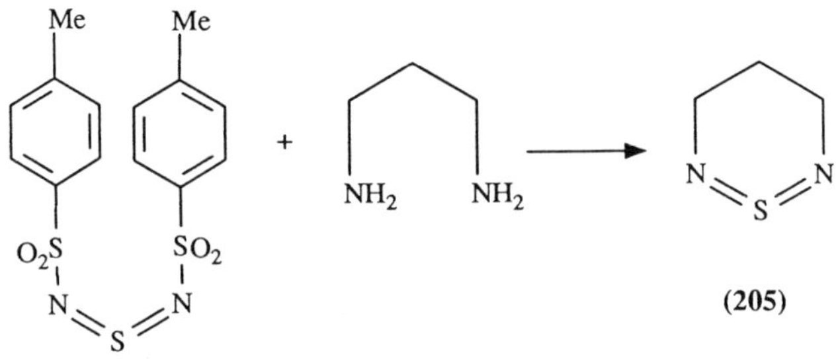

SCHEME 17

chloride to give 2,6-bis(piperidino-sulfenyl)-2,3,5,6-tetrahydro-4H-1,2,6-thiadiazine, which is the first example of a derivative of the fully saturated 1,2,6-thiadiazine ring system and, interestingly, includes an N(SN)$_3$ chain [84JCS(P1)2591].

    ii. *From [3 + 3] atom fragments.* The availability of sulfamides and sulfur diimides has made this the most common means of effecting 1,2,6-thiadiazine synthesis (79MI1; 88AHC81). Only synthetic methods involving sulfur diimides, however, are described here.

    Dialkylsulfur diimides **201** may be condensed with either unsubstituted or monosubstituted malonic esters (Scheme 16) to yield 4,5-dihydro-3,5-dioxo-3H-1λ^6,2,6-thiadiazines **203**. Although dialkylmalonates do not react with **201**, the 4,4-dialkylated compounds **204** can be made by alkylation of salts **202** [70AG(E)373]. The 4,5-dihydro derivative, **205,** of the parent compound is formed as a stable liquid by the treatment of sulfur bis(*p*-toluenesulfonylimide) with 1,3-diaminopropane (Scheme 17) (69TL4117). Substituted malonyl chlorides R$_2$CHCH(COCl)$_2$ react with di-*t*-butylsulfur diimide to give **206,** and S,S-dialkylsulfur diimides react with disubstituted malonyl chlorides to yield **207** (79MI1).

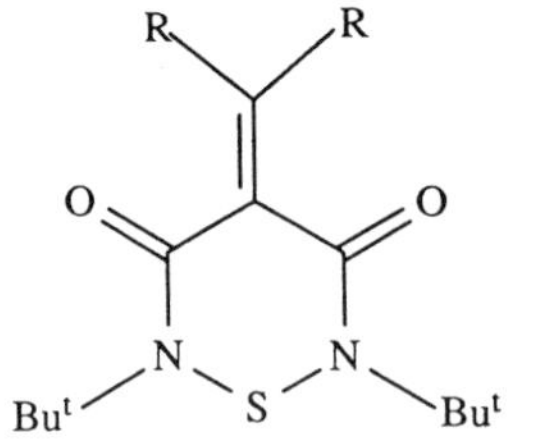

(206) 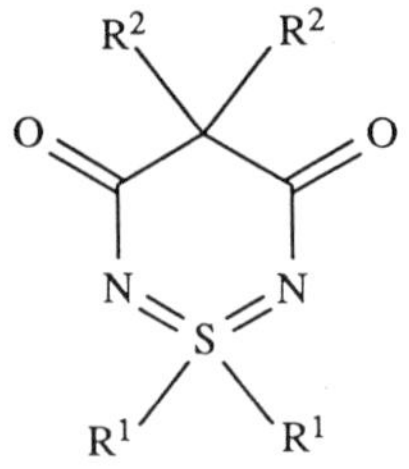   (207)

Another series of 1λ^6,2,6-thiadiazines can be prepared by reaction of dialkylsulfur diimides with substituted α,β-unsaturated nitriles and esters (Scheme 18) (Z = CN or COOR, X = OR, SR, CN). The main route to **209**

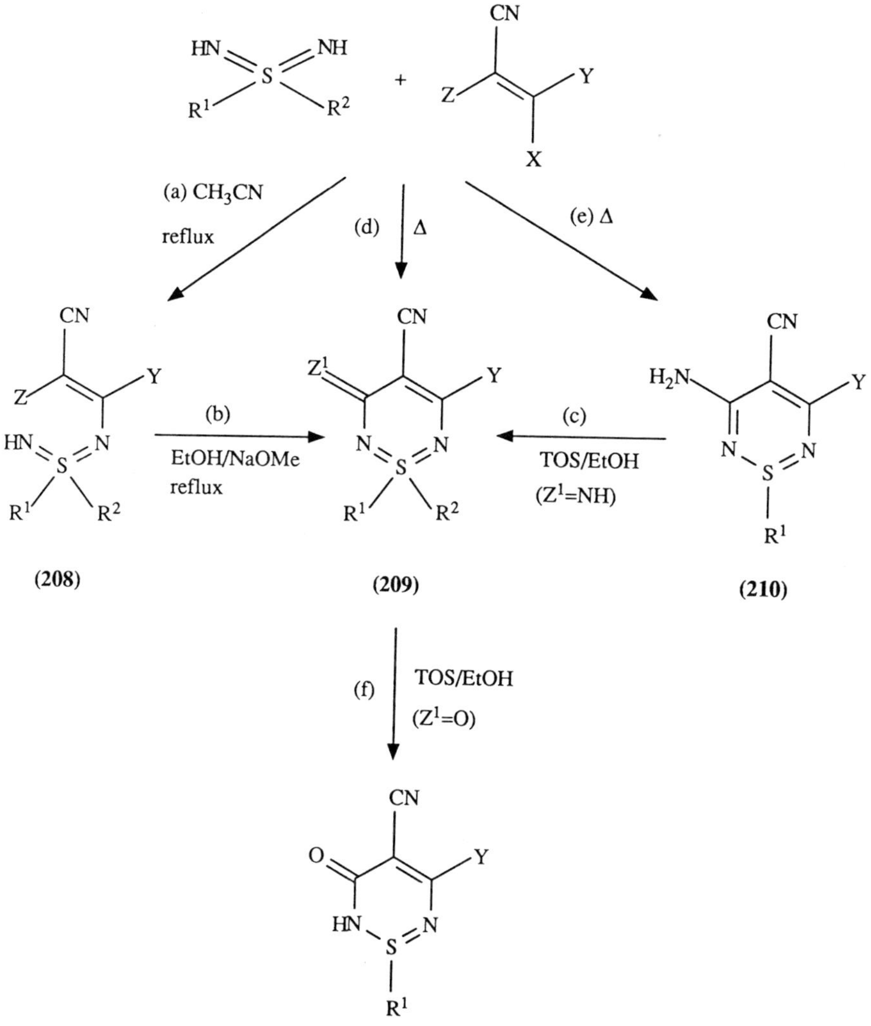

(208)

(209)

(210)

(211)

SCHEME 18

is through (a) and (b) via intermediate **208.** Route (d) is a more direct, though less efficient route to **209.** When R^1 or R^2 is benzyl, routes (c) and (f) may be taken beyond **209** to give thiadiazines **210** and **211.** Also, when R^1 and/or R^2 is benzyl, direct route (e) to **210** is possible in some cases (87CB1455).

Dialkylsulfur diimides condense with diketene to give, for instance, thiadiazines **212** (Scheme 19). Alternatively, **212** can be made by reaction with β-ketoesters (Scheme 19), although in lesser yields than

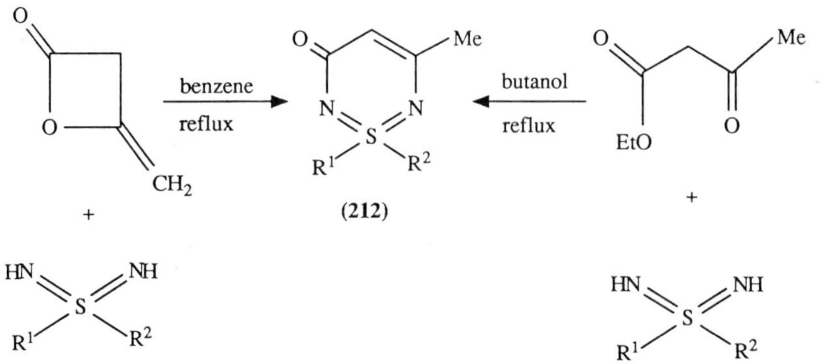

SCHEME 19

from diketene. The expected open-chain compound $MeCOCH_2$-$CON{=}S(R)^1)_2{=}NH$ was not found among the products (85CB2561).

Ketene mercaptals **213** (X = CN, COOMe; Y = OMe; R^1 = Me, benzyl) react with S,S-dialkysulfur diimides (Scheme 20) [R^2 = Me, Et, $(CH_2)_2$] to give colorless $1\lambda^6,2,6$-thiadiazines **214**. With $MeCCl{=}C$-$(COOEt)_2$, the same diimides yield thiadiazines **215** (86CB1745).

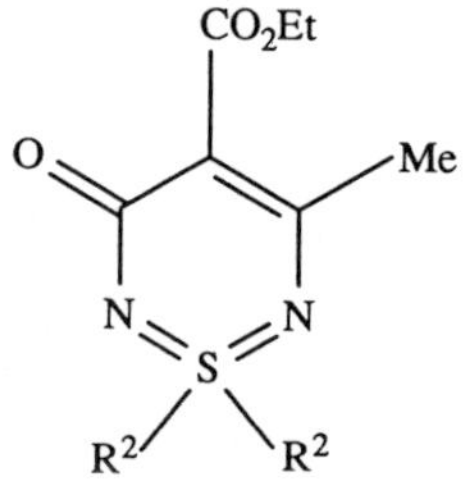

(215)

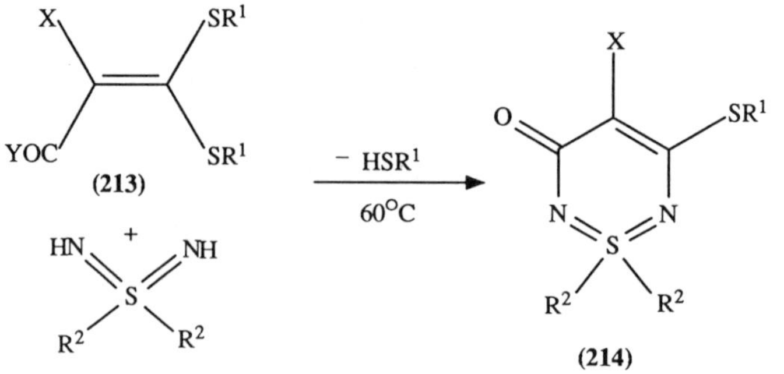

SCHEME 20

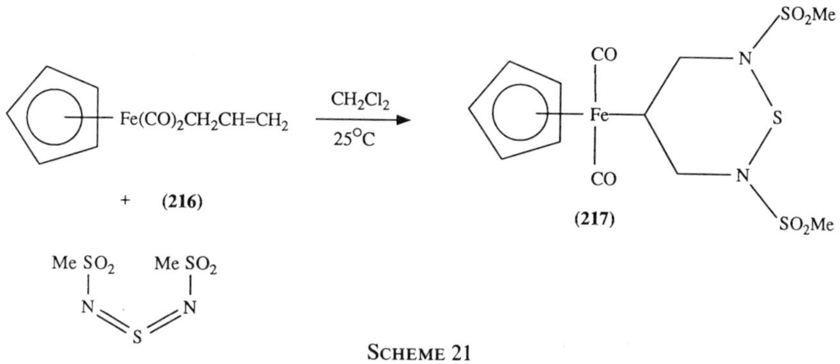

SCHEME 21

Ligated η^1-allyl compounds (Scheme 21) can react with sulfur diimides, e.g., **216** reacts with sulfur bis(methanesulfonylimide) to give the iron complex **217**. The [3 + 3] cycloadduct may possibly result from initial [3 + 2] cycloaddition followed by rearrangement (83MI3; 86MI4).

2. 2,1,3-Benzothiadiazines

Methods in use before the coverage of this review for making 2,1,3-benzothiadiazines include the reaction of (i) sulfonyl diamide with o-aminoaryl ketones or o-aminobenzylamines, (ii) sulfamoyl chlorides with o-aminobenzoic esters, (iii) o-aminobenzamides with N-sulfinyl-p-toluenesulfonamide or N,N'-bis(toluenesulfonyl)sulfur diimide or thionyl chloride (85MI2), and (iv) ring closure of o-methoxycarbonylphenylsulfonyl diamides with base, of carboxyphenylsulfonyl diamide with phosphoryl chloride, and of phenylsulfonyl diamide with formaldehyde (79MI1; 84MI1; 88AHC81). However, since 2,1,3-benzothiadiazine 2,2-dioxides have already been reviewed (88AHC81), only the synthetic method (iii) is described here.

The propensity of N-sulfinylanilines to cycloadd, which has already been used to prepare five- and six-membered heterocyclic sulfinamides, has now been applied to the synthesis of 2,1,3-benzothiadiazines, but with

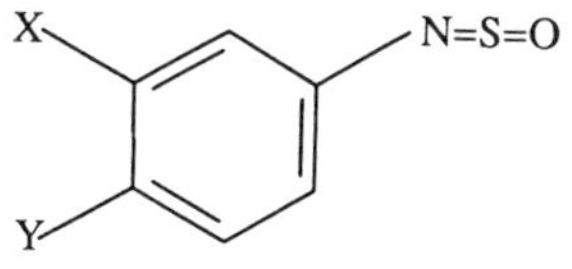

(218) 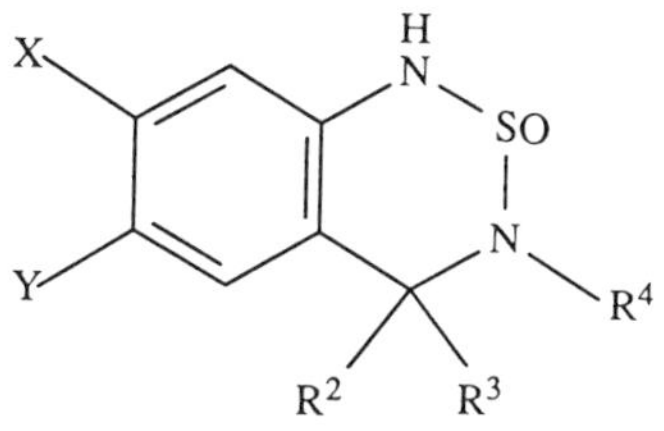   (219)

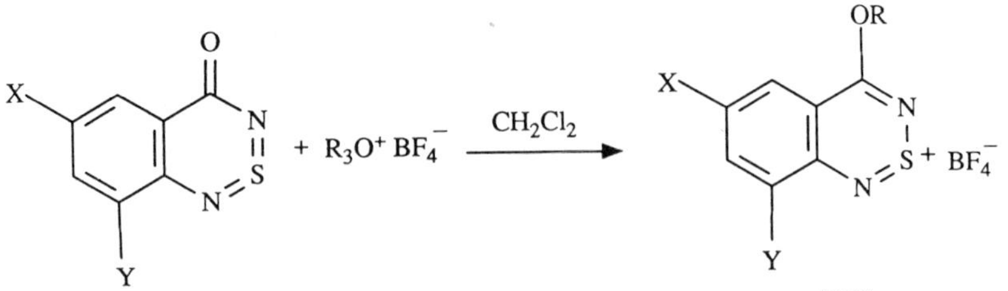

SCHEME 22

the variation that they behave as dienes rather than add across the N=S double bond. Thus, N-sulfinylanilines **218** (X = H, Y = H; X = Me, Y = H; X = H, Y = Me) react with benzalanilines $R^1R^2C=NR^4$ (R^4 = aryl) to afford the product 2-oxides **219**. In prior instances, only in their cycloaddition to norbornenes had N-sulfinylanilines behaved as dienes. Attempts to use Schiff bases with N-alkyl groups in this reaction yielded no thiadiazine (85TL2813). The action of N-sulfinyl-p-toluenesulfonamide on anthranilamide or 2-aminothiobenzamide gives 3,4-dihydro-4-oxo-1H-2,1,3-benzothiadiazine 2-oxide (71LA171).

Benzothiadiazinones **220** are formed by the reaction of o-aminobenzamides on sulfur bis(p-toluenesulfonylimide). These products provide a route to the heteroaromatic cation of 2,1,3-benzothiadiazinylium salts **222** by treating them (**220**) with oxonium salts **221** (Scheme 22) [77AG(E)780].

3. Miscellaneous Condensed 1,2,6-Thiadiazines

S,S-Dialkylsulfur diimides can be reacted with thiophene dioxide **223** in the presence of triethylamine to yield intermediate **224,** which may then be ring-closed with sodium methoxide to give **225** (84CB2779).

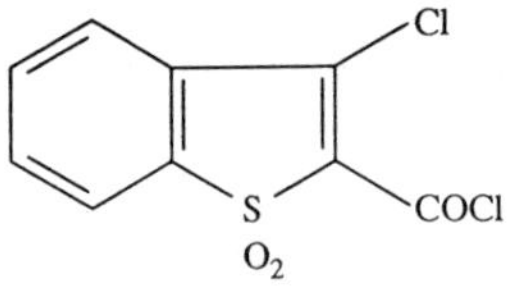

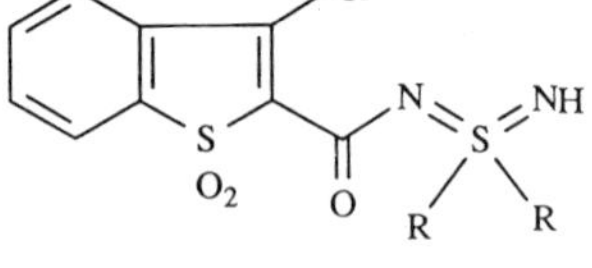

The reaction of *peri*-diamines with compounds of the type X=S=Y to yield naphtho[1,8-*cd*][1λ^4,2,6]thiadiazines, such as **229,** has been known for some time [67AG(E)149]. Piperidine- or morpholine-1-sulfenyl chloride have been shown to react with 1,8-diaminoaphthalenes and

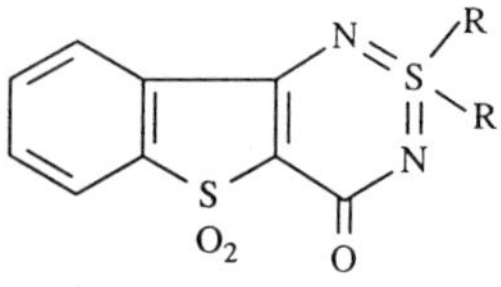

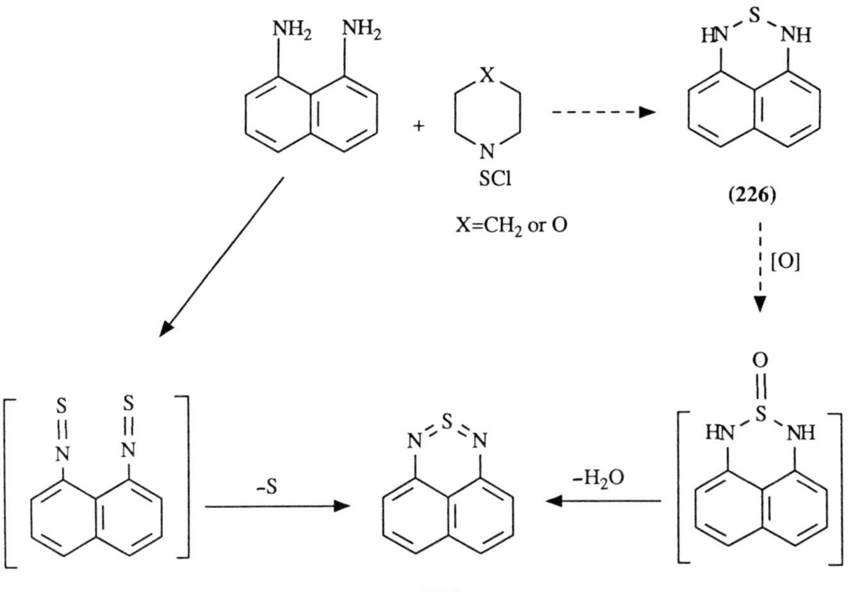

(228) (229) (227) (226)

SCHEME 23

1,4,5,8-tetraaminonaphthalene to afford **229** (Scheme 23) and **230** (Scheme 24), respectively. Piperidine-1-sulfenyl chloride gives better yields than the morpholine analog. A route for the reaction via intermediates **226** and **227** is unlikely on the grounds that **227** does not react with morpholine-1-sulfenyl chloride under preparation conditions to yield a thiadiazine [84JCS(P1)2591]. Compound **227** is known to be unstable and readily hydrolyzes to 1,8-diaminonaphthalene (79MI1). An alternative route is tentatively suggested which proceeds through the bis(thionitroso) compound, **228.** However, the thionitroso group is not detectable by means

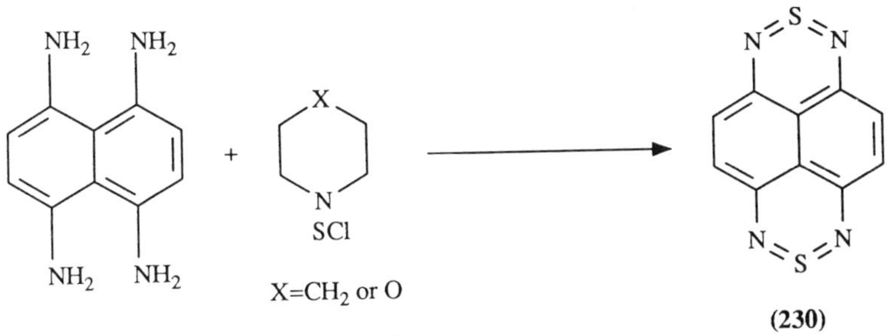

(230)

SCHEME 24

of the trapping agents, 2,3-dimethylbutadiene or cyclopentadiene [84JCS(P1)2591].

B. Structure

1. *Theoretical Methods*

The first effective claimant (**222**) of the 1,2,6-thiadiazine series to be heteroaromatic came to light in 1977. Evidence for the heteroaromaticity of this 2,1,3-benzothiadiazinylium salt (**222**) rests largely on the resonance energy calculated on an Hückel molecular orbital (HMO) basis to be 0.313β (cf. naphthalene 0.398β). Also, the ^{13}C NMR of these salts show strong deshielding of the C6 atom, due undoubtedly to the NSN group. A substituent increment of $+15.5$ ppm at this position has almost the same value as the diazonium group. Compound **222** (X = H, Y = H, R = Et) takes part in a facile Diels–Alder addition reaction with 1,3-dienes, such as 2,3-dimethylbutadiene, 1,3-pentadiene or isoprene. The double bond between the N1 and S atoms behaves as the dienophile in these reactions. Justification for the usefulness of Pariser–Parr–Pople (PPP) and CNDO S calculations is apparent in the application of these methods to explain the observed smoothness and regiospecificity of this Diels–Alder reaction. The calculations indicate that the LUMO, which is a π orbital, is almost completely localized on the N1 and S atoms. Such localization, together with the indicated nodal plane between these two atoms and the low energy of the LUMO, allow optimum reaction with the highest occupied molecular orbital (HOMO) of the diene [77AG(E)780].

The NSN group can be considered to be a four-electron, three-center π-system, which is isoelectronic with the allyl anion. Whereas the electron-releasing strength of the NSN group is weakened relative to the allyl anion because of the presence of the more electronegative nitrogen atoms, the electron-accepting capability is increased (77PS179). The likely outcome of this is that the NSN group should behave both as a donor or acceptor, and whichever of these roles it plays will depend on the nature of the bridging moiety in cyclic compounds.

The PPP configuration analysis technique and the molecules-in-molecules/localized configuration interaction (MIM/LCI) method can provide an explanation of the fact that some cyclic sulfur diimides are colorless, while others are not. Thus, while 1,2,5-thiadiazoles **231** and **232** are colorless, naphtho[1,8-*cd*][1λ^4,2,6]thiadiazine **229** is colored (low intensity absorption at 642 nm). The energy level of the color band appears to be largely determined by frontier orbital interactions. By considering the overlap between the two fragments, (i) the NSN group and (ii) the

(231)   (232)

remaining subunit in the structure, it is possible to show that relatively short wavelength absorption is expected if the HOMO of one fragment and the LUMO of the other (or vice versa) show the same symmetry with respect to the mirror plane. Otherwise, if either of the HOMO's has the same symmetry as the corresponding LUMO, long wavelength absorption is expected. In the case of **229,** by mixing in the naphthalene-to-NSN charge transfer (CT) configurations, the ground as well as the lowest excited state undergoes stabilization, but the excited state stabilization is much greater, and a bathochromic effect is observed. Sulfur diimides that are strongly stabilized in their ground states are less likely to be colored (79PS61).

2. Molecular Dimensions

The application of X-ray crystallography in determining the structure of 1,2,6-thiadiazine 1,1-dioxides confirms the ring is not aromatic. In general, the ring is nonplanar and either exists in an envelope form with the sulfur atom at the flap (79JOC4191; 88AHC81) or in a boat form with one of the oxygen atoms attached to the ring sulfur atom situated close to the C4 hydrogen atom. However, 1,2,4-thiadiazines that are not cyclic sulfamides have not been as frequently submitted to X-ray crystallographic analysis as their S,S-dioxide counterparts. There remains, therefore, more uncertainty about their conformational structure. X-ray crystallographic analysis has been applied to a study of the structure of 3,5-dichloro-4-oxo-4H-1,2,6-thiadiazine **200** [78AX(B)2927]. The ring frame **233** of **200** is almost planar, and the terminal oxygen and chlorine atoms deviate from the plane by small, though significant amounts (O −4.4 pm, Cl1 9.8 pm, Cl2 3.3 pm). The ring bond-angles centered on S1 and C4 are less than 120°, while the other ring angles are greater than 120° (**233**). The S—N bond lengths (161.5 and 161.9 pm) are greater than the value in the sulfur diimine group (154 pm) and suggest there is some conjugation with the remainder of the molecule. The bond lengths in the C—C—C segment (147.7 and 147.4 pm) also suggest some conjugation. However, the C—N bond lengths (127.2 and 127.6 pm) are typical for a C=N double bond. These bond lengths have been used to calculate Bird's aromaticity index (I_6). The value of 54.2

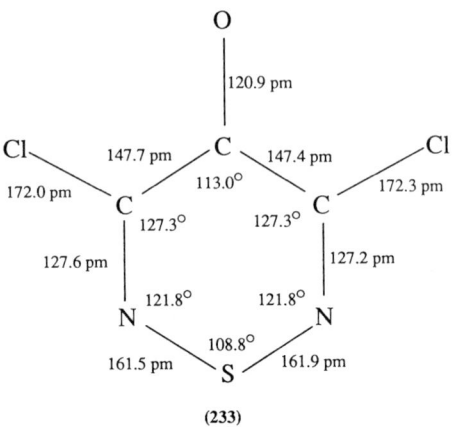

(233)

(cf. benzene 100, pyrylium cation 65.8, pyran-4-thione 47.9, and 1,3,5-thiadiazinium cation 53.1) resulted, which suggests moderate-to-low aromaticity for this compound (86T89).

3. Molecular Spectra

Ever since the polymer $(SN)_x$ was shown to be metallic and to represent the first example of a polymeric superconductor (78JA1235), compounds that contain the NSN group have come under close scrutiny. For instance, the compound naphtho[1,8-cd;4,5-c'd']bis[1λ^4,2,6]thiadiazine (**230**) is a metallic-green compound that is stable to the atmosphere and melts at about 290°C. It is difficultly soluble in organic solvents. However, despite this apparent stability, the half-wave polarographic potentials of the compound clearly show it is readily oxidized and reduced in acetonitrile.

The electron spin resonance (ESR) spectra of the radical ions of **230** indicate there are no large deviations from the free-electron g value that would have been expected had the $3d$ orbitals of the sulfur atom played an important part in influencing the spin density of the molecule. Consequently, structure **230** may not be the main contributor to the electronic structure of the compound. Such stability in this compound could be attributed to the inertness of the NSN group and the presence of the aromatic naphthalene ring. However, the ^{1}H-NMR chemical shifts (δ = 4.45 ppm) suggest the compound is antiaromatic. The compound is therefore referred to as an ambiguous aromatic compound (78JA1235).

Photoelectron spectroscopy studies indicate trimethylenesulfur diimide **205** may possibly take up chair and envelope (C4 at the flap) conformations (78ZN284). The benzo[4,5]thieno[3,2-c][1λ^6,2,6]thiadiazine-4-one 5,5-dioxides (**225**) exhibit blue fluorescence in long wavelength UV radiation (84CB2779).

The IR absorption maxima corresponding to the N=S=N asymmetric and symmetric stretching vibrations in trimethylenesulfur diimide **205** appear at 1135 and 1020 cm^{-1}, respectively (69TL4417). The 4,5-dioxo compounds **203** and **204** show N=S=N absorptions at 1330 and 1050 cm^{-1} [70AG(E)373]. The presence of one or more N—S groups in 2,6-bis (piperidino-sulfenyl)-2,3,5,6-tetrahydro-4H-1,2,6-thiadiazine is suggested by the presence of the absorption at 758 cm^{-1}.

The mass spectrum of **205** differs from other compounds containing the N=S=N group. The main fragmentations involve losses of NS, CH$_2$N and aziridine fragments (76CB2442; 78ACH275).

C. REACTIVITY OF THE RING

The presence of an N=S=N group can contribute a modicum of stability to some 1,2,6-thiadiazines such as **230,** although even in this case electrochemical reduction and oxidation can be quite facile (78JA1235). The presence of other polar groups, such as carbonyl, can modify the properties of 1,2,6-thiadiazines. Thus, the colorless 4,5-dioxo compounds **203** and **204** decompose at their melting points and many are hygroscopic, easily water-soluble substances [70AG(E)373]. The 3-oxo compounds **212** are readily soluble in protic solvents, and though stable towards bases, they are decomposed on exposure for a short time to acid (85CB2561). The unconjugated compound, trimethylenesulfur diimide **205,** is thermally unstable, and although it is quite stable to cold water, the addition of acid causes hydrolysis to the diamine precursor (69TL4117).

In contrast to the facile oxidation and reduction of 1,2-thiazine 1-oxides to the corresponding 1,1-dioxides and thiazines, respectively, the oxidation of 2,1,3-benzothiadiazine 2-oxides **219** to the 2,2-dioxides can be difficult with either hydrogen peroxide or m-chloroperbenzoic acid. Reduction of the benzothiadiazines with lithium aluminium hydride similarly leads to ill-defined oily products. The reaction of 2,1,3-benzothiadiazine 2-oxides **219** (R^4 = Ar) with either acetic anhydride or methanesulfonyl chloride leads to the corresponding acetanilides (R^{4}NHCOMe) and methanesulfonamides (R^{4}NHSO$_2$Me), respectively, in stoichiometric yields (85TL2813). Oxidation of 3,4-dihydro-4-oxo-1H-2,1,3-benzothiadiazine 2-oxide to the 2,2-dioxide can be accomplished with hydrogen peroxide at 20°C, but at 100°C, extrusion of sulfur dioxide occurs with the formation of 2,3-dihydro-3-oxo-1H-indazole (71LA171).

The extrusion of SO from some 1,2,6-thiadiazine 1-oxides, such as **198,** to give the corresponding pyrazole can be achieved by heating in toluene [79CC891; 81JCS(P1)1891; 83JCS(P1)2273]. The reaction of bromine/ dichloromethane with 3-oxo 1,2,5-thiadiazine **212** (R^1 = Me, R^2 = Me) in

the presence of boron trifluoride etherate results in bromination at the
4-position (85CB2561).

D. Reactivity of Substituents

What little is known of substituent reactivity in 1,2,6-thiadiazines that
do not include the 1,1-dioxide group concerns almost exclusively nucleophilic replacement of halogen. Thus, in 3,5-dichloro-4-oxo-4H-1,2,6-
thiadiazine **200,** the two chlorine atoms can be successively replaced by
N-nucleophiles. An excess of ammonia/ether gives the monoamino derivative, whereas heating with ammonia to 75°C in a Carius tube leads to
replacement of the second chlorine atom. Primary or secondary amines
replace one chlorine atom. Heating above 100°C is required to bring about
disubstitution. Alcohols do not react, but alkoxides do: one equivalent of
sodium methoxide gives the monomethoxy derivative, and a second
equivalent effects disubstitution. With S-nucleophiles, such as the very
nulceophilic thiolate anions, mono- and disubstitution can occur below
0°C. Thiols in the presence of triethylamine are used to introduce one SR
group and RSNa to supply the second. Interestingly, replacement of a
chlorine atom by an amino group causes a significant change (increase in
wavelength) in the UV spectra of these compounds, which can be put
down to an increase in conjugation. Additional evidence for the contribution of conjugated forms such as **234** is provided by the shifts of the C=O

X = NH$_2$, NHR; Y = Cl, NH$_2$, NHR

(234)

stretching vibration-absorption in the IR spectra to lower frequencies. The
likelihood of some conjugation in **200** itself is apparent from the X-ray
crystallographic data mentioned previously in Section V,B,2. The effect
on UV and IR data is less marked when either OR or SR replaces a chlorine
atom in **200,** which is not surprising in view of the lower conjugating ability
of oxygen and sulfur atoms (74RTC270).

E. Applications

Monosubstitution products of 3,5-dichloro-4-oxo-1,2,6-thiadiazine **235,** in which a chlorine atom is replaced by reaction with, for instance, *p*-hydroxybenzyl alcohol, followed by treatment with phosphorus tribromide, are featured as fungicides in a patent (78USP4097594). 2,1,3-

(235)   (236)   (237)

Benzothiadiazinone **236** (R = H) can be converted with BrCN to the 1-cyano derivative **236** (R = CN) which, along with a number of related compounds, has herbicidal properties (78GEP2656289). Furthermore, **236** (R = CH$_2$Cl), when stirred with sodium azide in DMF, is said to lead to **236** (R = CH$_2$N$_3$), a representive of another series of herbicides (78GEP2656290). The 2-oxides (m = 1) and 2,2-dioxides (m = 2) **237** (Z = O or S) have also been patented as herbicides (79USP4155746).

References

1883M131	R. Andreasch, *Monatsh Chem.* **4,** 131 (1883).
04M682	H. Wolfbauer, *Monatsh Chem.* **25,** 682 (1904).
14M137	F. Kucera, *Monatsh Chem.* **35,** 137 (1914).
14M151	F. Kucera, *Monatsh Chem.* **35,** 151 (1914).
17JPR180	E. Schrader, *J. Prakt. Chem.* **96,** 180 (1917).
47JA1393	R. Winterbottom, J. W. Clapp, W. H. Miller, J. P. English, and R. O. Roblin Jr., *J. Am. Chem. Soc.* 69, 1393 (1947).
49JOC946	R. A. Donia, J. A. Shotton, L. O. Bentz, and G. E. P. Smith, *J. Org. Chem.* **14,** 946 (1949).
50USP2514200	J. Shotton and R. A. Donia (Firestone Tire and Rubber Co.) U.S. Pat. 2,514,200 (1950) [*CA* **45,** 384d].
51USP2537633	G. E. P. Smith Jr. and J. A. Shotton, U.S. Pat. 2,537,633 (1951) [*CA* **45,** 5727a].
58E463	G. de Stevens, L. H. Werner, A. Holomandaris and S. Ricca, Jr., *Experientia* **14,** 463 (1958).
59JOC1983	B. E. Hoogenboom, R. Abbott, L. Locatell, and R. L. Hinman, *J. Org. Chem.* **24,** 1983 (1959).

60JOC970	F. C. Novello, S. C. Bell, E. L. A. Abrams, C. Ziegler, and J. M. Sprague, *J. Org. Chem.* **25**, 970 (1960).
61JOC3461	R. L. Hinman and B. E. Hoogenboom, *J. Org. Chem.* **26**, 3461 (1961).
61MI1	G. W. Stacey, *in* "Heterocyclic Compounds" (R. C. Elderfield, ed.) Vol. 7, p. 815. John Wiley, New York, 1961.
62AG(E)235	E. Schittler, G. De Stevens and L. Werner, *Angew. Chem. Int. Ed. Engl.* **1**, 235 (1962).
62BEP615374	(Ciba Ltd.), Belg. Pat. 615,374 (1962) [*CA* **60**, 1778h].
62DIS65	R. L. Abbott, *Diss. Abstr.* **23**, 65 (1962).
62FES320,331	L. Raffa, M. Di Bella, M. Melegari, and G. Vampa, *Farmaco, Ed. Sci.* **17**, 320, 331 (1962).
62HCA996	P. Schmidt, K. Eichenberger and M. Wilhelm, *Helv. Chim. Acta* **45**, 996 (1962).
62JOC1703	B. Loev and M. Kormendy, *J. Org. Chem.* **27**, 1703 (1962).
62JPR56	B. Helferich, R. Hoffmann, and H. Mylenbusch, *J. Prakt. Chem.* **19**, 56 (1962).
63FRP2166	M. Goudal, A. Goudal, P. Vernadeau, and J. Vernadeau, Fr. Pat. M2166 (1963) [*CA* **60**, 8048f].
63JMC603	S. Wawzonek and R. L. Abbott, *J. Med. Chem.* **6**, 603 (1963).
64LA189	A. W. Wagner and G. Reinoehl, *Justus Liebigs Ann. Chem.* **675**, 189 (1964).
66FRP1458701	Geistlich, Ed., Sohne A. -G. (fuer Chemische Industrie) Fr. Pat. 1,458,701 (1966) [*CA* **69**, 36190].
67AG(E)149	G. Kresze and W. Wucherpfennig, *Angew. Chem. Int. Ed. Engl.* **6**(2), 149 (1967).
67LA96	R. W. Hoffmann and W. Sieber, *Justus Liebigs Ann. Chem.* **703**, 96 (1967).
68JHC453	J. B. Wright, *J. Heterocycl. Chem.* **5**, 453 (1968).
68LA223	G. Kresze, C. Seyfried, and A. Trede, *Justus Liebigs Ann. Chem.* **715**, 223 (1968).
69CC33	J. K. King, A. Hanson, D. Deaken, and J. Komery, *J. C. S., Chem. Commun.* 33 (1969).
69TL4117	G. Kresze and H. Grill, *Tetrahedron Lett.* 4117 (1969).
70AG(E)373	M. Haake, *Angew. Chem., Int. Ed. Engl.* **9**(5), 373 (1970).
70CRV593	A. Lawson and R. B. Tinkler, *Chem. Rev.* **70**, 593 (1970).
70MI1	C. Kawasaki and K. Masaomi, *Bitamin* **41**(3), 207 (1970).
70MI2	A. J. Wohl, *Mol. Pharmacol.* **6**, 189 (1970).
70SST473	D. H. Reid, *Org. Compd. Sulfur, Selenium, Tellurium* **1**, 473 (1970).
71GEP2108461	J. B. Adams Jr., Ger. Pat. 2,108,461 (1971) [*CA* **76**, 14596].
71LA171	H. Grill and G. Kresze, *Justus Liebigs Ann. Chem.* **749**, 171 (1971).
72AX(B)2340	L. Dupont and O. Dideberg, *Acta Crystallogr.*, Part B **28**, 2340 (1972).
72BCJ1893	K. Hasegawa and S. Hirooka, *Bull. Chem. Soc. Jpn.* **45**, 1893 (1972).
72CB757	E. Cohen and J. Mahnke, *Chem. Ber.* **105**, 757 (1972).
72GEP2215733	A. C. Barker, P. Doyle, R. G. Foster, and J. R. Hadfield, Ger. Pat. 2,215,733 (1972) [*CA* **78**, 16198].
72JA8505	A. T. Fanning Jr., G. R. Bickford, and T. D. Roberts, *J. Am. Chem. Soc.* **94**, 8505 (1972).
73BSF985	A. Etienne, A. Le Berre, and J.-P. Giorgetti, *Bull. Soc. Chim. Fr.* 985 (1973).

73BSF2361   A. Etienne, A. Le Berre and J. P. Giorgetti, *Bull. Soc. Chim. Fr.*
                 2361 (1973).
73S225   G. Entenmann, *Synthesis* **4,** 225 (1973).
73TL4489   H. Kristinsson, *Tetrahedron Lett.* 4489 (1973).
73ZOR2038   E. A. Darmokhval, E. S. Levchenko, and L. N. Morkovskii, *Zh.*
                 Org. Khim. **9,** 2038 (1973).
74BRP1350277   A. C. Barker and R. Gregory, Brit. Pat. 1,350,277 (1974) [*CA* **81,**
                 25669].
74BSF1395   A. Etienne, A. Le Berre, and J. -P. Giorgetti, *Bull. Soc. Chim. Fr.*
                 1395 (1974).
74BSF1917   A. Etienne, A. Le Berre, and J. -P. Giorgetti, *Bull. Soc. Chim. Fr.*
                 1917 (1974).
74GEP2321786   E. Cohnen, Ger. Pat. 2,321,786 (1974) [*CA* **82,** 73035].
74JMC549   W. L. Matier and W. T. Comer, *J. Med. Chem.* **17,** 549 (1974).
74RTC270   J. Geevers and W. P. Trompen, *Recl. Trav. Chim. Pays-Bas* **93**(9–
                 10), 270 (1974).
75CSR189   P. D. Kennewell and J. B. Taylor, *Chem. Soc. Rev.* **4,** 189
                 (1975).
75GEP2451566   F. A. Sowinski and B. R. Vogt, Ger. Pat. 2,451,566 (1975) [*CA* **83,**
                 97399].
75OMS579   G. Entenmann, *Org. Mass. Spectrom.* **10**(8), 579 (1975).
76CB2442   R. Appel, J. -R. Lundehn, and E. Lassmann, *Chem. Ber.* **109,** 2442
                 (1976).
76GEP2438099   H. Koelling, H. Thomas, A. Widding, and H. Wollweber, Ger. Pat.
                 2,438,099 (1976) [*CA* **85,** 21484].
76KGS479   V. P. Chernykh, V. I. Gridasov, and P. A. Petyunin, *Khim. Getero-*
                 sikl. Soedin. 479 (1976) [*CA* **85,** 46602].
76PS309   S. L. Huang and D. Swern, *Phosphorus Sulfur* **1,** 309 (1976).
77AG(E)780   W. Kosbahn and H. Schaeffer, *Angew. Chem. Int. Ed. Engl.* **16**(11),
                 780 (1977).
77GEP2541742   H. Loewe, J. Urbanietz, D. Duewel, and R. Kirsch, Ger. Pat.
                 2,541,742 (1977) [*CA* **87,** 39545].
77GEP2541751   H. Heinz, J. Urbanietz, D. Duewel, and R. Kirsch, Ger. Pat.
                 2,541,751 (1977) [*CA* **87,** 5976].
77JOC952   K. Schaffner-Sabba, H. Tomaselli, B. Henrici, and H. B. Renfrone,
                 J. Org. Chem. **42,** 952 (1977).
77MI1   W. E. Truce, T. C. Klinger, and W. W. Brand, *in* "Organic Chem-
                 istry of Sulfur" (S. Oae, ed.), p. 588. Plenum Press, New York,
                 1977.
77PS179   J. Fabian and R. Mayer, *Phosphorus Sulfur,* **3,** 179 (1977).
78ACH275   I. Lengyel, G. Kresze, M. Berger, W. Kosbahn, and H. Schaeffer,
                 Acta Chim. Acad. Sci. Hung. **96,** 275 (1978).
78AX(B)2927   S. Harkema, *Acta Crystallogr.* Part B, **34,** 2927 (1978).
78GEP2656289   G. Stubenrauch, G. Hamprecht, B. Wuerzer, and G. Relzleff, Ger.
                 Pat. 2,656,289 (1978) [*CA* **89,** 129498].
78GEP2656290   G. Hamprecht, G. Stubenrauch, H. Urbach, and B. Wuerzer, Ger.
                 Pat. 2,656,290 (1978) [*CA* **89,** 129548].
78JA1235   R. C. Haddon, M. L. Kapland, and J. H. Marshall, *J. Amer. Chem.*
                 Soc. **100,** 1235 (1978).
78USP4097594   B. L. Davidson, N. W. Harnish, and C. J. Peake, U.S. Pat. 4,097,594
                 (1978) [*CA* **90,** 23124].

78ZN284	B. Solouki, H. Bock, and D. Glemser, *Z. Naturforsch.* **33B,** 284 (1978).
79CC891	J. Barluenga, J. F. Lopes-Ortiz, and V. Gotor, *J. C. S. Chem. Commun.* 891 (1979).
79JCS(P1)1043	Y. Girard, J. G. Atkinson and J. Rokach, *J. Chem. Soc., Perkin Trans.* 1, 1043 (1979).
79JOC4191	H. A. Albrecht, J. F. Blount, F. M. Konzelmann, and J. T. Plati, *J. Org. Chem.* **44,** 4191 (1979).
79MI1	J. K. Landquist, *in* "Comprehensive Organic Chemistry" (D. Barton and W. D. Ollis) Vol. 4, *Heterocyclic Compounds* (P. G. Sammes, ed.), p. 1108. Pergamon, Oxford, 1979.
79PS61	J. Fabian, R. Mayer, and S. Bleisch, *Phosphorus Sulfur* **7**(1), 61 (1979).
79T2151	P. Jakobsen and S. Treppendahl, *Tetrahedron* **35,** 2151 (1979).
79TL1281	J. Rokach, P. Hamel, J. Girard, and G. Reader, *Tetrahedron Lett.* 1281 (1979).
79USP4155746	L. H. McKendry and W. P. Bland, U.S. Pat. 4,155,746 (1979) [*CA* **91,** 103730].
80CSR477	P. D. Kennewell and J. B. Taylor, *Chem. Soc. Rev.* **9,** 477 (1980).
81FES905	M. G. Andrisano, M. Di Bella, P. Ferrari, L. Raffa, and S. G. Baggio, *Farmaco, Ed. Sci.* **36**(11), 905 (1981).
81JCS(P1)1891)	J. Barluenga, J. F. Lopes-Ortiz, M. Tomas, and V. Gotor, *J. Chem. Soc. Perkin Trans. 1* 1891 (1981).
81JCS(P1)2322	B. F. Bonini, G. Maccagnani, G. Mazzanti, and E. Foresti, *J. Chem. Soc. Perkin Trans. 1* 2322 (1981).
81ZOR2619	L. N. Markowskii, V. S. Talanov, O. M. Polumbrik, and Y. G. Shermolovich, *Zh. Org. Khim.* **17**(12), 2619 (1981) [*CA* **96,** 142810].
82CS248	K. G. Jensen and E. B. Pedersen, *Chem. Scr.* **20**(5), 248 (1982).
82JMC113	Y. Girard, J. G. Atkinson, D. R. Haubrich, M. Williams, and G. G. Yarborough, *J. Med. Chem.* **25,** 113 (1982).
82MI1	F. Erb, N. Febvay, and M. Imbenotte, *Talanta* **29,** 953 (1982).
82USP4338435	R. D. Haugwitz, U.S. Pat. 4,338,435 (1982) [*CA* **97,** 182465].
83AF688	Y. Orita, A. Ando, S. Yamabe, T. Nakanishi, Y. Arakawa, and H. Abe, *Arzneim.-Forsch.* **33**(51), 688 (1983).
83EUP88350	M. E. Smith, J. T. Witkowski, R. J. Doll, E. H. Gold, B. R. Neustadt, and A. S. Yehaskel, Eur. Pat. 88,350 (1983) [*CA* **100,** 175294].
83EUP95584	R. Haugwitz and P. W. Sprague, Eur. Pat. 95,584 (1983) [*CA* **100,** 175285].
83FES466	M. Di Bella, G. Gamberini, A. Tait, U. Fabio, and G. P. Quaglio. *Farmaco, Ed. Sci.* **38**(7), 466 (1983).
83FES738	C. Parenti, M. Di Bella, L. Raffa, G. G. Baggio, and S. Guarini, *Farmaco, Ed. Sci.* **38**(10), 738 (1983).
83GEP3208187	G. Staehler, W. Knauf, A. Waltersdorfer, and B. Sachse, Ger. Pat. 3,208,187 (1983) [*CA* **100,** 68526].
83JAP(K)58111943	Konishiroku Photo Industry Co. Ltd., Jpn. Kokai 58,-111,943 (1983) [*CA* **101,** 181055].
83JAP(K)58117542	Konishiroku Photo Industry Co. Ltd., Jpn. Kokai 58,-117,542 (1983) [*CA* **99,** 166957].

83JCS(P1)49   T. L. Gilchrist, C. W. Rees, and D. Vaughan, *J. Chem. Soc. Perkin Trans. 1*, 49 (1983).
83JCS(P1)55   T. L. Gilchrist, C. W. Rees, and D. Vaughan, *J. Chem. Soc. Perkin Trans. 1*, 55 (1983).
83JCS(P1)2273   J. Barluenga, M. Tomas, J. F. Lopez-Ortiz, and V. Gotor, *J. Chem. Soc. Perkin Trans. 1* 2273 (1983).
83MI1   V. I. Gridasov, L. D. Khaleeva and I. M. Timasheva, *Farm. Zh. (Kiev)*, (2), 62 (1983) [*CA* **99**, 38439].
83MI2   A. A. El-Maghraby and H. A. Eyada, *Egypt. J. Chem.* **26**(4), 355 (1983).
83MI3   T. W. Leung, G. G. Christoph, and A. Wojcicki, *Inorg. Chim. Acta* **76**(5–6), L281 (1983).
83MI4   A. A. El-Maghraby and H. A. Eyada, *Egypt. J. Chem.* **24**(1–4), 177 (1983) [*CA* **106**, 18503].
83S851   Y. Imai, A. Mochizuki, and M. Kakimoto, *Synthesis* (10), 851 (1983).
83T2073   D. Hellwinkel, R. Lenz, and F. Laemmerzahl, *Tetrahedron*, **39**, 2073 (1983).
84C316   M. Schlaepfer-Daehler, J. H. Bieri, and H. Heimgartner, *Chimia*, **38**(9), 316 (1984).
84CB2779   W. Ried and R. Pauli, *Chem. Ber.* **117**, 2779 (1984).
84EUP105732   F. D. King, C. S. Fake, and G. Burrell, Eur. Pat. 105,732 (1984) [*CA* **101**, 151835].
84EUP121830   J. J. Piwinski, J. T. Suh, R. P. Menard, H. Jones, and E. S. Neiss, Eur. Pat. 121,830 (1984) [*CA* **102**, 149792].
84GEP3321969   W. Dompert, T. Glaser, H. Horstmann, T. Schuurman, P. R. Seidel, and J. Traber, Ger. Pat. 3,321,969 (1984) [*CA* **102**, 220896].
84H1667   M. Schlaepfer-Daehler, R. Prewo, and H. Heimgartner, *Heterocycles* **22**, 1667 (1984).
84JCS(P1)2591   M. R. Bryce, *J. Chem. Soc. Perkin Trans. 1* 2591 (1984).
84JMC228   C. H. Levenson and R. B. Meyer Jr., *J. Med. Chem.* **27**, 228 (1984).
84KGS907   V. S. Fedenko, Z. F. Solomko, and V. I. Avramenko, *Khim. Geterotsikl. Soedin* (7), 907 (1984) [*CA* **101**, 191860].
84LA904   G. Kresze, A. Schwoebel, A. Hatjiisaak, K. Ackermann, and T. Minami, *Justus Liebigs Ann. Chem.* 904 (1984).
84MI1   C. J. Moody *in* "Comprehensive Heterocyclic Chemistry" (A. Katritzky and C. W. Rees, eds.), Vol. 3 (J. Boulton and A. McKillop, eds.) p. 1039. Pergamon, Oxford, 1984.
84MI2   A. A. El-Maghraby, F. M. Aly, and H. A. Eyada, *Curr. Sci.* **53**(10), 517 (1984).
84MI3   V. S. Fedenko, Z. F. Solomko, V. I. Avramenko, and M. P. Khmel, USSR Pat. 1,074,873 (1984); from *Orkrytiya, Izobret., Prom. Obraztsy, Tovarnye Znaki*. (7), 88 (1984) [*CA* **101**, 90989].
84USP4468396   C. V. Magatti, U.S. Pat. 4,468,396 (1984) [*CA* **102**, 24650].
84ZOR196   E. V. Levchenko, G. S. Borovikova, E. I. Borovik, and V. N. Kalinin, *Zh. Org. Khim.* **20**(1), 196 (1984) [*CA* **100**, 191836].
84ZOR589   Z. F. Solomko, V. S. Fedenko, and V. I. Avramenko, *Zh. Org. Khim.* **20**(3), 589 (1984) [*CA* **101**, 110453].
85AP903   C. Parenti, L. Costantino, M. Di Bella, L. Raffa, G. G. Baggio, and P. Zanoli, *Arch. Pharm. (Weinheim, Ger.).* **318**(10), 903 (1985).

85CB2561   W. Ried and R. Pauli, *Chem. Ber.* **118,** 2561 (1985).
85EUP153755   J. T. Suh, J. J. Piwinski, H. Jones, and E. S. Neiss, Eur. Pat. 153,755 (1985) [*CA* **104,** 110168].
85EUP161143   S. D. Jones, P. D. Kennewell, and R. W. Tully, Eur. Pat. 161,143 (1985) [*CA* **104,** 88615].
85EUP161498   H. Fukami, R. Kikumoto, K. Nakao, I. Nitta, and S. Inoue, Eur. Pat. 161,498 (1985) [*CA* **104,** 148911].
85IJC(B)1295   A. V. N. Reddy, A. Kamal and P. B. Sattur, *Indian J. Chem. Sect. B,* **24B**(12), 1295 (1985).
85JAP(K)6025984   Tokyo Institute of Technology, Jpn. Kokai 60-25,984 (1985) [*CA* **103,** 87910].
85JAP(K)6072868   Fujisawa Pharmaceutical Co. Ltd., Jpn. Kokai 60-72,868 (1985) [*CA* **103,** 142004].
85JAP(K)6078990   Hokko Chemical Co. Ltd., Jpn. Kokai 60-78,990 (1985) [*CA* **103,** 142022].
85JAP(K)6092278   Hokuriku Pharmaceutical Co. Ltd., Jpn. Kokai 60-92,278 (1985) [*CA* **103,** 215323].
85JAP(K)6097970   Hokko Chemical Industry Co. Ltd., Jpn. Kokai 60-97,970 (1985) [*CA* **103,** 142023].
85JAP(K)60112781   SmithKline Beckman Corp., Jpn. Kokai 60-112,781 (1985) [*CA* **103,** 196130].
85JAP(K)60156680   Y. Ito, S. Senda, H. Kato, N. Ogawa, E. Etsuchu, and H. Nishino, Jpn. Kokai 60-156,680 (1985) [*CA* **104,** 148920].
85LA579   A. Bender, D. Guenther and R. Wingen. *Justus Liebigs Ann. Chem.,* 579 (1985).
85MI1   I. G. Vinnichenko, V. S. Fedenko, Z. F. Solomko, and V. I. Avramenko, *Vopr. Khim. Khim. Tekhnol.* **77,** 13 (1985) [*CA* **106,** 155814].
85MI2   M. E. Suh, *Yakhak Hoechi* **29**(2), 103 (1985).
85T2625   E. Gomez, C. Avendano, and A. McKillop, *Tetrahedron* **42,** 2625 (1986).
85TL2813   U. Zoller and P. Rona, *Tetrahedron Lett.* **26,** 2813 (1985).
85USP4536501   J. E. Sundeen, R. D. Haugwitz, and P. W. Sprague, U.S. Pat. 4,536,501 (1985) [*CA* **104,** 6186].
85USP4556655   D. R. Andrews and F. C. A. Gaeta, U.S. Pat. 4,556,655 (1985) [*CA* **107,** 97126].
85USP4559340   B. R. Neustadt, D. R. Andrews, and P. E. McNamara, U.S. Pat. 4,559,340 (1985) [*CA* **105,** 6825].
86BSF871   J. Greiner, V. Bayer, R. Pastor, and A. Cambon, *Bull. Soc. Chim. Fr.* (6), 871 (1986).
86CB1745   W. Ried and M. A. Jakobi, *Chem. Ber.* **119,** 1745 (1986).
86GEP3427309   H. Koeppe, W. Abele, F. Esser, W. Gaida, and W. Hoefke, Ger. Pat. 3,427,309 (1986) [*CA* **104,** 224918].
86GEP3433037   H. Barth, J. Hartenstein, G. Satzinger, E. Fritschi, H. Osswalt, and G. Bartoszyk, Ger. Pat. 3,433,037 (1986) [*CA* **105,** 72690].
86JMC44   C. R. Young, G. J. Durant, J. C. Emmett, C. R. Ganellin, M. J. Graham, R. C. Mitchell, H. D. Prain, and M. L. Roantree, *J. Med. Chem.* **29**(1), 44 (1986).
86JOC1967   C. V. Kumar, K. R. Gopidas, K. Bhattacharyya, P. K. Das, and M. V. George, *J. Org. Chem.* **51,** 1967 (1986).

86MI1   V. P. Chernikh, V. F. Konev, V. I. Stepanenko, and V. I. Gridasov, *Farm. Zh.* (Kiev) (4), 59 (1986) [*CA* **107**, 96683].

86MI2   A. V. N. Reddy, A. Kamal, A. B. Rao, and P. B. Sattur, *Eur. J. Med. Chem.–Chim. Ther.* **21**(2), 177 (1986).

86MI3   A. S. Khan and F. F. Cantwell, *Talanta* **33**(2), 119 (1986).

86MI4   T. W. Leung, C. G. Christoph, J. Gallucci, and A. Wojcicka, *Organometallics* **5**(5), 864 (1986).

86SAP00083   B. R. Neustadt, E. M. Smith, C. V. Magatti, and E. H. Gold, S. Afr. Pat. 86-00,083 (1986) [*CA* **107**, 40330].

86S864   A. V. N. Reddy, A. Kamal, and P. B. Sattur, *Synthesis* **107**, 864 (1986).

86T89   C. W. Bird, *Tetrahedron* **42**, 89 (1986).

86TH1   S. D. Jones, Ph.D. Thesis, p. 45. Univ. Leeds, (1986).

86TL5703   D. Jacob, H. Peter-Niedermann, and H. Meier, *Tetrahedron Lett.* **27**, 5703 (1986).

86USP4616012   B. R. Neustadt, D. R. Andrews, P. E. McNamara, and R. W. Watkins U.S. Pat. 4,616,012 (1986) [*CA* **106**, 156866].

86USP4584285   R. J. Doll, B. R. Neustadt, E. M. Smith, C. V. Magatti, and E. H. Gold, U.S. Pat. 4,584,285 (1986) [*CA* **105**, 79362].

87CB1455   W. Ried and M. A. Jakobi, *Chem. Ber.* **120**, 1455 (1987).

87HCA2045   H. Balli, M. Huys-Francotte, and F. Schmidlin, *Helv. Chim. Acta* **70**, 2045 (1987).

87JAP(K)62195371   N. I. Kawa and A. Takaoka, Jpn. Kokai 62-195,371 (1987) [*CA* **108**, 94597].

87JHC1531   N. P. Peet, S. Sundar, R. J. Barbuch, E. W. Huber, and E. M. Bargar, *J. Heterocycl. Chem.* **24**, 1531 (1987).

87KGS1280   T. I. Orlova, S. P. Epshtein, A. F. Rukasov, N. S. Magomedova, V. K. Bel'skii, V. P. Tashchi, and Y. G. Putsykin, *Khim. Geterotikl. Soedin* (9), 1280 (1987) [*CA* **108**, 167366].

87MI1   R. C. Young, M. J. Graham, and L. Roantree, *Pharmacochem. Libr.* **10**(QSAR Drug Des. Toxicol.), 91 (1987).

87MI2   R. C. Young, *J. Pharm. Pharmacol.* **39**(10), 861 (1987).

87MI3   A. F. Casy, *J. Pharm. Biomed. Anal.* **5**(B). 247 (1987).

87MI4   A. V. N. Reddy, A. Kamal, A. B. Rao, and P. B. Sattur, *Eur. J. Med. Chem.* **22**(2), 157 (1987).

87MI5   C. Parenti, L. Costantino, M. Di Bella, L. Raffa and G. G. Baggio, *Boll. Chim. Farm.* **126**(6), 239 (1987).

87PS41   G. Kresze and A. Hatjiisaak, *Phosphorus Sulfur* **29**(1), 41 (1987).

87TL2641   S. N. Mazumdar, M. Sharma and M. M. Pal, *Tetrahedron Lett.* **28**, 2641 (1987).

87USP4634689   J. T. Witkowski and M. F. Czarniecki, U.S. Pat. 4,634,689 (1987) [*CA* **106** 120067].

87USP4634698   D. R. Andrews and F. C. A. Gaeta, U.S. Pat. 4,634,698 (1987) [*CA* **108**, 22287].

87USP4666906   J. J. Piwinski, J. T. Suh, P. Menard, and H. Jones, U.S. Pat. 4,666,906 (1987) [*CA* **108**, 38433).

88AHC81   V. J. Aran, P. Goya, and C. Ochoa, *Adv. Heterocycl. Chem.* **44**, 81 (1988).

88CB1689   W. Ried and M. A. Jacobi, *Chem. Ber.* **121**, 1689 (1988).

88EUP268990   T. Oku, E. Todo, C. Kasahara, K. Nakamura, H. Kayakiri, and M. Kasahara, Eur. Pat. 268,990 (1988) [*CA* **110**, 173265].

88GEP3632544   H. Koeppe, F. Esser, W. Gaida, and W. Hoefke, Ger. Pat. 3,632,544 (1988) [*CA* **109**, 93076].

88IJC109   U. V. Nabar, M. S. Mayadeo, and K. D. Deodhar, *Indian J. Chem.* **27B**(2), 109 (1988).

88S521   W. Ried, G. Beller, and J. W. Bats, *Synthesis* (7), 521 (1988).